Lecture Notes in Mathematics

Volume 2341

This series reports on new developments in all areas of mathematics and their applications - quickly, informally and at a high level. Mathematical texts analysing new developments in modelling and numerical simulation are welcome. The type of material considered for publication includes:

1. Research monographs
2. Lectures on a new field or presentations of a new angle in a classical field
3. Summer schools and intensive courses on topics of current research.

Texts which are out of print but still in demand may also be considered if they fall within these categories. The timeliness of a manuscript is sometimes more important than its form, which may be preliminary or tentative. Please visit the LNM Editorial Policy (https://drive.google.com/file/d/19XzCzDXr0FyfcV-nwVojWYTIIhCeo2LN/view?usp=sharing)

Titles from this series are indexed by Scopus, Web of Science, Mathematical Reviews, and zbMATH.

Chuchu Chen • Jialin Hong • Lihai Ji

Numerical Approximations of Stochastic Maxwell Equations

via Structure-Preserving Algorithms

 Springer

Chuchu Chen
LSEC, ICMSEC, Academy of Mathematics
and Systems Sciences
Chinese Academy of Sciences
Beijing, China

Jialin Hong
LSEC, ICMSEC, Academy of Mathematics
and Systems Sciences
Chinese Academy of Sciences
Beijing, China

Lihai Ji
Laboratory of Computational Physics
Institute of Applied Physics
and Computational Mathematics
Beijing, China

ISSN 0075-8434   ISSN 1617-9692   (electronic)
Lecture Notes in Mathematics
ISBN 978-981-99-6685-1   ISBN 978-981-99-6686-8   (eBook)
https://doi.org/10.1007/978-981-99-6686-8

Mathematics Subject Classification: 65-XX, 60-XX, 30-XX

This Springer imprint is published by the registered company Springer Nature Singapore Pte Ltd.
The registered company address is: 152 Beach Road, #21-01/04 Gateway East, Singapore 189721,
Singapore

Paper in this product is recyclable.

Preface

Since the pioneering works of Ampère, Faraday, and Maxwell in the early days, electromagnetism has become a fascinating area of physics, engineering, and mathematics. Its mathematical description is provided by the Maxwell equations, which form a system of partial differential equations expressed in terms of physical quantities such as the electric field, the magnetic field, and the current density. With the development of technology, certain electromagnetic noises, caused by the chaotic thermal motion of charged micro-particles [81, 149], the radiating sources [10, 124, 125], and the unpredictability of the environments or incomplete knowledge of the systems [3, 4, 13], now can be observed from various electric devices. To enhance the correspondence between real-life situations and theoretical results, researchers introduce randomness into the Maxwell equations, making them stochastic in nature. These equations are known as the stochastic Maxwell equations.

The study of the stochastic Maxwell equations has attracted great attention from researchers across many disciplines. There is currently an enormous effort to come up with various ways to study problems related to the stochastic Maxwell equations in fields like fluctuational electrodynamics, statistical radiophysics, integrated circuits, and stochastic inverse problems. The theoretical analysis of the stochastic Maxwell equations, including well-posedness, controllability, and homogenization, has been established in [148]. Since it is usually too complex to solve the governing stochastic Maxwell equations in an analytic form, different numerical algorithms are proposed to discretize these equations at grid points. However, no monograph treating this field has appeared. Our purpose in this monograph is to take a step toward filling this gap by discussing the developments in structure-preserving algorithms, which have good performance in long-term computations, for the stochastic Maxwell equations.

This monograph, in particular, brings a self-contained account of the numerical analysis of the stochastic Maxwell equations. Precisely, it includes the construction and analysis of structure-preserving algorithms with an emphasis on the preservation of geometric structures, physical properties, and asymptotic behaviors of the original equations. The objects considered here are related to several fascinating mathematical fields: numerical analysis, stochastic analysis, (multi-)symplectic geometry, large deviations principle, ergodic theory, partial differential equation, probability theory, etc.

This monograph consists of six chapters. Chapter 1 starts with a brief review of the deterministic Maxwell equations. Then we turn to the stochastic Maxwell equations. Here we introduce the origins of stochasticity, mathematical descriptions of some commonly used noises, constitutive relations, and the perfectly electrically conducting boundary condition. The governing equations for two fundamental polarizations, transverse electric and transverse magnetic polarizations, and those for the time-harmonic case are derived. At the end of this chapter, applications of the stochastic Maxwell equations in inverse random source problems and thermal radiation are discussed.

The solution theory of the stochastic Maxwell equations, which is crucial in the numerical analysis of algorithms, is presented in Chap. 2. We begin with formulating the stochastic Maxwell equations as the equivalent stochastic evolution equation to show the well-posedness in both the globally Lipschitz and non-globally Lipschitz continuous cases. Furthermore, regularities in the $\mathscr{D}(M^k)$-norm and the H^k-norm, as well as the differentiability of the solution on the initial datum, are given provided that more assumptions on coefficients, initial fields, and noises are employed.

A prerequisite for constructing the structure-preserving algorithms is to identify the intrinsic properties of the continuous system. This proceeds in Chap. 3 for the stochastic Maxwell equations, with the analysis of geometric structures (infinite-dimensional stochastic symplectic structure and stochastic multi-symplectic structure), physical properties (stochastic energy and divergence evolution laws), asymptotic behaviors (large deviations and ergodicity), etc.

With the preparatory work in Chap. 3, we turn to consider the construction of stochastic algorithms in Chap. 4 which preserve the intrinsic structures of the stochastic Maxwell equations. More precisely, we first present several temporal semi-discretizations for the stochastic Maxwell equations, including the stochastic midpoint method, stochastic symplectic Runge–Kutta methods, and exponential-type methods. *A priori* estimates and intrinsic discrete structures of these temporal semi-discretizations are analyzed. Then we further discretize those temporal semi-discretizations in the spatial direction to construct fully discrete algorithms via the finite difference method, the wavelet collocation method, and the discontinuous Galerkin method, respectively. Moreover, the splitting technique is introduced to reduce computational costs and improve the efficiency of implementing these algorithms. After discussing the well-posedness and regularity of stochastic structure-preserving algorithms in Chap. 4, we move on to their rigorous convergence analyses in Chap. 5.

The last chapter is dedicated to numerical experiments which verify the theoretical results obtained in this monograph. Not only a friendly introduction to the simulation of the noise but also a practical route into the simulations of the stochastic Maxwell equations with some structure-preserving algorithms are provided using MATLAB for the reader's convenience. In the appendix, we collect some results on commonly used identities and inequalities, semigroups, Sobolev spaces, differential calculus, estimates related to Maxwell operators, and stochastic partial differential equations.

We hope that our monograph could benefit a wide variety of readers comprising researchers, engineers, and graduate students in computational mathematics, stochastic analysis, applied physics, electrical engineering, telecommunications, etc. It is intended to review recent developments in the numerical analysis of the stochastic Maxwell equations. It is also intended to provide those who are interested in using and applying results of numerical algorithms and MATLAB codes without worrying about mathematical details or proofs.

Beijing, China   Chuchu Chen
November 2022   Jialin Hong
                                                                    Lihai Ji

Acknowledgements

Without the help and encouragement of many people, this monograph would not have been completed. We would like to use this opportunity to thank them. We thank our families for their kind love and constant support. Our thanks also go to Daniel Wang and Springer professionals for their help and assistance in finalizing the book. It is our great pleasure to thank our colleagues: Prof. Xu Wang and Prof. Liying Sun; Ph.D. students: Derui Sheng, Tonghe Dang, Tau Zhou, Ge Liang, and Guanlin Yang; and Postdocs: Xiaojing Zhang, Xinjie Dai, Qiang Li, Meng Cai, and Guoting Song for the suggestions and lists of mistakes which helped us to improve the presentation in a considerable way. We would also like to thank all the referees, whose comments and suggestions are of great help in improving this monograph. In closing, financial support from the National Natural Science Foundation of China (No. 12022118, No. 12031020, No. 11971470, No. 11871068, No. 12171047, No. 11971458) and the Youth Innovation Promotion Association CAS is gratefully acknowledged.

Contents

Notation and Symbols

$\mathbb{R}^d$	d-Dimensional Euclidean space
$\mathbf{x}$	Space variable in $\mathbb{R}^d$ ($d \geq 1$) with $\mathbf{x} = (x_1, \ldots, x_d)^\top$, specially $\mathbf{x} = (x, y, z)^\top$ when $d = 3$
$|\mathbf{x}|$	Euclidean norm in $\mathbb{R}^d$ for $\mathbf{x} \in \mathbb{R}^d$
$\mathbf{x} \cdot \mathbf{y}$	Inner product in $\mathbb{R}^d$ for $\mathbf{x}, \mathbf{y} \in \mathbb{R}^d$
$D \subset \mathbb{R}^d$	Open, bounded, and Lipschitz domain in $\mathbb{R}^d$
∂D	Boundary of D
$\min\{s, t\}$ (resp. $\max\{s, t\}$)	Minimum (resp. maximum) of $s, t \in \mathbb{R}$
$\mathbb{N}$	Set of all nonnegative integers
$\mathbb{N}_+$	Set of all positive integers
$\mathbb{Z}$	Set of all integers
$\otimes$	Kronecker inner product
$(\Omega, \mathscr{F}, \{\mathscr{F}_t\}, \mathbb{P})$	Filtered probability space
$N(0, 1)$	Standard normal distribution
$\mathbb{E}(u)$	Expectation of the random variable u
$\mathbb{E}(u|\mathscr{G})$	Conditional expectation of u with respect to a σ-algebra $\mathscr{G} \subset \mathscr{F}$
$\mathscr{B}(\mathbb{H})$	Borel σ-algebra on $\mathbb{H}$
$\mathscr{G}_1 \times \mathscr{G}_2$	Product σ-algebra of two σ-algebras $\mathscr{G}_1$ and $\mathscr{G}_2$
$\mathbf{E}$	Electric field $\mathbf{E} = (E_1, E_2, E_3)^\top$
$\mathbf{H}$	Magnetic field $\mathbf{H} = (H_1, H_2, H_3)^\top$
M	Maxwell operator
$\mathscr{D}(M)$	Domain of the Maxwell operator M
$\mathrm{ran}(B)$	Range of the operator B
$\mathrm{Tr}(B)$	Trace of the operator B
$(\mathbb{H}, \langle \cdot, \cdot \rangle_{\mathbb{H}}, \| \cdot \|_{\mathbb{H}})$	Hilbert space $\mathbb{H} = L^2(D)^3 \times L^2(D)^3$
$(U, \langle \cdot, \cdot \rangle_U, \| \cdot \|_U)$	Hilbert space $U = L^2(D)$
$\mathscr{L}(U, \mathbb{H})$	Space of all bounded linear operators from U to $\mathbb{H}$, denoted by $\mathscr{L}(\mathbb{H})$ if $U = \mathbb{H}$

$HS(U, \mathbb{H})$	Space of all Hilbert–Schmidt operators from U to $\mathbb{H}$
$L^p(D)^d$	Lebesgue space of p-integrable vector functions $f :$ $D \to \mathbb{R}^d$ with d components
$W^{k,p}(D)$	Sobolev spaces consisting of all functions $f : D \to \mathbb{R}$ whose derivatives up to order k are functions in $L^p(D)$
$H^k(D)$	Sobolev spaces of all functions $f : D \to \mathbb{R}$ whose derivatives up to order k are square integrable
$H(\mathrm{curl}, D)$	Space $\{v \in L^2(D)^3 : \nabla \times v \in L^2(D)^3\}$
$H_0(\mathrm{curl}, D)$	Space $\{v \in H(\mathrm{curl}, D) : \mathbf{n} \times v|_{\partial D} = 0\}$
$H(\mathrm{div}, D)$	Space $\{v \in L^2(D)^3 : \nabla \cdot v \in L^2(D)\}$
$H_0(\mathrm{div}, D)$	Space $\{v \in H(\mathrm{div}, D) : \mathbf{n} \cdot v|_{\partial D} = 0\}$
$C^k(D)$	Space of continuous functions with continuous derivatives up to order k
$C^{0,\alpha}(D)$	Space of Hölder continuous functions of order α
$C^{k,\alpha}(D)$	Space of functions of $C^k(D)$ whose derivatives of order k belong to $C^{0,\alpha}(D)$
$C^\infty(D)$	Space of infinitely differentiable functions in D
$C_0^\infty(D)$	Space of infinitely differentiable functions with compact support in D
$C_b^k(\mathbb{H})$	Space of all smooth functionals with bounded derivatives of order up to k on $\mathbb{H}$
$L^p(\Omega, \mathbb{H})$	Space of all $\mathbb{H}$-valued functions which are p-integrable with respect to $\mathbb{P}$
$w_1 \wedge w_2$	2-Form generated by the exterior product of two 1-forms
df	Differential 1-form
$df \wedge dg$	Differential 2-form
diag	Diagonal matrix
Id	Identity operator or identity matrix
$\mathbb{I}$	Rate function
$\mathbf{1}_A(\cdot)$	Indicator function of the set A
$\mathbb{J}$	Standard symplectic matrix $\mathbb{J} = \begin{bmatrix} 0 & Id \\ -Id & 0 \end{bmatrix}$
∇^2	Hessian matrix operator
∇	Gradient operator, i.e., $\nabla f = (\partial_x f, \partial_y f, \partial_z f)^\top$ for a scalar function f
$\nabla\times$	Curl operator, i.e., $\nabla \times \mathbf{v} = (\partial_y v_3 - \partial_z v_2, \partial_z v_1 - \partial_x v_3, \partial_x v_2 - \partial_y v_1)^\top$ for $\mathbf{v} = (v_1, v_2, v_3)^\top \in \mathbb{R}^3$
$\nabla\cdot$	Divergence operator, i.e., $\nabla \cdot \mathbf{v} = \partial_x v_1 + \partial_y v_2 + \partial_z v_3$ for $\mathbf{v} = (v_1, v_2, v_3)^\top \in \mathbb{R}^3$
FDTD	Finite-difference time-domain
dG	Discontinuous Galerkin
FE	Finite element

FV	Finite volume
TE	Transverse electric
TM	Transverse magnetic
PEC	Perfectly electrically conducting

Chapter 1
Introduction

In this chapter, we start with a brief introduction to the deterministic Maxwell equations including boundary and interface conditions, intrinsic properties, and numerical algorithms. With the development of technology, certain electromagnetic noises can be observed from various electric devices. Hence, the electromagnetic fields may not be deterministic but should be modeled by random fields, namely by the solution of the stochastic Maxwell equations. We then present the origin of stochasticity in the Maxwell equations and summarize some commonly used noises which drive the stochastic system. Two polarizations of the electromagnetic fields and the time-harmonic stochastic Maxwell equations are also discussed. Finally, we list some applications of the stochastic Maxwell equations in areas such as inverse random source problems and thermal radiation. For more details on the electromagnetic theory, we refer to [11, 140, 162] for the deterministic case and to [148, 149] for the stochastic case.

1.1 Deterministic Maxwell Equations

The Maxwell equations form the fundamentals of classical electrodynamics and optics. They were completely formulated by James Clerk Maxwell in the period from 1861 to 1865. They consist of a set of partial differential equations which describe the propagation of electromagnetic waves through media. The Maxwell equations in the differential form read as follows

Faraday's law of induction: $\quad \partial_t \mathbf{B}(t, \mathbf{x}) + \nabla \times \mathbf{E}(t, \mathbf{x}) = 0, \quad\quad (1.1)$

Ampère's circuital law: $\quad \partial_t \mathbf{D}(t, \mathbf{x}) - \nabla \times \mathbf{H}(t, \mathbf{x}) = -\mathbf{J}_e(t, \mathbf{x}), \quad (1.2)$

Gauss's law: $\quad \nabla \cdot \mathbf{D}(t, \mathbf{x}) = \rho_e, \quad\quad (1.3)$

© The Author(s), under exclusive license to Springer Nature Singapore Pte Ltd. 2023
C. Chen et al., *Numerical Approximations of Stochastic Maxwell Equations*,
Lecture Notes in Mathematics 2341, https://doi.org/10.1007/978-981-99-6686-8_1

Gauss's law for magnetism: $\nabla \cdot \mathbf{B}(t, \mathbf{x}) = 0,$ (1.4)

where $\mathbf{D}$ is the electric displacement, $\mathbf{E}$ is the electric field, $\mathbf{B}$ is the magnetic induction, $\mathbf{H}$ is the magnetic field intensity, $\mathbf{J}_e$ is the electric current density, and ρ_e is the electric charge density. After differentiating (1.3) with respect to the time variable and using (1.2), we derive the continuity equation

$$\partial_t \rho_e + \nabla \cdot \mathbf{J}_e = 0,$$

which expresses the conservation of charge.

Notice that the number of equations in (1.1)–(1.4) is less than the number of unknowns. To guarantee the well-posedness of the Maxwell equations (1.1)–(1.4), the constitutive relations describing macroscopic properties of the medium need to be introduced. Generally, the relations are complicated and depend strongly on the material in which the electromagnetic wave propagates. We start with the following typical representations of the form

$$\mathbf{D} = \varepsilon_0 \mathbf{E} + \mathbf{P}, \quad \mathbf{B} = \mu_0(\mathbf{H} + \mathbf{M}),$$ (1.5)

where ε_0 and μ_0 are the vacuum permittivity and permeability respectively, and $\mathbf{P}$ and $\mathbf{M}$ are electric polarization and magnetization respectively. Note that $\mathbf{P}$ and $\mathbf{M}$ are caused by the impinging fields, which can influence the organization of electrical charges and magnetic dipoles in a medium and vanish in the free space.

If one ignores ferroelectric and ferromagnetic media and if the fields are relatively small, the constitutive relations can be modeled by the following linear form

$$\mathbf{D} = \varepsilon \mathbf{E}, \quad \mathbf{B} = \mu \mathbf{H}$$ (1.6)

with ε and μ being the electric permittivity and magnetic permeability, respectively. In this case, the medium is called linear. For an anisotropic linear medium, parameters ε and μ are tensors, i.e., $\varepsilon, \mu : \mathbb{R}^3 \to \mathbb{R}^{3 \times 3}$. While in the isotropic case, parameters ε and μ are scalar functions, i.e., $\varepsilon, \mu : \mathbb{R}^3 \to \mathbb{R}$.

Moreover, if the medium is conductive, a further constitutive relation should be given by the fact that the electromagnetic field induces an electric current. In a linear approximation, this is described by Ohm's law

$$\mathbf{J}_e = \sigma \mathbf{E},$$ (1.7)

where σ is called the electrical conductivity. The parameter σ is a tensor for the anisotropic medium but a scalar for the isotropic medium. Particularly, a medium is called lossy if $\sigma > 0$, whereas it is lossless if $\sigma \equiv 0$. A medium is said to be temporally dispersive if parameters ε, μ, and σ depend also on the time variable. In this monograph, we constrict ourselves on the non-dispersive media case, i.e., the parameters ε, μ, and σ are independent of time.

1.1.1   Boundary and Interface Conditions

Electromagnetic fields generally exist in the whole space and are generated by charges and currents. In order to determine solutions to boundary value problems, we need to introduce some boundary conditions which apply to both the deterministic and stochastic Maxwell equations.

Let S be a surface. The permittivity and permeability parameters ε and μ are piecewise constants with jumps on S. As displayed in Fig. 1.1, for any subset $\Gamma \subset S$, we consider the cylindrical domain $\Theta = \Gamma \times (-\iota, \iota)$ for $\iota > 0$, which is separated by the surface S into two parts Θ_1 and Θ_2. Let $\mathbf{n}$ be the unit outward normal vector on $\partial\Theta$. Denote by Γ_1 and Γ_2 the part of surface $\partial\Theta$ in Material 1 and Material 2, respectively. It is clear to note that $\partial\Theta_1 = \Gamma_1 \cup \Gamma$ and $\partial\Theta_2 = \Gamma_2 \cup \Gamma$. To derive the interface conditions, we denote by $\mathbf{n}_S$ the unit normal vector on S pointing from Material 2 to Material 1, and by $\mathbf{E}_j$, $\mathbf{D}_j$, $\mathbf{H}_j$, $\mathbf{B}_j$, ε_j, μ_j ($j = 1, 2$) the restrictions of the respective functions to Material 1 and Material 2, respectively.

By (1.3) and the divergence theorem, we obtain

$$\int_\Theta \rho_e \, d\mathbf{x} = \int_\Theta \nabla \cdot \mathbf{D} \, d\mathbf{x} = \oint_{\partial\Theta} \mathbf{n} \cdot \mathbf{D} \, ds = \int_{\Gamma_1} \mathbf{n} \cdot \mathbf{D}_1 \, ds + \int_{\Gamma_2} \mathbf{n} \cdot \mathbf{D}_2 \, ds.$$

It follows from the divergence theorem again that

$$\int_\Theta \rho_e \, d\mathbf{x} = \int_\Theta \nabla \cdot \mathbf{D} \, d\mathbf{x} = \int_{\Theta_1} \nabla \cdot \mathbf{D}_1 \, d\mathbf{x} + \int_{\Theta_2} \nabla \cdot \mathbf{D}_2 \, d\mathbf{x}$$

$$= \left(\int_{\Gamma_1} \mathbf{n} \cdot \mathbf{D}_1 \, ds - \int_\Gamma \mathbf{n}_S \cdot \mathbf{D}_1 \, ds \right) + \left(\int_{\Gamma_2} \mathbf{n} \cdot \mathbf{D}_2 \, ds + \int_\Gamma \mathbf{n}_S \cdot \mathbf{D}_2 \, ds \right).$$

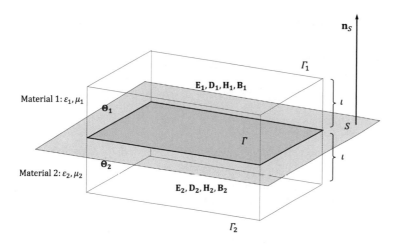

Fig. 1.1 Interface conditions: neighbourhood of the interface S

Combining the above two equations yields

$$\int_\Gamma \mathbf{n}_S \cdot (\mathbf{D}_1 - \mathbf{D}_2)\, ds = 0 \quad \forall\, \Gamma \subset S,$$

which implies that the normal component $\mathbf{n}_S \cdot \mathbf{D}$ has to be continuous, i.e.,

$$\mathbf{n}_S \cdot (\mathbf{D}_1 - \mathbf{D}_2)\, |_S = 0. \tag{1.8}$$

Using (1.4) and the divergence theorem, a similar argument holds for the magnetic field, that is, the normal component $\mathbf{n}_S \cdot \mathbf{B}$ is continuous across the surface

$$\mathbf{n}_S \cdot (\mathbf{B}_1 - \mathbf{B}_2)\, |_S = 0. \tag{1.9}$$

For the continuity conditions for the electromagnetic fields, it follows from (1.1), (1.4), and the divergence theorem that

$$\int_{\partial\Theta} \mathbf{n} \cdot \nabla \times \mathbf{E}\, ds = - \int_{\partial\Theta} \mathbf{n} \cdot \partial_t \mathbf{B}\, ds = -\partial_t \int_\Theta \nabla \cdot \mathbf{B}\, dx = 0.$$

Moreover, from (1.1), with the help of the Stokes theorem, one can deduce

$$\int_{\partial\Theta} \mathbf{n} \cdot \nabla \times \mathbf{E}\, ds = \int_{\Gamma_1} \mathbf{n} \cdot \nabla \times \mathbf{E}_1\, ds + \int_{\Gamma_2} \mathbf{n} \cdot \nabla \times \mathbf{E}_2\, ds = \oint_{\partial\Gamma} \hat{\mathbf{n}} \cdot (\mathbf{E}_1 - \mathbf{E}_2)\, dl,$$

where $\hat{\mathbf{n}}$ is the unit tangential vector on the contour $\partial\Gamma$. Combining the above two equations yields

$$\mathbf{n}_S \times (\mathbf{E}_1 - \mathbf{E}_2)\, |_S = 0 \tag{1.10}$$

due to the fact that the curve $\partial\Gamma$ is arbitrary on S. This demonstrates that the tangential trace of the electric field is continuous across the surface. Similarly, from (1.2) and (1.3) one can obtain

$$\mathbf{n}_S \times (\mathbf{H}_1 - \mathbf{H}_2)\, |_S = 0, \tag{1.11}$$

which means that the tangential trace of the magnetic field is continuous across the surface.

For the special case that one side of S is occupied by a perfect conductor, a popular boundary condition, named the perfectly electrically conducting (PEC) boundary condition, has drawn a lot of attention. Since the perfect conductor prevents the electric energy from entering, the interface condition (1.10) implies

$$\mathbf{n}_S \times \mathbf{E}|_S = 0. \tag{1.12}$$

The PEC boundary condition (1.12) is commonly used when the conductivity of a conductor is sufficiently large in realistic situations.

1.1.2 Intrinsic Properties

The deterministic Maxwell equations possess several intrinsic properties, including energy conservation properties, symplectic and multi-symplectic geometric structures. Let $D \subset \mathbb{R}^3$ be an open, bounded, and Lipschitz domain with boundary ∂D. Recall the source-free Maxwell equations

$$\begin{cases} \varepsilon \partial_t \mathbf{E} = \nabla \times \mathbf{H}, \quad \mu \partial_t \mathbf{H} = -\nabla \times \mathbf{E}, & \text{in } (0, T] \times D, \\ \mathbf{E}(0, \mathbf{x}) = \mathbf{E}_0(\mathbf{x}), \quad \mathbf{H}(0, \mathbf{x}) = \mathbf{H}_0(\mathbf{x}), & \text{in } D, \\ \mathbf{n} \times \mathbf{E} = 0, & \text{on } [0, T] \times \partial D, \end{cases} \tag{1.13}$$

where $\mathbf{E} = (E_1, E_2, E_3)^\top$, $\mathbf{H} = (H_1, H_2, H_3)^\top$, and $\mathbf{n}$ is the unit outward normal vector on ∂D.

(i) Physical conservation laws

We first present the energy conservation properties of (1.13). The electromagnetic energy is given by

$$\mathscr{E}(t) := \int_D \left(\varepsilon |\mathbf{E}(t, \mathbf{x})|^2 + \mu |\mathbf{H}(t, \mathbf{x})|^2 \right) d\mathbf{x}.$$

Recall that the Poynting vector $\mathbf{F} := \mathbf{E} \times \mathbf{H}$ models the energy flux density. Using (1.13) and the Green formula, we obtain the Poynting theorem

$$\frac{d\mathscr{E}}{dt} = -2 \int_D \nabla \cdot \mathbf{F} d\mathbf{x} = -2 \oint_{\partial D} \mathbf{n} \cdot (\mathbf{E} \times \mathbf{H}) ds = -2 \oint_{\partial D} \mathbf{H} \cdot (\mathbf{n} \times \mathbf{E}) ds = 0,$$

i.e.,

$$\int_D \left(\varepsilon |\mathbf{E}(t, \mathbf{x})|^2 + \mu |\mathbf{H}(t, \mathbf{x})|^2 \right) d\mathbf{x} \equiv \text{Constant}, \tag{1.14}$$

which represents the conservation of electromagnetic energy in a lossless medium. Noting that $\mathbf{E}$ satisfies the PEC boundary condition, then

$$\mathbf{n} \times \partial_t \mathbf{E} = 0, \quad \text{on } [0, T] \times \partial D.$$

Taking the derivative with respect to t on both sides of the first equation of (1.13) and using the Green formula again, we derive the second energy conservation law

$$\int_D \left(\varepsilon |\partial_t \mathbf{E}(t, \mathbf{x})|^2 + \mu |\partial_t \mathbf{H}(t, \mathbf{x})|^2 \right) d\mathbf{x} \equiv \text{Constant.} \tag{1.15}$$

Similarly, if we assume that ε and μ are positive constants, then we can show that electromagnetic waves possess the following energy conservation laws

$$\int_D \left(\varepsilon |\partial_\alpha \mathbf{E}(t, \mathbf{x})|^2 + \mu |\partial_\alpha \mathbf{H}(t, \mathbf{x})|^2 \right) d\mathbf{x} \equiv \text{Constant} \tag{1.16}$$

and

$$\int_D \left(\varepsilon |\partial_t \partial_\alpha \mathbf{E}(t, \mathbf{x})|^2 + \mu |\partial_t \partial_\alpha \mathbf{H}(t, \mathbf{x})|^2 \right) d\mathbf{x} \equiv \text{Constant,} \tag{1.17}$$

where $\alpha = x$, y or z.

(ii)　Geometric structures

We now turn to the geometric structures for the deterministic Maxwell equations. It is known that (1.13) is an infinite-dimensional Hamiltonian system. Define a Hamiltonian $\mathscr{H} : L^2(D)^3 \times L^2(D)^3 \to \mathbb{R}$ as

$$\mathscr{H}(\mathbf{E}, \mathbf{H}) = -\int_D \left(\frac{1}{2\mu} \mathbf{E}(\mathbf{x}) \cdot \nabla \times \mathbf{E}(\mathbf{x}) + \frac{1}{2\varepsilon} \mathbf{H}(\mathbf{x}) \cdot \nabla \times \mathbf{H}(\mathbf{x}) \right) d\mathbf{x}.$$

We recall that the variation of the functional $\mathscr{H}(\mathbf{E}, \mathbf{H})$ is defined as

$$\delta \mathscr{H}(\mathbf{E}, \mathbf{H}) = \mathscr{H}(\mathbf{E} + \delta \mathbf{E}, \mathbf{H} + \delta \mathbf{H}) - \mathscr{H}(\mathbf{E}, \mathbf{H})$$

$$= \int_D \left[\frac{\delta \mathscr{H}}{\delta \mathbf{E}(\mathbf{x})} \delta \mathbf{E}(\mathbf{x}) + \frac{\delta \mathscr{H}}{\delta \mathbf{H}(\mathbf{x})} \delta \mathbf{H}(\mathbf{x}) \right] d\mathbf{x},$$

where $\frac{\delta \mathscr{H}}{\delta \mathbf{E}(\mathbf{x})}$ (resp. $\frac{\delta \mathscr{H}}{\delta \mathbf{H}(\mathbf{x})}$) is the functional derivative of the functional $\mathscr{H}(\mathbf{E}, \mathbf{H})$ with respect to the function $\mathbf{E}$ (resp. $\mathbf{H}$) at the space point $\mathbf{x}$. Hence, under the assumption that ε and μ are positive constants, we derive that

$$\frac{\delta \mathscr{H}}{\delta \mathbf{E}} = -\mu^{-1} \nabla \times \mathbf{E}, \qquad \frac{\delta \mathscr{H}}{\delta \mathbf{H}} = -\varepsilon^{-1} \nabla \times \mathbf{H}.$$

The above equations mean that (1.13) has the following Hamiltonian formulation

$$\frac{d}{dt} \begin{bmatrix} \mathbf{E} \\ \mathbf{H} \end{bmatrix} = \mathbb{J}^{-1} \begin{bmatrix} \frac{\delta \mathscr{H}}{\delta \mathbf{E}} \\ \frac{\delta \mathscr{H}}{\delta \mathbf{H}} \end{bmatrix} \tag{1.18}$$

with the standard symplectic matrix

$$\mathbb{J} = \begin{bmatrix} 0 & Id \\ -Id & 0 \end{bmatrix},$$

where Id is the identity operator on $L^2(D)^3$.

The following theorem shows that the phase flow of (1.13) preserves the symplectic structure.

Theorem 1.1 *Suppose that ε and μ are two positive constants. Under the homogeneous boundary condition, the phase flow of (1.13) preserves the symplectic structure*

$$\bar{\omega}(t) := \int_D d\mathbf{E}(t, \mathbf{x}) \wedge d\mathbf{H}(t, \mathbf{x})d\mathbf{x} = \int_D d\mathbf{E}_0(\mathbf{x}) \wedge d\mathbf{H}_0(\mathbf{x})d\mathbf{x} = \bar{\omega}(0) \qquad (1.19)$$

for all $t \geq 0$, which means that the spatial integral of the oriented areas of projections onto the coordinate plane $(\mathbf{E}_0, \mathbf{H}_0)$ is an invariant.

Proof We have

$$\bar{\omega}(t) = \int_D \left(\frac{\partial \mathbf{E}}{\partial \mathbf{E}_0} d\mathbf{E}_0 + \frac{\partial \mathbf{E}}{\partial \mathbf{H}_0} d\mathbf{H}_0 \right) \wedge \left(\frac{\partial \mathbf{H}}{\partial \mathbf{E}_0} d\mathbf{E}_0 + \frac{\partial \mathbf{H}}{\partial \mathbf{H}_0} d\mathbf{H}_0 \right) d\mathbf{x}$$

$$= \int_D \left(d\mathbf{E}_0 \wedge \left(\frac{\partial \mathbf{E}}{\partial \mathbf{E}_0} \right)^\top \frac{\partial \mathbf{H}}{\partial \mathbf{E}_0} d\mathbf{E}_0 \right) d\mathbf{x}$$

$$+ \int_D \left(d\mathbf{H}_0 \wedge \left(\frac{\partial \mathbf{E}}{\partial \mathbf{H}_0} \right)^\top \frac{\partial \mathbf{H}}{\partial \mathbf{H}_0} d\mathbf{H}_0 \right) d\mathbf{x}$$

$$+ \int_D \left[d\mathbf{E}_0 \wedge \left(\left(\frac{\partial \mathbf{E}}{\partial \mathbf{E}_0} \right)^\top \frac{\partial \mathbf{H}}{\partial \mathbf{H}_0} - \left(\frac{\partial \mathbf{H}}{\partial \mathbf{E}_0} \right)^\top \frac{\partial \mathbf{E}}{\partial \mathbf{H}_0} \right) d\mathbf{H}_0 \right] d\mathbf{x},$$

which implies

$$\frac{d\bar{\omega}(t)}{dt} = \int_D \left(d\mathbf{E}_0 \wedge \frac{d}{dt} \left(\left(\frac{\partial \mathbf{E}}{\partial \mathbf{E}_0} \right)^\top \frac{\partial \mathbf{H}}{\partial \mathbf{E}_0} \right) d\mathbf{E}_0 \right) d\mathbf{x}$$

$$+ \int_D \left(d\mathbf{H}_0 \wedge \frac{d}{dt} \left(\left(\frac{\partial \mathbf{E}}{\partial \mathbf{H}_0} \right)^\top \frac{\partial \mathbf{H}}{\partial \mathbf{H}_0} \right) d\mathbf{H}_0 \right) d\mathbf{x} \qquad (1.20)$$

$$+ \int_D \left[d\mathbf{E}_0 \wedge \frac{d}{dt} \left(\left(\frac{\partial \mathbf{E}}{\partial \mathbf{E}_0} \right)^\top \frac{\partial \mathbf{H}}{\partial \mathbf{H}_0} - \left(\frac{\partial \mathbf{H}}{\partial \mathbf{E}_0} \right)^\top \frac{\partial \mathbf{E}}{\partial \mathbf{H}_0} \right) d\mathbf{H}_0 \right] d\mathbf{x}.$$

It follows from (1.13) that

$$\frac{d}{dt}\left(\frac{\partial \mathbf{E}}{\partial \mathbf{E}_0}\right) = \varepsilon^{-1}\nabla\times\left(\frac{\partial \mathbf{H}}{\partial \mathbf{E}_0}\right),\quad \frac{d}{dt}\left(\frac{\partial \mathbf{H}}{\partial \mathbf{E}_0}\right) = -\mu^{-1}\nabla\times\left(\frac{\partial \mathbf{E}}{\partial \mathbf{E}_0}\right),$$

$$\frac{d}{dt}\left(\frac{\partial \mathbf{E}}{\partial \mathbf{H}_0}\right) = \varepsilon^{-1}\nabla\times\left(\frac{\partial \mathbf{H}}{\partial \mathbf{H}_0}\right),\quad \frac{d}{dt}\left(\frac{\partial \mathbf{H}}{\partial \mathbf{H}_0}\right) = -\mu^{-1}\nabla\times\left(\frac{\partial \mathbf{E}}{\partial \mathbf{H}_0}\right). \tag{1.21}$$

Plugging (1.21) into (1.20) gives

$$\frac{d\bar{\varpi}(t)}{dt} = \int_D d\mathbf{E}_0 \wedge \left[\varepsilon^{-1}\left(\nabla\times\left(\frac{\partial \mathbf{H}}{\partial \mathbf{E}_0}\right)\right)^\top \frac{\partial \mathbf{H}}{\partial \mathbf{E}_0} - \mu^{-1}\left(\frac{\partial \mathbf{E}}{\partial \mathbf{E}_0}\right)^\top \nabla\times\left(\frac{\partial \mathbf{E}}{\partial \mathbf{E}_0}\right)\right]d\mathbf{E}_0 d\mathbf{x}$$

$$+ \int_D d\mathbf{H}_0 \wedge \left[\varepsilon^{-1}\left(\nabla\times\left(\frac{\partial \mathbf{H}}{\partial \mathbf{H}_0}\right)\right)^\top \frac{\partial \mathbf{H}}{\partial \mathbf{H}_0} - \mu^{-1}\left(\frac{\partial \mathbf{E}}{\partial \mathbf{H}_0}\right)^\top \nabla\times\left(\frac{\partial \mathbf{E}}{\partial \mathbf{H}_0}\right)\right]d\mathbf{H}_0 d\mathbf{x}$$

$$+ \int_D d\mathbf{E}_0 \wedge \left[\varepsilon^{-1}\left(\nabla\times\left(\frac{\partial \mathbf{H}}{\partial \mathbf{E}_0}\right)\right)^\top \frac{\partial \mathbf{H}}{\partial \mathbf{H}_0} - \mu^{-1}\left(\frac{\partial \mathbf{E}}{\partial \mathbf{E}_0}\right)^\top \nabla\times\left(\frac{\partial \mathbf{E}}{\partial \mathbf{H}_0}\right)\right]d\mathbf{H}_0 d\mathbf{x}$$

$$+ \int_D d\mathbf{E}_0 \wedge \left[\mu^{-1}\left(\nabla\times\left(\frac{\partial \mathbf{E}}{\partial \mathbf{E}_0}\right)\right)^\top \frac{\partial \mathbf{E}}{\partial \mathbf{H}_0} - \varepsilon^{-1}\left(\frac{\partial \mathbf{H}}{\partial \mathbf{E}_0}\right)^\top \nabla\times\left(\frac{\partial \mathbf{H}}{\partial \mathbf{H}_0}\right)\right]d\mathbf{H}_0 d\mathbf{x}.$$

Therefore, we have

$$\frac{d\bar{\varpi}(t)}{dt} = \int_D \left[\varepsilon^{-1}d(\nabla\times\mathbf{H})\wedge d\mathbf{H} + \mu^{-1}d(\nabla\times\mathbf{E})\wedge d\mathbf{E}\right]d\mathbf{x}$$

$$= \int_D \varepsilon^{-1}\left[\partial_x(dH_2\wedge dH_3) + \partial_y(dH_3\wedge dH_1) + \partial_z(dH_1\wedge dH_2)\right]d\mathbf{x}$$

$$+ \int_D \mu^{-1}\left[\partial_x(dE_2\wedge dE_3) + \partial_y(dE_3\wedge dE_1) + \partial_z(dE_1\wedge dE_2)\right]d\mathbf{x}$$

$$= 0$$

due to the homogeneous boundary condition.   □

Remark 1.1 To avoid ambiguities, we clarify that the notations 'd' in (1.18) and 'd' in (1.19) have different meanings. In (1.18), $\mathbf{E}$ and $\mathbf{H}$ are treated as functions of time, and $\mathbf{E}_0$ and $\mathbf{H}_0$ are fixed parameters, while exterior differentiation 'd' in (1.19) is made with respect to the initial datum $(\mathbf{E}_0, \mathbf{H}_0)$.

Note that if we consider the following modified Hamiltonian $\widetilde{\mathscr{H}} : L^2(D)^3 \times L^2(D)^3 \to \mathbb{R}$,

$$\widetilde{\mathscr{H}}(\mathbf{E}, \mathbf{H}) = \int_D \left(\frac{1}{2}\varepsilon|\mathbf{E}(\mathbf{x})|^2 + \frac{1}{2}\mu|\mathbf{H}(\mathbf{x})|^2\right)d\mathbf{x},$$

then the Maxwell equations (1.13) can be rewritten into the non-canonical Hamiltonian formulation

$$\frac{\partial}{\partial t}\begin{bmatrix} \mathbf{E} \\ \mathbf{H} \end{bmatrix} = \begin{bmatrix} 0 & (\varepsilon\mu)^{-1}\nabla\times \\ -(\varepsilon\mu)^{-1}\nabla\times & 0 \end{bmatrix}\begin{bmatrix} \frac{\delta\widetilde{\mathscr{H}}}{\delta\mathbf{E}} \\ \frac{\delta\widetilde{\mathscr{H}}}{\delta\mathbf{H}} \end{bmatrix}. \tag{1.22}$$

Taking the exterior derivative on both sides of (1.22) yields the preservation of the following symplectic structure.

Theorem 1.2 *Suppose that ε and μ are two positive constants. The phase flow of (1.13) preserves the symplectic conservation law*

$$\tilde{\omega}(t) := \int_D \left(d\mathbf{E}(t, \mathbf{x}) \wedge [(\varepsilon\mu)^{-1}\nabla\times]d\mathbf{H}(t, \mathbf{x})\right)d\mathbf{x} = \tilde{\omega}(0) \quad \forall\, t \geq 0.$$

Remark 1.2 When ε and μ depend on the space variable $\mathbf{x}$, (1.22) is not Hamiltonian. However, the modified Hamiltonian $\widetilde{\mathscr{H}}$ can still be preserved by introducing $\mathbf{X}(t) := \sqrt{\varepsilon}\mathbf{E}(t)$ and $\mathbf{Y}(t) := \sqrt{\mu}\mathbf{H}(t)$. In terms of fields $\mathbf{X}(t)$ and $\mathbf{Y}(t)$, the corresponding equations form an infinite-dimensional Hamiltonian system.

Next, we turn to investigate the geometric structure for (1.13) from another point of view by regarding (1.13) as a Hamiltonian partial differential equation.

Definition 1.1 Let $\mathbb{F}$ and $\mathbb{K}_i$, $i = 1, 2, \ldots, d$ be skew-symmetric matrices in $\mathbb{R}^{n\times n}$, and let $S : \mathbb{R}^n \to \mathbb{R}$ be a smooth function for $n \geq 3$. Then a partial differential equation is called a Hamiltonian partial differential equation if it possesses the form

$$\mathbb{F}\partial_t u + \sum_{i=1}^{d}\mathbb{K}_i\partial_{x_i}u = \nabla_u S(u),$$

where $u : [0, T] \times \mathbb{R}^d \to \mathbb{R}^n$.

Setting $u = (\mathbf{E}^\top, \mathbf{H}^\top)^\top$ and skew-symmetric matrices

$$
\mathbb{F} = \begin{bmatrix} & & & 1 & 0 & 0 \\ & \mathbf{0} & & 0 & 1 & 0 \\ & & & 0 & 0 & 1 \\ -1 & 0 & 0 & & & \\ 0 & -1 & 0 & & \mathbf{0} & \\ 0 & 0 & -1 & & & \end{bmatrix}, \quad
\mathbb{K}_1 = \begin{bmatrix} 0 & 0 & 0 & & & \\ 0 & 0 & -\frac{1}{\mu} & & \mathbf{0} & \\ 0 & \frac{1}{\mu} & 0 & & & \\ & & & 0 & 0 & 0 \\ & \mathbf{0} & & 0 & 0 & -\frac{1}{\varepsilon} \\ & & & 0 & \frac{1}{\varepsilon} & 0 \end{bmatrix},
$$

$$
\mathbb{K}_2 = \begin{bmatrix} 0 & 0 & \frac{1}{\mu} & & & \\ 0 & 0 & 0 & & \mathbf{0} & \\ -\frac{1}{\mu} & 0 & 0 & & & \\ & & & 0 & 0 & \frac{1}{\varepsilon} \\ & \mathbf{0} & & 0 & 0 & 0 \\ & & & -\frac{1}{\varepsilon} & 0 & 0 \end{bmatrix}, \quad
\mathbb{K}_3 = \begin{bmatrix} 0 & -\frac{1}{\mu} & 0 & & & \\ \frac{1}{\mu} & 0 & 0 & & \mathbf{0} & \\ 0 & 0 & 0 & & & \\ & & & 0 & -\frac{1}{\varepsilon} & 0 \\ & \mathbf{0} & & \frac{1}{\varepsilon} & 0 & 0 \\ & & & 0 & 0 & 0 \end{bmatrix},
$$

$$(1.23)$$

we obtain the Hamiltonian partial differential equation form for (1.13):

$$\mathbb{F}\partial_t u + \mathbb{K}_1 \partial_x u + \mathbb{K}_2 \partial_y u + \mathbb{K}_3 \partial_z u = 0. \tag{1.24}$$

Theorem 1.3 *Suppose that ε and μ are two positive constants. The Maxwell equations (1.13) preserve the multi-symplectic geometric conservation law*

$$\partial_t \varpi + \partial_x \kappa_1 + \partial_y \kappa_2 + \partial_z \kappa_3 = 0,$$

where $\varpi = \frac{1}{2} du \wedge \mathbb{F} du$ and $\kappa_p = \frac{1}{2} du \wedge \mathbb{K}_p du$, $p = 1, 2, 3$ are differential 2-forms. More precisely,

$$\partial_t \Big(dE_1 \wedge dH_1 + dE_2 \wedge dH_2 + dE_3 \wedge dH_3 \Big) + \partial_x \Big(\frac{1}{\varepsilon} dH_3 \wedge dH_2 + \frac{1}{\mu} dE_3 \wedge dE_2 \Big)$$

$$+ \partial_y \Big(\frac{1}{\varepsilon} dH_1 \wedge dH_3 + \frac{1}{\mu} dE_1 \wedge dE_3 \Big) + \partial_z \Big(\frac{1}{\varepsilon} dH_2 \wedge dH_1 + \frac{1}{\mu} dE_2 \wedge dE_1 \Big) = 0.$$

Proof Taking partial derivative of ϖ with respect to t yields

$$\partial_t \varpi = \frac{1}{2} \left(du_t \wedge \mathbb{F} du + du \wedge \mathbb{F} du_t \right) = du \wedge \mathbb{F} du_t$$

due to the skew-symmetry of $\mathbb{F}$. Similarly, we can obtain

$$\partial_x \kappa_1 = du \wedge \mathbb{K}_1 du_x, \quad \partial_y \kappa_2 = du \wedge \mathbb{K}_2 du_y, \quad \partial_z \kappa_3 = du \wedge \mathbb{K}_3 du_z.$$

Combining the above results gives the desired equality

$$\partial_t \varpi + \partial_x \kappa_1 + \partial_y \kappa_2 + \partial_z \kappa_3 = du \wedge d(\mathbb{F}u_t + \mathbb{K}_1 u_x + \mathbb{K}_2 u_y + \mathbb{K}_3 u_z) = 0.$$

The proof is thus completed. □

1.1.3 Numerical Algorithms

The numerical simulation of an electromagnetic wave is a fundamental and important electromagnetic subject. In the past few decades, many researchers have developed, analyzed, and tested various numerical algorithms for the Maxwell equations. In this section, we will give a brief overview of classical numerical algorithms for the Maxwell equations, among which the structure-preserving ones will be discussed in some detail.

(i) **Classical algorithms**

The finite-difference time-domain (FDTD) method is an efficient numerical algorithm in the field of computational electromagnetism. It was introduced in 1966 by Yee [178] and was further developed by Taflove [162] in the 1970s. The FDTD method uses second-order accurate central differences in time and space on a staggered grid. There are many excellent theoretical results on the FDTD method for solving the Maxwell equations, such as the energy-preserving splitting FDTD method [34, 36, 84], the high order FDTD method [118, 127] and the alternating direction implicit FDTD method [75], etc. This method can be efficiently implemented on vector computers, which makes it feasible to solve complex problems on early supercomputers. As an example, in 1987 SAAB (a company in Sweden) performed lightning analysis on the Swedish fighter aircraft Gripen on a grid with approximately $60 \times 30 \times 30$ cells. However, the major drawback of the FDTD method is the inability to represent curved boundaries and small geometrical details.

Another well-developed method for the Maxwell equations is the finite element (FE) method, including the edge element method, the discontinuous Galerkin (dG) method, etc. The FE method is based on a variational formulation of the partial differential equation in some suitable Hilbert space. Approximations to the solution are then sought in a finite-dimensional subspace. There is extensive literature on designing such numerical methods to the Maxwell equations. For instance, we refer to [32, 96, 113, 139, 140, 168, 172] for the edge element method, and to [50, 52, 123, 131] for the dG method.

Even though the FE method provides excellent tools for solving electromagnetic problems on geometrically complex domains, marching techniques applied to this method produce implicit schemes. In order to obtain an explicit scheme to solve the Maxwell equations on geometrically complex domains, the finite volume (FV) method is considered. The FV method was introduced to the Maxwell equations in [151] by exporting methods from computational fluid dynamics. For a detailed description of this method as well as its implementation for the Maxwell equations, we refer to [30, 47–49] and references therein.

(ii) **Structure-preserving algorithms**

We list several commonly used structure-preserving algorithms, including symplectic algorithms, multi-symplectic algorithms, and energy-preserving algorithms.

(a) **Splitting multi-symplectic algorithm:** we split the Hamiltonian partial differential equation (1.24) into three local one-dimensional subsystems

$$\frac{1}{3}\mathbb{F}\partial_t u + \mathbb{K}_1\partial_x u = 0, \quad \frac{1}{3}\mathbb{F}\partial_t u + \mathbb{K}_2\partial_y u = 0, \quad \frac{1}{3}\mathbb{F}\partial_t u + \mathbb{K}_3\partial_z u = 0. \quad (1.25)$$

It can be verified that the above three subsystems possess the multi-symplectic conservation laws

$$\frac{1}{3}\partial_t \varpi + \partial_x \kappa_1 = 0, \quad \frac{1}{3}\partial_t \varpi + \partial_y \kappa_2 = 0, \quad \frac{1}{3}\partial_t \varpi + \partial_z \kappa_3 = 0,$$

respectively. The central box scheme or the Preissman scheme is applied to discretize subsystems in (1.25) and thereby this gives the multi-symplectic algorithm for the deterministic Maxwell equations (1.13). Here, for example, we present the splitting multi-symplectic algorithm by utilizing the central box scheme. To avoid notational complications, we only discrete the first subsystem of (1.25) and obtain

$$\frac{1}{\tau}\left((E_2)^*_{i+\frac{1}{2},j,k} - (E_2)^n_{i+\frac{1}{2},j,k}\right) + \frac{1}{\varepsilon\Delta x}\left((H_3)^{n+\frac{1}{2}*}_{i+1,j,k} - (H_3)^{n+\frac{1}{2}*}_{i,j,k}\right) = 0,$$

$$\frac{1}{\tau}\left((E_3)^*_{i+\frac{1}{2},j,k} - (E_3)^n_{i+\frac{1}{2},j,k}\right) - \frac{1}{\varepsilon\Delta x}\left((H_2)^{n+\frac{1}{2}*}_{i+1,j,k} - (H_2)^{n+\frac{1}{2}*}_{i,j,k}\right) = 0,$$

$$\frac{1}{\tau}\left((H_2)^*_{i+\frac{1}{2},j,k} - (H_2)^n_{i+\frac{1}{2},j,k}\right) - \frac{1}{\mu\Delta x}\left((E_3)^{n+\frac{1}{2}*}_{i+1,j,k} - (E_3)^{n+\frac{1}{2}*}_{i,j,k}\right) = 0,$$

$$\frac{1}{\tau}\left((H_3)^*_{i+\frac{1}{2},j,k} - (H_3)^n_{i+\frac{1}{2},j,k}\right) + \frac{1}{\mu\Delta x}\left((E_2)^{n+\frac{1}{2}*}_{i+1,j,k} - (E_2)^{n+\frac{1}{2}*}_{i,j,k}\right) = 0,$$

where τ is the step size in the temporal direction, and Δx is the mesh size in x-direction; u^* is an intermediate value between u^n and u^{n+1}, and $u^{n+\frac{1}{2}*} = (u^n + u^*)/2$. We refer to [26, 117] and references therein for more details.

(b) **Averaged vector field algorithm:** we make a spatial semi-discretization for the infinite-dimensional Hamiltonian system (1.18)

$$\begin{bmatrix} d_t\mathbf{E} \\ d_t\mathbf{H} \end{bmatrix} = -\mathbb{J}^{-1}\begin{bmatrix} \mu^{-1}\widetilde{\mathrm{curl}}(\mathbf{E}) \\ \varepsilon^{-1}\widetilde{\mathrm{curl}}(\mathbf{H}) \end{bmatrix}, \quad (1.26)$$

where 'curl' is some discretization of the operator '$\nabla \times$', for example, $\widetilde{\mathrm{curl}}\,\mathbf{E} :=$
$(\bar{\delta}_y E_3 - \bar{\delta}_z E_2, \ \bar{\delta}_z E_1 - \bar{\delta}_x E_3, \ \bar{\delta}_x E_2 - \bar{\delta}_y E_1)^\top$ with $(\bar{\delta}_x E_3)_{i,j,k} = \frac{(E_3)_{i+1,j,k} - (E_3)_{i-1,j,k}}{2\Delta x}$
and other notations are defined similarly. Applying the second-order averaged vector
field algorithm to (1.26) yields

$$
\frac{1}{\tau}\begin{bmatrix} \mathbf{E}^{n+1} - \mathbf{E}^n \\ \mathbf{H}^{n+1} - \mathbf{H}^n \end{bmatrix} = -\mathbb{J}^{-1} \begin{bmatrix} \mu^{-1}\widetilde{\mathrm{curl}}\left(\int_0^1 [(1-\xi)\mathbf{E}^n + \xi \mathbf{E}^{n+1}]d\xi \right) \\ \varepsilon^{-1}\widetilde{\mathrm{curl}}\left(\int_0^1 [(1-\xi)\mathbf{H}^n + \xi \mathbf{H}^{n+1}]d\xi \right) \end{bmatrix}.
$$

Based on the variational equation, it can be proved that this algorithm preserves the
symplecticity and multi-symplecticity; see [27, 28, 158] and references therein for
more details.

(c) **Wavelet collocation algorithm:** based on the autocorrelation function θ of
the Daubechies scaling function (see e.g., [16]), we take the component E_1 as an
example and approximate it by the interpolation operator

$$
\mathscr{I} E_1(t, x, y, z) = \sum_{i=1}^{N_1}\sum_{j=1}^{N_2}\sum_{k=1}^{N_3} (E_1)_{i,j,k}\theta\left(\frac{N_1}{L_1}x - i\right)\theta\left(\frac{N_2}{L_2}y - j\right)\theta\left(\frac{N_3}{L_3}z - k\right),
$$

where N_1, N_2, and N_3 are the numbers of grid points of spatial domains $[0, L_1]$,
$[0, L_2]$, and $[0, L_3]$, respectively. Then plugging the interpolation operator into
(1.24), and integrating the semi-discrete system by the midpoint method in time,
we obtain the following multi-symplectic algorithm

$$
Id \otimes \mathbb{F}\frac{U^{n+1} - U^n}{\tau} + A_1 \otimes \mathbb{K}_1 U^{n+\frac{1}{2}} + A_2 \otimes \mathbb{K}_2 U^{n+\frac{1}{2}} + A_3 \otimes \mathbb{K}_3 U^{n+\frac{1}{2}} = 0,
$$

where

$$
U^n = \Big((E_1)^n_{1,1,1}, (E_1)^n_{2,1,1}, \ldots, (E_1)^n_{N_1,1,1}, (E_1)^n_{1,2,1}, \ldots, (E_1)^n_{N_1,2,1}, \ldots,
$$
$$
(E_1)^n_{N_1,N_2,N_3},
$$
$$
(E_2)^n_{1,1,1}, \ldots, (E_2)^n_{N_1,N_2,N_3}, \ldots, (E_3)^n_{N_1,N_2,N_3}, (H_1)^n_{1,1,1}, \ldots,
$$
$$
(H_3)^n_{N_1,N_2,N_3}\Big)^\top
$$

with $(E_s)^n_{i,j,k}$ and $(H_s)^n_{i,j,k}$, $s = 1, 2, 3$, $i = 1, \ldots, N_1$, $j = 1, \ldots, N_2$, $k = 1, \ldots, N_3$ being the approximations of $E_s(t^n, x_i, y_j, z_k)$ and $H_s(t^n, x_i, y_j, z_k)$,
respectively. Here $\mathbb{F}$, $\mathbb{K}_1$, $\mathbb{K}_2$, and $\mathbb{K}_3$ are skew-symmetric matrices given by (1.23);
A_1, A_2, and A_3 are three skew-symmetric first-order differentiation matrices in
x, y, z-directions, respectively. We refer to [182] and references therein for more
details of the symplectic and multi-symplectic wavelet collocation algorithms for
the deterministic Maxwell equations.

(d) Splitting FDTD algorithm: we rewrite the Maxwell equations (1.13) as

$$\frac{\partial}{\partial t} \begin{bmatrix} \varepsilon \mathbf{E} \\ \mu \mathbf{H} \end{bmatrix} = \begin{bmatrix} 0 & \nabla \times \\ -\nabla \times & 0 \end{bmatrix} \begin{bmatrix} \mathbf{E} \\ \mathbf{H} \end{bmatrix} =: A \begin{bmatrix} \mathbf{E} \\ \mathbf{H} \end{bmatrix}. \tag{1.27}$$

Then we split the operator A into

$$A = \underbrace{\begin{bmatrix} 0 & (\nabla \times)_+ \\ (\nabla \times)_+^\top & 0 \end{bmatrix}}_{A_+} + \underbrace{\begin{bmatrix} 0 & (\nabla \times)_- \\ (\nabla \times)_-^\top & 0 \end{bmatrix}}_{A_-}$$

with

$$(\nabla \times)_+ := \begin{bmatrix} 0 & 0 & \partial_y \\ \partial_z & 0 & 0 \\ 0 & \partial_x & 0 \end{bmatrix}, \quad (\nabla \times)_- := \begin{bmatrix} 0 & -\partial_z & 0 \\ 0 & 0 & -\partial_x \\ -\partial_y & 0 & 0 \end{bmatrix}.$$

Note that $(\nabla \times)_+^\top = -(\nabla \times)_-$ and $(\nabla \times)_-^\top = -(\nabla \times)_+$.

Another kind of splitting approach for the operator A is given by

$$A = \underbrace{\begin{bmatrix} 0 & (\nabla \times)_x \\ (\nabla \times)_x^\top & 0 \end{bmatrix}}_{A_x} + \underbrace{\begin{bmatrix} 0 & (\nabla \times)_y \\ (\nabla \times)_y^\top & 0 \end{bmatrix}}_{A_y} + \underbrace{\begin{bmatrix} 0 & (\nabla \times)_z \\ (\nabla \times)_z^\top & 0 \end{bmatrix}}_{A_z},$$

where

$$(\nabla \times)_x := \begin{bmatrix} 0 & 0 & 0 \\ 0 & 0 & -\partial_x \\ 0 & \partial_x & 0 \end{bmatrix}, \quad (\nabla \times)_y := \begin{bmatrix} 0 & 0 & \partial_y \\ 0 & 0 & 0 \\ -\partial_y & 0 & 0 \end{bmatrix}, \quad (\nabla \times)_z := \begin{bmatrix} 0 & -\partial_z & 0 \\ \partial_z & 0 & 0 \\ 0 & 0 & 0 \end{bmatrix}.$$

Note that $(\nabla \times)_\alpha^\top = -(\nabla \times)_\alpha, \alpha \in \{x, y, z\}$.

These two splitting formulas for the operator A lead to the following approximations of (1.27):

$$\begin{bmatrix} \sqrt{\varepsilon} \mathbf{E}(t_{n+1}) \\ \sqrt{\mu} \mathbf{H}(t_{n+1}) \end{bmatrix} \approx e^{cA_+\tau} e^{cA_-\tau} \begin{bmatrix} \sqrt{\varepsilon} \mathbf{E}(t_n) \\ \sqrt{\mu} \mathbf{H}(t_n) \end{bmatrix}$$

and

$$\begin{bmatrix} \sqrt{\varepsilon} \mathbf{E}(t_{n+1}) \\ \sqrt{\mu} \mathbf{H}(t_{n+1}) \end{bmatrix} \approx e^{cA_z\tau} e^{cA_y\tau} e^{cA_x\tau} \begin{bmatrix} \sqrt{\varepsilon} \mathbf{E}(t_n) \\ \sqrt{\mu} \mathbf{H}(t_n) \end{bmatrix},$$

where $c = 1/\sqrt{\varepsilon\mu}$. Using the Crank–Nicolson scheme to solve the one-dimensional subsystems which are obtained by the splitting of the operator A, one constructs two energy-preserving and unconditionally stable algorithms; for more details see [25, 26, 34, 36, 83, 181] and references therein.

(e) **Discontinuous Galerkin algorithm:** define a partition with cells denoted by $I_i \times J_j \times G_k = [x_{i-\frac{1}{2}}, x_{i+\frac{1}{2}}] \times [y_{j-\frac{1}{2}}, y_{j+\frac{1}{2}}] \times [z_{k-\frac{1}{2}}, z_{k+\frac{1}{2}}]$ of rectangular spatial domain D. We define the piecewise polynomial space $\mathbb{H}_{h,r}$ as follows

$$\mathbb{H}_{h,r} := \{v \in L^2(D) : v|_{I_i \times J_j \times G_k} \in P^r(I_i) \otimes P^r(J_j) \otimes P^r(G_k),$$

$$i = 1, \ldots, N_1; j = 1, \ldots, N_2; k = 1, \ldots, N_3\}^6$$

with $P^r(I)$ being the space of polynomials on I of degree up to $r \geq 1$ and h being the maximum diameter of I_i, J_j, and G_k for all i, j, k.

The dG algorithm for (1.24) is formulated as: finding a numerical solution $u_h \in \mathbb{H}_{h,r}$ such that for any test function $\boldsymbol{\varphi} \in \mathbb{H}_{h,r}$,

$$\int_{G_k} \int_{J_j} \int_{I_i} \mathbb{F}(u_h)_t \cdot \boldsymbol{\varphi} \mathrm{d}x \mathrm{d}y \mathrm{d}z - \int_{G_k} \int_{J_j} \left[\int_{I_i} \mathbb{K}_1 u_h \cdot \boldsymbol{\varphi}_x \mathrm{d}x - \left(\widehat{\mathbb{K}_1 u_h} \cdot \boldsymbol{\varphi}^- \right)_{i+\frac{1}{2},y,z} \right.$$

$$\left. + \left(\widehat{\mathbb{K}_1 u_h} \cdot \boldsymbol{\varphi}^+ \right)_{i-\frac{1}{2},y,z} \right] \mathrm{d}y \mathrm{d}z - \int_{G_k} \int_{I_i} \left[\int_{J_j} \mathbb{K}_2 u_h \cdot \boldsymbol{\varphi}_y \mathrm{d}y \right.$$

$$\left. - \left(\widehat{\mathbb{K}_2 u_h} \cdot \boldsymbol{\varphi}^- \right)_{x,j+\frac{1}{2},z} \right.$$

$$\left. + \left(\widehat{\mathbb{K}_2 u_h} \cdot \boldsymbol{\varphi}^+ \right)_{x,j-\frac{1}{2},z} \right] \mathrm{d}x \mathrm{d}z - \int_{J_j} \int_{I_i} \left[\int_{G_k} \mathbb{K}_3 u_h \cdot \boldsymbol{\varphi}_z \mathrm{d}z \right.$$

$$\left. - \left(\widehat{\mathbb{K}_3 u_h} \cdot \boldsymbol{\varphi}^- \right)_{x,y,k+\frac{1}{2}} \right.$$

$$\left. + \left(\widehat{\mathbb{K}_3 u_h} \cdot \boldsymbol{\varphi}^+ \right)_{x,y,k-\frac{1}{2}} \right] \mathrm{d}x \mathrm{d}y = 0,$$

where the hatted terms $\widehat{\mathbb{K}_\ell u_h}$, $\ell = 1, 2, 3$, are the numerical fluxes defined on the element interfaces, and φ^+, φ^- represent the function limit of φ from right and left, respectively. It can be shown that this method can simultaneously preserve the multi-symplectic structure and the energy conservation law; see [159] for dG algorithms of a general class of Hamiltonian partial differential equations.

(f) **Other algorithms:** in order to keep the symmetry in the temporal direction and improve the convergence order, [35] proposed a symmetric energy-preserving algorithm by distinguishing the time steps between even and odd time steps; [75] investigated an alternating direction implicit method which can preserve the energy of the Maxwell equations with currents, charges, and conductivity; [27] proposed two energy-preserving algorithms by discretizing the Maxwell equations with the Fourier pseudo-spectral method in the spatial direction and the averaged vector field algorithm in the temporal direction.

1.2 Stochastic Maxwell Equations

In practical circumstances and many applications, there is uncertainty concerning either the externally imposed sources or the nature of the medium under consideration, to quote Varadhan (see [169]), *"The world we live in has never been very predictable, and randomness has always been part of our lives"*. In this regard, the stochastic Maxwell equations are proposed to strengthen the correspondence between some theoretical results and real-life situations, to better understand the role of thermodynamic fluctuations presented in electromagnetic fields, and to gain a deeper insight regarding the propagation of electromagnetic waves in complex media.

Taking the thermal fluctuations into account, it is known that there are the following three entirely equivalent ways to describe the effects caused by the randomness (see [33]): (i) introducing the randomness directly in Newton's equation of motion, (ii) adding a random term to the displacement field $\mathbf{D}$, and (iii) adding a random source term to the original current density $\mathbf{J}_e$, in the Maxwell equations.

We start with the first way by introducing randomness into Newton's equation of motion. Consider a particle of mass m and charge e immersed in a fluid of temperature T, under the action of an additional external electric field $\mathbf{E}$. According to the classical Newton's second law of motion, the equation which characterizes the dynamics of the particle can be written as

$$\frac{d^2\mathbf{x}}{dt^2} + \alpha\frac{d\mathbf{x}}{dt} + \omega_0^2\mathbf{x} = e\frac{\mathbf{E}}{m}, \tag{1.28}$$

where $\mathbf{x}$ is the position of the particle, α is the frictional coefficient expressing the energy loss in the system, and ω_0 is the natural resonant frequency of the system. Converting (1.28) to polarization via an electric polarization $\mathbf{P} = ne\mathbf{x}$, we have

$$\frac{d^2\mathbf{P}}{dt^2} + \alpha\frac{d\mathbf{P}}{dt} + \omega_0^2\mathbf{P} = v\mathbf{E}, \tag{1.29}$$

where $v = ne^2/m$ with n being the density of the positive charge.

Following the Langevin approach, we introduce a random term $\mathbf{K} : (t, \mathbf{x}) \mapsto \mathbf{K}(t, \mathbf{x})$ representing thermal fluctuations on the electromagnetic field to the right-hand side of (1.29). The response of the system then reads as

$$\frac{d^2\mathbf{P}}{dt^2} + \alpha\frac{d\mathbf{P}}{dt} + \omega_0^2\mathbf{P} = v\mathbf{E} + \mathbf{K}. \tag{1.30}$$

Using a time-harmonic variation of e^{-iwt} with an excitation angular frequency w, it follows from (1.30) that the corresponding expression for the polarization P reads

$$\mathbf{P}(w, \mathbf{x}) = \frac{v\mathbf{E}(w, \mathbf{x})}{\omega_0^2 - w^2 - i\alpha w} + \frac{\mathbf{K}(w, \mathbf{x})}{\omega_0^2 - w^2 - i\alpha w}.$$

It follows from the constitutive relation (1.5) that the electric displacement $\mathbf{D}$ consists of an external field $\varepsilon_0 \mathbf{E}$, a deterministic polarization-induced component $v\mathbf{E}(w, \mathbf{x})/(\omega_0^2 - w^2 - i\alpha w)$, and a random component $\mathbf{Q}_e(w, \mathbf{x}) = \mathbf{K}(w, \mathbf{x})/(\omega_0^2 - w^2 - i\alpha w)$.

Therefore, transforming $\mathbf{P}$ and $\mathbf{Q}_e$ into the time domain, Ampère's circuital law (1.2) becomes

$$\nabla \times \mathbf{H}(t, \mathbf{x}) = \mathbf{J}_e(t, \mathbf{x}) + \partial_t \Big(\mathbf{D}(t, \mathbf{x}) + \mathbf{Q}_e(t, \mathbf{x}) \Big). \tag{1.31}$$

Alternatively, instead of introducing randomness to the electric displacement $\mathbf{D}$, we could add a random term $\mathfrak{J}_e^r = \partial_t \mathbf{Q}_e$ to the current density $\mathbf{J}_e$, and the end result would be completely same. Thus, we have shown that the three ways (i), (ii), and (iii) are entirely equivalent. In the following chapters, we adopt the third approach to analyze the effect of randomness on electromagnetic fields.

Mathematical symmetry and beauty have drawn important considerations in twentieth-century physics, both in creating new physical theories and in elegantly connecting symmetry with conservation laws (see [18]). Taking this symmetry into account, we will focus on the following stochastic Maxwell equations

$$\partial_t \mathbf{D}(t, \mathbf{x}) - \nabla \times \mathbf{H}(t, \mathbf{x}) = -\mathbf{J}_e(t, \mathbf{x}) - \mathbf{J}_e^r(t, \mathbf{x})\gamma, \tag{1.32}$$

$$\partial_t \mathbf{B}(t, \mathbf{x}) + \nabla \times \mathbf{E}(t, \mathbf{x}) = -\mathbf{J}_m(t, \mathbf{x}) - \mathbf{J}_m^r(t, \mathbf{x})\gamma, \tag{1.33}$$

together with the constitutive relations (1.6). Here, $\mathbf{J}_m$ is the magnetic current density, γ denotes the noise (see Sect. 1.2.1 for details), $\mathbf{J}_e^r(t, \mathbf{x})\gamma$ corresponds to the random term $\partial_t \mathbf{Q}_e$ in (1.31), and $\mathbf{J}_m^r(t, \mathbf{x})\gamma$ is the symmetric part in the magnetic term. We say that the stochastic Maxwell equations are driven by additive noise if $\mathbf{J}_e^r$ and $\mathbf{J}_m^r$ do not depend on the electromagnetic field, while the equations are driven by multiplicative noise if $\mathbf{J}_e^r$ and $\mathbf{J}_m^r$ are some functions of the electromagnetic field (linear or nonlinear).

1.2.1 Formulation of Noises

This section summarizes and provides the mathematical description of several kinds of noises γ. We classify γ into colored noise or white noise based on the covariance function

$$\mathbb{E}[\gamma(t, \mathbf{x})\gamma(s, \mathbf{y})] = B(t, s)K(\mathbf{x}, \mathbf{y}), \quad t, s \geq 0, \quad \mathbf{x}, \mathbf{y} \in D,$$

where $D \subset \mathbb{R}^d$ $(d \geq 1)$ is connected and bounded with a Lipschitz boundary ∂D. When B is a Hilbert–Schmidt kernel, that is, $\int_0^{+\infty} \int_0^{+\infty} |B(t, s)|^2 dt ds < \infty$ and has the finite trace, the noise is said to be colored in time. When $B(t, s) = \delta(t - s)$ with $\delta(\cdot)$ being the Dirac delta function, the noise is said to be white in time. The situation is similar for the spatial kernel K. We refer to e.g., [88] for more details about noises.

(i) **Spatial and temporal colored noise**

The generalized infinite-dimensional temporal colored noise is given by

$$\gamma(t, \mathbf{x}) = \sum_{k \in \mathbb{N}} \sqrt{\eta_k} e_k(\mathbf{x}) z_k(t), \quad t \geq 0, \quad \mathbf{x} \in D. \tag{1.34}$$

Here $\{e_k, \eta_k\}_{k \in \mathbb{N}}$ is the orthonormal eigenpairs induced by the covariance operator Q,

$$(Q e_k)(\mathbf{x}) = \int_D K(\mathbf{x}, \mathbf{y}) e_k(\mathbf{y}) d\mathbf{y} = \eta_k e_k(\mathbf{x}), \quad \mathbf{x} \in D, \quad k \in \mathbb{N} \tag{1.35}$$

with the decreasing ordered eigenvalues $\eta_1 \geq \eta_2 \geq \ldots > 0$ and $\sum_{k \in \mathbb{N}} \eta_k < \infty$. And $\{z_k\}_{k \in \mathbb{N}}$ is a sequence of independent stochastic processes with $\mathbb{E}[z_k(t)] = 0$ and $\mathbb{E}[z_k(t)z_k(s)] = B(t, s)$, defined by

$$dz_k(t) = a_k(z_k, t)dt + b_k(z_k, t)d\beta_k(t), \quad t \geq 0, \quad k \in \mathbb{N},$$

where $\{\beta_k\}_{k \in \mathbb{N}}$ is a sequence of independent standard Brownian motions, and $\{a_k(\cdot, t)\}_{k \in \mathbb{N}}$ and $\{b_k(\cdot, t)\}_{k \in \mathbb{N}}$ are Lipschitz continuous for any $t \geq 0$, which guarantees the existence and uniqueness of the solution.

The colored noise in space is determined by the spatial kernel function $K(\cdot, \cdot)$. Note that when Q is of finite trace, i.e., $\text{Tr}(Q) = \sum_{k \in \mathbb{N}} \eta_k < \infty$, it follows from (1.34) that the covariance function of the random field $\gamma(t, \mathbf{x})$ reads

$$\begin{aligned}
\mathbb{E}[\gamma(t, \mathbf{x})\gamma(s, \mathbf{y})] &= \mathbb{E}\left[\left(\sum_{k \in \mathbb{N}} \sqrt{\eta_k} e_k(\mathbf{x}) z_k(t)\right)\left(\sum_{k \in \mathbb{N}} \sqrt{\eta_k} e_k(\mathbf{y}) z_k(s)\right)\right] \\
&= \mathbb{E}\left[\sum_{k \in \mathbb{N}} \eta_k e_k(\mathbf{x}) e_k(\mathbf{y}) z_k(t) z_k(s)\right] \\
&= \sum_{k \in \mathbb{N}} \eta_k e_k(\mathbf{x}) e_k(\mathbf{y}) \mathbb{E}\left[z_k(t) z_k(s)\right] \\
&= B(t, s) \sum_{k \in \mathbb{N}} \eta_k e_k(\mathbf{x}) e_k(\mathbf{y}) = B(t, s) K(\mathbf{x}, \mathbf{y})
\end{aligned}$$

due to Mercer's theorem (see e.g., [138]). In this case, the representation of (1.34) is a spatial and temporal colored noise.

(ii) Spatial white and temporal colored noise

When $\mathrm{Tr}(Q) = \infty$, there exists an orthonormal basis $\{e_k\}_{k \in \mathbb{N}}$ such that $K(\mathbf{x}, \mathbf{y}) = \sum_{k \in \mathbb{N}} \eta_k e_k(\mathbf{x}) e_k(\mathbf{y}) = \delta(\mathbf{x} - \mathbf{y})$, then

$$\mathbb{E}[\gamma(t, \mathbf{x})\gamma(s, \mathbf{y})] = B(t, s)\delta(\mathbf{x} - \mathbf{y}), \quad t, s \geq 0, \quad \mathbf{x}, \mathbf{y} \in D.$$

In this case, the spatial white and temporal colored noise is given by

$$\gamma(t, \mathbf{x}) = \sum_{k \in \mathbb{N}} e_k(\mathbf{x}) z_k(t), \quad t \geq 0, \quad \mathbf{x} \in D. \tag{1.36}$$

(iii) Spatial colored and temporal white noise

If the kernel function $K(\cdot, \cdot)$ of γ is a symmetric Hilbert–Schmidt kernel with a finite trace, then the covariance operator Q defined by (1.35) is a compact, semi-positive, and self-adjoint operator with $\mathrm{Tr}(Q) < \infty$. Hence the generalized spatial colored and temporal white noise can be represented as

$$\gamma(t, \mathbf{x}) = \frac{\mathrm{d}}{\mathrm{d}t} W(t, \mathbf{x}), \quad t \geq 0, \quad \mathbf{x} \in D, \tag{1.37}$$

where W has the Karhunen–Loève expansion

$$W(t, \mathbf{x}) = \sum_{k \in \mathbb{N}} \sqrt{\eta_k} e_k(\mathbf{x}) \beta_k(t), \quad t \geq 0, \quad \mathbf{x} \in D.$$

In fact, for a.e. ω, the sample path $t \mapsto W(t, \omega)$ is nowhere differentiable. Thus γ does not really exist and we only make use of the above symbolic expression. We would like to stress that the expression of the random source in equations (1.32) and (1.33) is purely formal.

(iv) Spatial and temporal white noise

Similar to the Hilbert–Schmidt kernel function, for the spatial and temporal white noise we can introduce the identity operator Id by the Dirac delta function which has the following fundamental property in the distributional sense

$$\big(Id\, e_k\big)(\mathbf{x}) := \int_D \delta(\mathbf{x} - \mathbf{y}) e_k(\mathbf{y}) \mathrm{d}\mathbf{y} = e_k(\mathbf{x}), \quad e_k \in L^2(D), \quad \mathbf{x} \in D, \quad k \in \mathbb{N}.$$

This kind of noise can be formally given by

$$\gamma(t, \mathbf{x}) = \frac{d}{dt} W(t, \mathbf{x}), \quad t \geq 0, \quad \mathbf{x} \in D, \tag{1.38}$$

where W is the cylindrical Wiener process

$$W(t, \mathbf{x}) = \sum_{k \in \mathbb{N}} e_k(\mathbf{x}) \beta_k(t), \quad t \geq 0, \quad \mathbf{x} \in D.$$

1.2.2 Two Polarizations

The polarization of electromagnetic waves is defined by the orientation of their electromagnetic fields relative to the plane of incidence. If we assume that the incident waves lie in the (x, y)-plane, that is the stochastic Maxwell equations are homogeneous in z-direction, then all z-derivatives in (1.32) and (1.33) will vanish. This gives us two decoupled two-dimensional stochastic Maxwell equations.

The first one is transverse electric (TE) polarization where the associated equations contain components E_1, E_2, H_3, i.e.,

$$\mathbf{E} = (E_1(t, x, y), E_2(t, x, y), 0)^\top, \quad \mathbf{H} = (0, 0, H_3(t, x, y))^\top.$$

And we have $\mathbf{J}_e = (J_{e1}, J_{e2}, 0)^\top$, $\mathbf{J}_e^r = (J_{e1}^r, J_{e2}^r, 0)^\top$, $\mathbf{J}_m = (0, 0, J_{m3})^\top$, and $\mathbf{J}_m^r = (0, 0, J_{m3}^r)^\top$. It describes the propagation of the electromagnetic waves where the electric field lies in the plane of propagation. Therefore, the stochastic Maxwell equations (1.32)–(1.33) with constitutive relations (1.6) become

$$\begin{cases} \varepsilon \partial_t E_1 = \partial_y H_3 - J_{e1} - J_{e1}^r \gamma, \\ \varepsilon \partial_t E_2 = -\partial_x H_3 - J_{e2} - J_{e2}^r \gamma, \\ \mu \partial_t H_3 = \partial_y E_1 - \partial_x E_2 - J_{m3} - J_{m3}^r \gamma. \end{cases} \tag{1.39}$$

In this case, the PEC boundary condition (1.12) on the boundary ∂D of the rectangle domain $D = (a_1^-, a_1^+) \times (a_2^-, a_2^+)$ can be recast as

$$E_1 = E_2 = 0, \quad \text{on} \quad \partial D.$$

The second case is transverse magnetic (TM) polarization. The set of stochastic Maxwell equations contains components H_1, H_2, E_3, i.e.,

$$\mathbf{E} = (0, 0, E_3(t, x, y))^\top, \quad \mathbf{H} = (H_1(t, x, y), H_2(t, x, y), 0)^\top.$$

And we have $\mathbf{J}_e = (0, 0, J_{e3})^\top$, $\mathbf{J}_e^r = (0, 0, J_{e3}^r)^\top$, $\mathbf{J}_m = (J_{m1}, J_{m2}, 0)^\top$, and $\mathbf{J}_m^r = (J_{m1}^r, J_{m2}^r, 0)^\top$. It describes the propagation of the electromagnetic waves where the electric field is perpendicular to the plane of propagation. Then (1.32)–(1.33) with constitutive relations (1.6) become

$$
\begin{cases}
\varepsilon \partial_t E_3 = \partial_x H_2 - \partial_y H_1 - J_{e3} - J_{e3}^r \gamma, \\
\mu \partial_t H_1 = -\partial_y E_3 - J_{m1} - J_{m1}^r \gamma, \\
\mu \partial_t H_2 = \partial_x E_3 - J_{m2} - J_{m2}^r \gamma.
\end{cases}
\tag{1.40}
$$

In this case, the PEC boundary condition (1.12) on the boundary ∂D of the rectangle domain $D = (a_1^-, a_1^+) \times (a_2^-, a_2^+)$ can be recast as

$$
H_1 = H_2 = 0, \quad \text{on} \quad \partial D.
$$

Example 1.1 In [38, 102], the authors studied the following two-dimensional stochastic Maxwell equations with additive noise in a lossless medium

$$
\begin{cases}
\partial_t E_3 = \partial_x H_2 - \partial_y H_1 - \lambda_1 \gamma, \\
\partial_t H_1 = -\partial_y E_3 + \lambda_2 \gamma, \\
\partial_t H_2 = \partial_x E_3 + \lambda_2 \gamma,
\end{cases}
$$

where $\lambda_1, \lambda_2 > 0$ are real numbers representing the scales of the noise, and γ is a spatial and temporal colored noise.

1.2.3   Time-Harmonic Stochastic Maxwell Equations

In the case of a monochromatic wave, assume that all fields are of the form

$$
\mathbf{E}(t, \mathbf{x}) = \mathbf{E}(\mathbf{x})e^{-ikt}, \quad \mathbf{H}(t, \mathbf{x}) = \mathbf{H}(\mathbf{x})e^{-ikt},
\tag{1.41}
$$

where $k > 0$ is the wavenumber and i is the imaginary unit. In addition, we assume that $\mathbf{J}_e(t, \mathbf{x}) = \mathbf{J}_e(\mathbf{x})e^{-ikt}$, $(\mathbf{J}_e^r \gamma)(t, \mathbf{x}) = \mathbf{J}_e^r(\mathbf{x})\gamma(\mathbf{x})e^{-ikt}$, $\mathbf{J}_m(t, \mathbf{x}) = \mathbf{J}_m(\mathbf{x})e^{-ikt}$, and $(\mathbf{J}_m^r \gamma)(t, \mathbf{x}) = \mathbf{J}_m^r(\mathbf{x})\gamma(\mathbf{x})e^{-ikt}$. Plugging (1.41) into (1.32)–(1.33) with constitutive relations (1.6), we have the time-harmonic stochastic Maxwell equations

$$
\begin{aligned}
-ik\varepsilon \mathbf{E}(\mathbf{x}) - \nabla \times \mathbf{H}(\mathbf{x}) &= -\mathbf{J}_e(\mathbf{x}) - \mathbf{J}_e^r(\mathbf{x})\gamma(\mathbf{x}), \\
-ik\mu \mathbf{H}(\mathbf{x}) + \nabla \times \mathbf{E}(\mathbf{x}) &= -\mathbf{J}_m(\mathbf{x}) - \mathbf{J}_m^r(\mathbf{x})\gamma(\mathbf{x}).
\end{aligned}
\tag{1.42}
$$

Eliminating $\mathbf{H}$ yields

$$\nabla \times \left(\mu^{-1}\nabla \times \mathbf{E}\right) - k^2 \varepsilon \mathbf{E} = \mathrm{i}k \left(\mathbf{J}_e + \mathbf{J}_e^r \gamma\right) - \nabla \times \left(\mu^{-1}\mathbf{J}_m + \mu^{-1}\mathbf{J}_m^r \gamma\right),$$

while eliminating $\mathbf{E}$ gives analogously

$$\nabla \times \left(\varepsilon^{-1}\nabla \times \mathbf{H}\right) - k^2 \mu \mathbf{H} = \mathrm{i}k\left(\mathbf{J}_m + \mathbf{J}_m^r \gamma\right) + \nabla \times \left(\varepsilon^{-1}\mathbf{J}_e + \varepsilon^{-1}\mathbf{J}_e^r \gamma\right).$$

Example 1.2 In [179, Sect. 3.1], the author investigated the following time-harmonic stochastic Maxwell equations with colored noise in a simple medium

$$\begin{cases} \nabla \times (\nabla \times \mathbf{E}) - k^2 \mathbf{E} = \mathrm{i}k \left(\mathbf{J}_e + \gamma_2\right), & \text{in } D, \\ \nabla \cdot \mathbf{E} = \rho_e + \gamma_1, & \text{in } D, \\ \mathbf{n} \times \mathbf{E} = 0, & \text{on } \partial D, \end{cases}$$

where γ_1 and γ_2 denote spatial colored noises.

1.3 Applications of Stochastic Maxwell Equations

Since 1979, there has been a tremendously growing interest in the study of the stochastic Maxwell equations and their applications in areas ranging from the inverse random source problem, thermal radiation, integrated circuit technology, metamaterial, and optical communication. At the end of this chapter, we present two related research on the stochastic Maxwell equations.

1.3.1 Inverse Random Source Problems

Inverse source problems are to infer the information of the radiating sources by using the measured wave fields generated by unknown sources. These problems have significant applications in many scientific areas such as antenna synthesis and design, biomedical engineering, medical imaging, and optical tomography. In particular, the inverse source problem modeled by the Maxwell equations is an important research subject.

In many practical situations, due to the unpredictability of the surrounding environment, the source of the system, in general, randomly varies in time and space. These random fluctuations are important in a variety of practical applications. Geophysicists are interested in the use of electromagnetic wave random fluctuations that occur due to the propagation through planetary atmospheres in order to remotely determine their dynamic characteristics. Physicians use electric or magnetic measurements with random fluctuation on the surface of the human head to infer the source currents in the brain which produce these measured fields. Radar engineers need to concern themselves with clutter echoes that follow some probability distribution produced by storms, rain, snow, or hail so that the radar target can be detected and identified accurately. In recent years, there has been rapid progress in the theoretical understanding and the numerical treatment of electromagnetic inverse random source problems; see [12, 82, 124, 125] and references therein for relevant studies.

More precisely, all of these inverse random source problems of determining a source are characterized by the microcorrelation strength or the statistical properties of the random source based on the stochastic Maxwell equations. For example, in geological prospecting, the random source problem can be depicted by the following time-harmonic stochastic Maxwell equations

$$\nabla \times \mathbf{E} = ik\mathbf{H}, \quad \nabla \times \mathbf{H} = -ik\mathbf{E} + \mathfrak{J}^r, \quad \text{in } \mathbb{R}^d \setminus D,$$

where $\mathfrak{J}^r$ is a random field and the impenetrable material D is a polyhedral scatterer in $\mathbb{R}^d$ ($d = 2, 3$). In medical imaging, the stochastic Helmholtz equation

$$\Delta \mathbf{E} + (k^2 + ik\beta)\mathbf{E} = \mathfrak{J}^r, \quad \text{in } D \subset \mathbb{R}^3$$

can be used to diagnose the body or the extent of the pathological tissue, where the attenuation coefficient $\beta \geq 0$ describes the electrical conductivity of the intracellular current (dendrite).

The solution theory and the structure-preserving algorithms for the stochastic Maxwell equations discussed in this monograph may provide some tools for the establishment of the well-posedness theory for inverse random source problems, the construction of highly efficient and stable reconstruction algorithms, etc.

1.3.2 Thermal Radiation

Thermal radiation, as a ubiquitous physical phenomenon, plays an important role in various research fields of science and engineering. The traditional understanding of radiation heat transfer inside a semitransparent medium relies on the radiative transfer equation, considering emission, absorption, and scattering. However, these phenomenological approaches do not fully account for the origin of thermal emission and break down when the interference and diffraction roles of waves become increasingly important. According to the fluctuation-dissipation theorem, thermal emission originates from the fluctuating currents induced by the random thermal motion of charges, known as thermally induced dipoles. In order to predict and understand thermal radiation, fluctuational electrodynamics needs to be used, leading to the stochastic Maxwell equations.

For example, if an object is at temperature T which is greater than absolute zero, thermal agitation causes a chaotic motion of charged particles inside the body. The random thermal motion of the charges generates in turn a fluctuating electromagnetic field. On a macroscopic level, the field fluctuations are due to space-time thermal fluctuations of charges and currents in the physically infinitesimal volume elements of bodies. In other words, two random extraneous current and charge density terms $\mathfrak{J}_e^r$ and $\mathfrak{J}_m^r$, which cause thermal fluctuations of the field, should be introduced in Faraday's law and Ampère's law:

$$\nabla \times \mathbf{H} = \frac{1}{c}\partial_t \mathbf{D} + \frac{4\pi}{c}\mathbf{J}_e + \frac{4\pi}{c}\mathfrak{J}_e^r, \quad \nabla \times \mathbf{E} = -\frac{1}{c}\partial_t \mathbf{B} - \frac{4\pi}{c}\mathfrak{J}_m^r$$

with the corresponding constitutive relations. Here, c is the speed of propagation of the electromagnetic wave. Particularly, as nanotechnology advances rapidly, there is an increasing demand for understanding the mechanism of radiation heat transfer in the processing and diagnostics of nanomaterials. The stochastic Maxwell equations provide a proper model for solving the radiation heat transfer problem. By investigating the fluctuation-dissipation theorem of the stochastic Maxwell equations, the bridge between the strength of the fluctuation of the current density $\mathfrak{J}_e^r$ and the local temperature of the emitting body could be built. There is extensive literature on this topic. We refer to [81, 144, 149, 155, 173] and references therein.

By applying some structure-preserving algorithms developed in this monograph to the considered model, we can give a better interpretation of some interesting physical phenomena and understand the physical mechanisms of thermal fluctuation more clearly.

Summary and Outlook

This chapter gives a brief introduction to the deterministic Maxwell equations and the origin of the stochastic Maxwell equations. In addition, several mathematical formulations of noise which provide preliminaries to theoretically and numerically analyze the stochastic Maxwell equations are discussed. Finally, some applications of the stochastic Maxwell equations are presented.

As is known, the deterministic Maxwell equations, as an important Hamiltonian system, possess certain intrinsic structures and properties. More precisely, the phase flow of the deterministic Maxwell equations preserves both the symplectic and multi-symplectic structures. In addition, the solution of the deterministic Maxwell equations satisfies the charge and divergence conservation laws. Various structure-preserving algorithms have been developed, analyzed, and tested in order to inherit the structures and properties of the original equations. Taking stochasticity into account, [112] investigates the theory of stochastic Hamiltonian partial differential equations, whose phase flow preserves the stochastic multi-symplectic structure.

And a stochastic multi-symplectic algorithm is developed in [112]. For the study of the stochastic Maxwell equations, some natural and important questions arise:

(i) What are the intrinsic properties, in the aspects of the geometric structure, evolution laws of physical quantities, and dynamical properties, of the stochastic Maxwell equations?
(ii) How to construct and analyze numerical algorithms which can preserve those intrinsic properties of the stochastic Maxwell equations?

We intend to answer these problems in the following chapters.

In recent years, the stochastic Maxwell equations have been applied to many other areas and there is also a lot of work related to their application and numerical analysis. For example, in the integrated circuit area, a technique based on the model order reduction to the stochastic Maxwell equations has been proposed for the simulation of a coplanar waveguide with dielectric overlay (see e.g., [3, 4, 13]); in the metamaterial area, researchers proposed some innovative methodologies based on the stochastic Maxwell equations to generate and optimize random metamaterial configurations, including the stochastic collocation method (see [126]) and the multi-element probabilistic collocation method (see [166]); in the optical soliton communication area, [122] investigated the propagation of ultra-short short solitons in a cubic nonlinear medium modeled by the nonlinear Maxwell equations with stochastic variations of media, and [132] developed an accurate coupled local-mode equation for ultra-short optical pulses based on the stochastic Maxwell equations.

Chapter 2
Solution Theory of Stochastic Maxwell Equations

This chapter is devoted to the solution theory of the stochastic Maxwell equations, including the well-posedness, the regularity of the solution, as well as the regular dependence of the solution on the initial datum.

The outline of this chapter is as follows. In Sect. 2.1, we provide a succinct introduction to function spaces and the Maxwell operator associated with the stochastic Maxwell equations and rewrite the equations in the formulation of the stochastic evolution equation for convenience. Then in Sect. 2.2, we show the existence and uniqueness of the solution for the stochastic Maxwell equations with the drift term being either globally Lipschitz or non-globally Lipschitz continuous. After that, we investigate the regularity of the solution in Sect. 2.3. Finally, in Sect. 2.4, the differentiability with respect to the initial datum of the solution for the stochastic Maxwell equations is discussed.

2.1 Preliminaries

In this section, we introduce several function spaces and present the properties of the Maxwell operator associated with the stochastic Maxwell equations, and give the formulation of the stochastic evolution equation by introducing two Nemytskij operators.

© The Author(s), under exclusive license to Springer Nature Singapore Pte Ltd. 2023
C. Chen et al., *Numerical Approximations of Stochastic Maxwell Equations*,
Lecture Notes in Mathematics 2341, https://doi.org/10.1007/978-981-99-6686-8_2

We focus on the study of the following stochastic Maxwell equations in the Itô sense:

$$\begin{cases} \varepsilon d\mathbf{E} = \left[\nabla \times \mathbf{H} - \mathbf{J}_e(t, \mathbf{x}, \mathbf{E}, \mathbf{H})\right]dt - \mathbf{J}_e^r(t, \mathbf{x}, \mathbf{E}, \mathbf{H})dW, & (t, \mathbf{x}) \in (0, T] \times D, \\ \mu d\mathbf{H} = \left[-\nabla \times \mathbf{E} - \mathbf{J}_m(t, \mathbf{x}, \mathbf{E}, \mathbf{H})\right]dt - \mathbf{J}_m^r(t, \mathbf{x}, \mathbf{E}, \mathbf{H})dW, & (t, \mathbf{x}) \in (0, T] \times D, \\ \mathbf{E}(0, \mathbf{x}) = \mathbf{E}_0(\mathbf{x}), \ \mathbf{H}(0, \mathbf{x}) = \mathbf{H}_0(\mathbf{x}), & \mathbf{x} \in D, \\ \mathbf{n} \times \mathbf{E} = 0, & t \in [0, T], \ \mathbf{x} \in \partial D, \end{cases}$$
$$(2.1)$$

where $D \subset \mathbb{R}^3$ is an open, bounded, and Lipschitz domain with boundary ∂D, and $\mathbf{n}$ is the unit outward normal vector on ∂D. Suppose that the medium is isotropic, which implies that ε and μ are real-valued scalar functions, i.e., ε, μ : $D \to \mathbb{R}$. Here, W is a Q-Wiener process defined on a filtered probability space $(\Omega, \mathscr{F}, \{\mathscr{F}_t\}_{0 \le t \le T}, \mathbb{P})$, where Q is a symmetric, nonnegative operator with a finite trace on $U := L^2(D)$. More precisely, W has the following Karhunen–Loève expansion

$$W(t, \mathbf{x}) = \sum_{j \in \mathbb{N}} Q^{\frac{1}{2}} e_j(\mathbf{x}) \beta_j(t), \quad t \in [0, T], \quad \mathbf{x} \in D,$$

where $\{\beta_j\}_{j \in \mathbb{N}}$ is a family of independent standard Brownian motions and $\{e_j\}_{j \in \mathbb{N}}$ is an orthonormal basis of U.

Throughout this monograph, we assume that the coefficients ε and μ satisfy the following assumption.

Assumption 2.1 *Assume that the electric permittivity $\varepsilon : D \to \mathbb{R}$ and the magnetic permeability $\mu : D \to \mathbb{R}$ satisfy*

$$\varepsilon, \ \mu \in L^\infty(D), \quad \varepsilon, \ \mu \ge \delta > 0 \tag{2.2}$$

with δ being a constant.

The basic Hilbert space we work with is $\mathbb{H} := L^2(D)^3 \times L^2(D)^3$ with the weighted inner product defined by

$$\left\langle \begin{bmatrix} \mathbf{E}_1 \\ \mathbf{H}_1 \end{bmatrix}, \begin{bmatrix} \mathbf{E}_2 \\ \mathbf{H}_2 \end{bmatrix} \right\rangle_{\mathbb{H}} := \int_D (\varepsilon \mathbf{E}_1 \cdot \mathbf{E}_2 + \mu \mathbf{H}_1 \cdot \mathbf{H}_2) d\mathbf{x},$$

which is equivalent to the standard inner product in $\mathbb{H}$ according to Assumption 2.1. The norm

$$\left\| \begin{bmatrix} \mathbf{E} \\ \mathbf{H} \end{bmatrix} \right\|_{\mathbb{H}} = \left(\int_D (\varepsilon |\mathbf{E}|^2 + \mu |\mathbf{H}|^2) d\mathbf{x} \right)^{1/2},$$

induced by the above inner product, corresponds to the electromagnetic energy of the physical system.

In addition, definitions and properties of important function spaces related to the divergence and curl operators are briefly introduced below, which play a fundamental role in the regularity analysis of the stochastic Maxwell equations.

The space $H(\text{div}, D)$ is defined by

$$H(\text{div}, D) := \left\{ v \in L^2(D)^3 : \nabla \cdot v \in U \right\},$$

which is a Hilbert space under the inner product

$$\langle u, v \rangle_{H(\text{div},D)} := \langle u, v \rangle_{L^2(D)^3} + \langle \nabla \cdot u, \nabla \cdot v \rangle_U$$

and the induced norm

$$\|u\|_{H(\text{div},D)} := \left(\|u\|^2_{L^2(D)^3} + \|\nabla \cdot u\|^2_U \right)^{1/2}.$$

The space $H(\text{div}, D)$ can be characterized as the closure of $C^\infty(\overline{D})^3$ with respect to $\| \cdot \|_{H(\text{div},D)}$. Define the subspace $H_0(\text{div}, D)$ of $H(\text{div}, D)$ as the closure of $C_0^\infty(\overline{D})^3$ with respect to $\| \cdot \|_{H(\text{div},D)}$, which can also be expressed as

$$H_0(\text{div}, D) = \left\{ v \in H(\text{div}, D) : \mathbf{n} \cdot v|_{\partial D} = 0 \right\}.$$

The space $H(\text{curl}, D)$ is defined by

$$H(\text{curl}, D) := \left\{ v \in L^2(D)^3 : \nabla \times v \in L^2(D)^3 \right\},$$

which is a Hilbert space under the inner product

$$\langle u, v \rangle_{H(\text{curl},D)} := \langle u, v \rangle_{L^2(D)^3} + \langle \nabla \times u, \nabla \times v \rangle_{L^2(D)^3}$$

and the norm

$$\|u\|_{H(\text{curl},D)} := \left(\|u\|^2_{L^2(D)^3} + \|\nabla \times u\|^2_{L^2(D)^3} \right)^{1/2}.$$

This space can be characterized as the closure of $C^\infty(\overline{D})^3$ with respect to $\| \cdot \|_{H(\text{curl},D)}$. Define the subspace $H_0(\text{curl}, D)$ of $H(\text{curl}, D)$ as the closure of $C_0^\infty(\overline{D})^3$ with respect to $\| \cdot \|_{H(\text{curl},D)}$, which can be expressed as

$$H_0(\text{curl}, D) = \left\{ v \in H(\text{curl}, D) : \mathbf{n} \times v|_{\partial D} = 0 \right\}.$$

Note that we have the following integration by parts formula:

$$\int_D (\nabla \times u) \cdot v \, d\mathbf{x} = \int_D u \cdot (\nabla \times v) \, d\mathbf{x} \tag{2.3}$$

for all $u \in H(\text{curl}, D)$ and $v \in H_0(\text{curl}, D)$.

Remark 2.1 In general, $u \in L^2(D)^3$ is not differentiable and does not possess a 'classical curl'. We clarify that $\nabla \times u$ always denotes the following variational curl of u, i.e., there exists $v \in L^2(D)^3$ such that

$$\int_D u \cdot \nabla \times \phi \, d\mathbf{x} = \int_D v \cdot \phi \, d\mathbf{x} \quad \forall \phi \in C_0^\infty(D)^3.$$

In this case, we write $\nabla \times u = v$. Similarly, we denote by $\nabla \cdot u$ the variational divergence of u, i.e., there exists $w \in U$ such that

$$\int_D u \cdot \nabla \psi \, d\mathbf{x} = - \int_D w \psi \, d\mathbf{x} \quad \forall \psi \in C_0^\infty(D).$$

In this case, we write $\nabla \cdot u = w$.

Based on the above preliminaries on function spaces, we now introduce the Maxwell operator

$$M = \begin{bmatrix} 0 & \varepsilon^{-1}\nabla\times \\ -\mu^{-1}\nabla\times & 0 \end{bmatrix} \tag{2.4}$$

with domain

$$\mathscr{D}(M) := \left\{ \begin{bmatrix} \mathbf{E} \\ \mathbf{H} \end{bmatrix} \in \mathbb{H} : M \begin{bmatrix} \mathbf{E} \\ \mathbf{H} \end{bmatrix} = \begin{bmatrix} \varepsilon^{-1}\nabla \times \mathbf{H} \\ -\mu^{-1}\nabla \times \mathbf{E} \end{bmatrix} \in \mathbb{H}, \ \mathbf{n} \times \mathbf{E}\big|_{\partial D} = 0 \right\}$$

$$= H_0(\text{curl}, D) \times H(\text{curl}, D).$$

The corresponding norm is defined as

$$\|u\|_{\mathscr{D}(M)} := \left(\|u\|_{\mathbb{H}}^2 + \|Mu\|_{\mathbb{H}}^2 \right)^{1/2}.$$

Theorem 2.1 *If ε, μ satisfy Assumption 2.1, then the Maxwell operator M : $\mathscr{D}(M) \to \mathbb{H}$ is closed and skew-adjoint, and thus generates a unitary C_0-semigroup $\{S(t) := e^{tM}, \ t \geq 0\}$ on $\mathbb{H}$.*

Proof The closedness of M follows from the closedness of the operator $\nabla\times$. To prove the skew-adjointness of M, it suffices to show that M is a skew-symmetric operator and that $Id \pm M$ have dense ranges.

Taking $\psi = (u^\top, v^\top)^\top, \tilde\psi = (\tilde u^\top, \tilde v^\top)^\top \in \mathscr{D}(M)$ and using (2.3) give that

$$\left\langle M\psi, \tilde\psi \right\rangle_{\mathbb{H}} = \int_D \left(\nabla \times v \cdot \tilde u - \nabla \times u \cdot \tilde v \right) dx$$

$$= \int_D \left(v \cdot \nabla \times \tilde u - u \cdot \nabla \times \tilde v \right) dx$$

$$= -\left\langle \psi, M\tilde\psi \right\rangle_{\mathbb{H}},$$

i.e., M is skew-symmetric.

By the standard spectral theory given in Lemma B.1, the Maxwell operator M is skew-adjoint if

$$\overline{\operatorname{ran}(Id \pm M)} = \mathbb{H}. \tag{2.5}$$

Skew-adjointness then implies the assertion in view of the Stone theorem (see Theorem B.1). Since $C^\infty(D)^6$ is dense in $\mathbb{H}$, the proof of (2.5) is equivalent to showing that for every $\mathbf{f} = (\mathbf{f}_1^\top, \mathbf{f}_2^\top)^\top \in C^\infty(D)^6$, there exists $\mathbf{g} = (\mathbf{E}^\top, \mathbf{H}^\top)^\top \in \mathscr{D}(M)$ such that

$$(Id \pm M)\mathbf{g} = \mathbf{f},$$

or equivalently,

$$\mathbf{E} \pm \varepsilon^{-1}\nabla \times \mathbf{H} = \mathbf{f}_1, \quad \mathbf{H} \mp \mu^{-1}\nabla \times \mathbf{E} = \mathbf{f}_2. \tag{2.6}$$

Plugging the second equation in (2.6) into the first one, we obtain

$$\varepsilon\mathbf{E} + \nabla \times (\mu^{-1}\nabla \times \mathbf{E}) = \varepsilon\mathbf{f}_1 \mp (\nabla \times \mathbf{f}_2) =: \mathbf{h}. \tag{2.7}$$

Then $\mathbf{h} \in L^2(D)^3$ due to the denseness of $H(\mathrm{curl}, D)$ in $L^2(D)^3$. For the well-posedness of these equations, we introduce the following symmetric bilinear form

$$a(u, v) = \int_D \left(\varepsilon u \cdot v + \mu^{-1}(\nabla \times u) \cdot (\nabla \times v) \right) dx$$

on $H(\mathrm{curl}, D)$. It can be seen that a is continuous and coercive under Assumption 2.1. Using the Lax–Milgram lemma, one obtains the existence of $\mathbf{E} \in H(\mathrm{curl}, D)$ satisfying

$$\int_D \left(\varepsilon\mathbf{E} \cdot v + \mu^{-1}\nabla \times \mathbf{E} \cdot \nabla \times v \right) dx = \int_D \mathbf{h} \cdot v \, dx \quad \forall v \in H(\mathrm{curl}, D).$$

It follows from $\mathbf{h} - \varepsilon\mathbf{E} \in L^2(D)^3$ that $\nabla \times (\mu^{-1}\nabla \times \mathbf{E}) \in L^2(D)^3$ and $\mathbf{E}$ satisfies (2.7). If we define $\mathbf{H} \in H(\text{curl}, D)$ by the second equation in (2.6), we obtain a solution $\mathbf{g} = (\mathbf{E}^\top, \mathbf{H}^\top)^\top \in \mathscr{D}(M)$ of (2.6), as asserted. $\qquad\square$

Remark 2.2 For the two-dimensional case in TE polarization (1.39) with $D = (a_1^-, a_1^+) \times (a_2^-, a_2^+) \subset \mathbb{R}^2$, the Maxwell operator is defined as

$$M^{TE} : \mathscr{D}(M^{TE}) \to L^2(D)^3, \quad \begin{bmatrix} u \\ v \\ w \end{bmatrix} \mapsto \begin{bmatrix} \varepsilon^{-1}\partial_y w \\ -\varepsilon^{-1}\partial_x w \\ \mu^{-1}\partial_y u - \mu^{-1}\partial_x v \end{bmatrix},$$

where the domain of M^{TE} is given by

$$\mathscr{D}(M^{TE}) := \left\{ \begin{bmatrix} u \\ v \\ w \end{bmatrix} \in L^2(D)^3 : M^{TE}\begin{bmatrix} u \\ v \\ w \end{bmatrix} \in L^2(D)^3, \; u|_{\partial D} = 0, \; v|_{\partial D} = 0 \right\}.$$

Under the weighted inner product

$$\left\langle \begin{bmatrix} u_1 \\ v_1 \\ w_1 \end{bmatrix}, \begin{bmatrix} u_2 \\ v_2 \\ w_2 \end{bmatrix} \right\rangle_{\varepsilon,\mu} := \int_D \left(\varepsilon u_1 u_2 + \varepsilon v_1 v_2 + \mu w_1 w_2 \right) \mathrm{d}x\mathrm{d}y$$

and Assumption 2.1, it can be verified that M^{TE} is skew-adjoint on $L^2(D)^3$, and thus generates a unitary C_0-semigroup $\{S^{TE}(t) := e^{tM^{TE}}, t \geq 0\}$ on $L^2(D)^3$. In a similar manner, we can define the Maxwell operator M^{TM} for the TM polarization (1.40).

To study the solution theory of (2.1), we always rewrite it in the following equivalent form of the infinite-dimensional stochastic evolution equation:

$$\begin{cases} \mathrm{d}u(t) = [Mu(t) + F(t, u(t))]\mathrm{d}t + B(t, u(t))\mathrm{d}W(t), & t \in (0, T], \\ u(0) = u_0, \end{cases} \tag{2.8}$$

where $u = (\mathbf{E}^\top, \mathbf{H}^\top)^\top$, $u_0 = (\mathbf{E}_0^\top, \mathbf{H}_0^\top)^\top$, M is the Maxwell operator defined in (2.4), and F, B are defined below.

- The drift term $F : [0, T] \times \mathbb{H} \to \mathbb{H}$ is a Nemytskij operator associated with $\mathbf{J}_e$ and $\mathbf{J}_m$, which is defined by

$$F(t, u)(\mathbf{x}) := \begin{bmatrix} -\varepsilon^{-1}\mathbf{J}_e(t, \mathbf{x}, \mathbf{E}(t, \mathbf{x}), \mathbf{H}(t, \mathbf{x})) \\ -\mu^{-1}\mathbf{J}_m(t, \mathbf{x}, \mathbf{E}(t, \mathbf{x}), \mathbf{H}(t, \mathbf{x})) \end{bmatrix}, \quad t \in [0, T], \; \mathbf{x} \in D$$

for $u = (\mathbf{E}^\top, \mathbf{H}^\top)^\top \in \mathbb{H}$.

- The diffusion term $B : [0, T] \times \mathbb{H} \to HS(U_0, \mathbb{H})$ is a Nemytskij operator associated with $\mathbf{J}_e^r$ and $\mathbf{J}_m^r$, which is defined by

$$(B(t, u)v)(\mathbf{x}) := \begin{bmatrix} -\varepsilon^{-1} \mathbf{J}_e^r(t, \mathbf{x}, \mathbf{E}(t, \mathbf{x}), \mathbf{H}(t, \mathbf{x}))v(\mathbf{x}) \\ -\mu^{-1} \mathbf{J}_m^r(t, \mathbf{x}, \mathbf{E}(t, \mathbf{x}), \mathbf{H}(t, \mathbf{x}))v(\mathbf{x}) \end{bmatrix}, \quad t \in [0, T], \mathbf{x} \in D$$

for $u = (\mathbf{E}^\top, \mathbf{H}^\top)^\top \in \mathbb{H}$ and $v \in U_0 := Q^{\frac{1}{2}} U$. See Appendix D.3 for more details on the notation U_0.

2.2 Well-Posedness

This section presents the well-posedness of (2.8) with either globally Lipschitz or non-globally Lipschitz continuous drift term.

2.2.1 Globally Lipschitz Continuous Case

Consider the stochastic Maxwell equations (2.8) with globally Lipschitz continuous coefficients. More precisely, assumptions on F and B are given as follows.

Assumption 2.2 *Assume that $F : [0, T] \times \mathbb{H} \to \mathbb{H}$ is measurable from $([0, T] \times \mathbb{H}, \mathscr{B}([0, T]) \times \mathscr{B}(\mathbb{H}))$ into $(\mathbb{H}, \mathscr{B}(\mathbb{H}))$, and there is a positive constant C such that*

$$\|F(t, u)\|_{\mathbb{H}} \leq C(1 + \|u\|_{\mathbb{H}}), \tag{2.9}$$

$$\|F(t, u) - F(s, v)\|_{\mathbb{H}} \leq C(|t - s| + \|u - v\|_{\mathbb{H}}) \tag{2.10}$$

for almost every $t, s \in [0, T]$ and $u, v \in \mathbb{H}$.

Assumption 2.3 *Assume that $B : [0, T] \times \mathbb{H} \to HS(U_0, \mathbb{H})$ is measurable from $([0, T] \times \mathbb{H}, \mathscr{B}([0, T]) \times \mathscr{B}(\mathbb{H}))$ into $(HS(U_0, \mathbb{H}), \mathscr{B}(HS(U_0, \mathbb{H})))$, and there is a positive constant C such that*

$$\|B(t, u)\|_{HS(U_0, \mathbb{H})} \leq C(1 + \|u\|_{\mathbb{H}}^2)^{1/2}, \tag{2.11}$$

$$\|B(t, u) - B(s, v)\|_{HS(U_0, \mathbb{H})} \leq C(|t - s| + \|u - v\|_{\mathbb{H}}) \tag{2.12}$$

for almost every $t, s \in [0, T]$ and $u, v \in \mathbb{H}$.

Remark 2.3 Assumptions 2.2 and 2.3 can be guaranteed by certain conditions on coefficients in (2.1). More precisely, suppose that there exists a positive constant L such that every $\mathbf{J} \in \{\mathbf{J}_e, \mathbf{J}_m, \mathbf{J}_e^r, \mathbf{J}_m^r\}$ satisfies the globally Lipschitz continuous condition:

$$|\mathbf{J}(t, \mathbf{x}, u_1, v_1) - \mathbf{J}(s, \mathbf{x}, u_2, v_2)| \leq L(|t - s| + |u_1 - u_2| + |v_1 - v_2|), \quad (2.13)$$

where $t, s \in [0, T]$, $\mathbf{x} \in D$, and $u_1, v_1, u_2, v_2 \in \mathbb{R}^d$. Let ε, μ satisfy Assumption 2.1. In addition, suppose that $Q^{1/2} \in HS(U, H^\gamma(D))$ for $\gamma > 3/2$. Then conditions (2.9)–(2.12) hold.

In fact, thanks to (2.13), we derive that

$$\|F(t, u)\|_{\mathbb{H}} = \left(\int_D \left(\varepsilon |\varepsilon^{-1} \mathbf{J}_e|^2 + \mu |\mu^{-1} \mathbf{J}_m|^2 \right) d\mathbf{x} \right)^{1/2}$$

$$\leq C\delta^{-1/2} \left(\int_D (1 + |\mathbf{E}|^2 + |\mathbf{H}|^2) d\mathbf{x} \right)^{1/2}$$

$$\leq C\delta^{-1/2} \left[|D|^{1/2} + \delta^{-1/2} \left(\int_D (\varepsilon |\mathbf{E}|^2 + \mu |\mathbf{H}|^2) d\mathbf{x} \right)^{1/2} \right]$$

$$\leq C(1 + \|u\|_{\mathbb{H}})$$

and

$$\|B(t, u)\|_{HS(U_0, \mathbb{H})}^2 = \|B(t, u) Q^{\frac{1}{2}}\|_{HS(U, \mathbb{H})}^2 = \sum_{j \in \mathbb{N}} \|B(t, u) Q^{\frac{1}{2}} e_j\|_{\mathbb{H}}^2$$

$$= \sum_{j \in \mathbb{N}} \int_D \left(\varepsilon^{-1} |\mathbf{J}_e^r Q^{\frac{1}{2}} e_j(\mathbf{x})|^2 + \mu^{-1} |\mathbf{J}_m^r Q^{\frac{1}{2}} e_j(\mathbf{x})|^2 \right) d\mathbf{x}$$

$$\leq C\delta^{-1} \sum_{j \in \mathbb{N}} \|Q^{\frac{1}{2}} e_j\|_{L^\infty(D)}^2 \int_D \left(1 + |\mathbf{E}|^2 + |\mathbf{H}|^2 \right) d\mathbf{x}$$

$$\leq C\delta^{-1} \|Q^{\frac{1}{2}}\|_{HS(U, H^\gamma(D))}^2 \left(|D| + \delta^{-1} \int_D (\varepsilon |\mathbf{E}|^2 + \mu |\mathbf{H}|^2) d\mathbf{x} \right)$$

$$\leq C(1 + \|u\|_{\mathbb{H}}^2),$$

where we used the Sobolev embedding $H^\gamma(D) \hookrightarrow L^\infty(D)$ with $\gamma > 3/2$ (see Theorem B.2). Thus we obtain (2.9) and (2.11). Proofs of (2.10) and (2.12) can be obtained similarly.

Theorem 2.2 *Assume that u_0 is an $\mathscr{F}_0$-measurable $\mathbb{H}$-valued random variable satisfying $\|u_0\|_{L^p(\Omega,\mathbb{H})} < \infty$ for some $p \geq 2$. Let Assumptions 2.1, 2.2, and 2.3 hold. Then* (2.8) *has a unique mild solution given by*

$$u(t) = S(t)u_0 + \int_0^t S(t-s)F(s, u(s))\mathrm{d}s$$

$$+ \int_0^t S(t-s)B(s, u(s))\mathrm{d}W(s), \quad \mathbb{P}\text{-}a.s. \tag{2.14}$$

for each $t \in [0, T]$. Moreover, there exists a positive constant $C = C(p, T, F, B, Q)$ such that

$$\mathbb{E}\Big[\sup_{t\in[0,T]} \|u(t)\|_{\mathbb{H}}^p\Big] \leq C\Big(1 + \mathbb{E}\big[\|u_0\|_{\mathbb{H}}^p\big]\Big). \tag{2.15}$$

Proof
Step 1: Existence and uniqueness. We first prove the existence and uniqueness of the solution based on the fixed point theorem. For $p \geq 2$ and $\beta > 0$, denote by $\mathscr{H}_{p,\beta}$ the Banach space of all $\mathbb{H}$-valued predictable stochastic processes $\{Y(t) : t \in [0, T]\}$ such that

$$\|Y\|_{p,\beta} := \sup_{t\in[0,T]} e^{-\beta t}\Big(\mathbb{E}\big[\|Y(t)\|_{\mathbb{H}}^p\big]\Big)^{1/p} < \infty.$$

Define a mapping $\mathscr{K}$ on $\mathscr{H}_{p,\beta}$ by

$$\mathscr{K}(Y)(t) := S(t)u_0 + \int_0^t S(t-s)F(s, Y(s))\mathrm{d}s$$

$$+ \int_0^t S(t-s)B(s, Y(s))\mathrm{d}W(s), \quad \mathbb{P}\text{-}a.s.$$

for all $t \in [0, T]$ and $Y \in \mathscr{H}_{p,\beta}$.

Now we show that the mapping $\mathscr{K} : \mathscr{H}_{p,\beta} \to \mathscr{H}_{p,\beta}$ is well-defined. In fact, taking into account Assumptions 2.2 and 2.3, and using Proposition D.4 (ii), one obtains that

$\|\mathscr{K}(Y)\|_{p,\beta}$

$$\leq \sup_{t\in[0,T]} e^{-\beta t}\Big[\Big(\mathbb{E}\big[\|S(t)u_0\|_{\mathbb{H}}^p\big]\Big)^{1/p} + \Big(\mathbb{E}\Big[\Big\|\int_0^t S(t-s)F(s, Y(s))\mathrm{d}s\Big\|_{\mathbb{H}}^p\Big]\Big)^{1/p}$$

$$+ \Big(\mathbb{E}\Big[\Big\|\int_0^t S(t-s)B(s, Y(s))\mathrm{d}W(s)\Big\|_{\mathbb{H}}^p\Big]\Big)^{1/p}\Big]$$

$$\leq \|u_0\|_{L^p(\Omega,\mathbb{H})} + \sup_{t\in[0,T]} e^{-\beta t} \int_0^t \|F(s,Y(s))\|_{L^p(\Omega,\mathbb{H})} ds$$

$$+ C \sup_{t\in[0,T]} e^{-\beta t} \left(\int_0^t \left(\mathbb{E}\left[\|B(s,Y(s))\|_{HS(U_0,\mathbb{H})}^p \right] \right)^{2/p} ds \right)^{1/2}$$

$$\leq \|u_0\|_{L^p(\Omega,\mathbb{H})} + C \int_0^T \left[1 + \left(\mathbb{E}\left[\|Y(s)\|_{\mathbb{H}}^p \right] \right)^{1/p} \right] ds$$

$$+ C \left(\int_0^T \left[1 + \left(\mathbb{E}\left[\|Y(s)\|_{\mathbb{H}}^p \right] \right)^{2/p} \right] ds \right)^{1/2}$$

$$\leq \|u_0\|_{L^p(\Omega,\mathbb{H})} + CT + C \left(\int_0^T e^{\beta s} ds + \left(\int_0^T e^{2\beta s} ds \right)^{1/2} \right) \|\!|Y|\!\|_{p,\beta} < \infty.$$

Let Y_1 and Y_2 be two arbitrary processes in $\mathscr{H}_{p,\beta}$. Then

$$\|\mathscr{K}(Y_1)(t) - \mathscr{K}(Y_2)(t)\|_{L^p(\Omega,\mathbb{H})}$$

$$\leq \left\| \int_0^t S(t-s)(F(s,Y_1(s)) - F(s,Y_2(s))) ds \right\|_{L^p(\Omega,\mathbb{H})}$$

$$+ \left\| \int_0^t S(t-s)(B(s,Y_1(s)) - B(s,Y_2(s))) dW(s) \right\|_{L^p(\Omega,\mathbb{H})}$$

$$=: I_1 + I_2.$$

For the term I_1, Assumption 2.2 implies that

$$I_1 \leq \int_0^t \|F(s,Y_1(s)) - F(s,Y_2(s))\|_{L^p(\Omega,\mathbb{H})} ds$$

$$\leq C \int_0^t \left(\mathbb{E}\left[\|Y_1(s) - Y_2(s)\|_{\mathbb{H}}^p \right] \right)^{1/p} ds$$

$$= C \int_0^t e^{\beta s} e^{-\beta s} \left(\mathbb{E}\left[\|Y_1(s) - Y_2(s)\|_{\mathbb{H}}^p \right] \right)^{1/p} ds$$

$$\leq C \int_0^t e^{\beta s} ds \, \|\!|Y_1 - Y_2|\!\|_{p,\beta} = \frac{C(e^{\beta t}-1)}{\beta} \|\!|Y_1 - Y_2|\!\|_{p,\beta}$$

for all $t \in [0,T]$. For the term I_2, it follows from Proposition D.4 (i) and Assumption 2.3 that

$$I_2 \leq C \left(\int_0^t \left(\mathbb{E}\left[\|B(s,Y_1(s)) - B(s,Y_2(s))\|_{HS(U_0,\mathbb{H})}^p \right] \right)^{2/p} ds \right)^{1/2}$$

$$\leq C\left(\int_0^t \left(\mathbb{E}\big[\|Y_1(s) - Y_2(s)\|_{\mathbb{H}}^p\big]\right)^{2/p} ds\right)^{1/2}$$

$$= C\left(\int_0^t e^{2\beta s} e^{-2\beta s} \left(\mathbb{E}\big[\|Y_1(s) - Y_2(s)\|_{\mathbb{H}}^p\big]\right)^{2/p} ds\right)^{1/2}$$

$$\leq C\left(\int_0^t e^{2\beta s} ds\right)^{1/2} \||Y_1 - Y_2\||_{p,\beta} = C\left(\frac{e^{2\beta t} - 1}{2\beta}\right)^{1/2} \||Y_1 - Y_2\||_{p,\beta}$$

for all $t \in [0, T]$. Therefore,

$$\||\mathscr{K}(Y_1) - \mathscr{K}(Y_2)\||_{p,\beta} = \sup_{t \in [0,T]} e^{-\beta t} \|\mathscr{K}(Y_1)(t) - \mathscr{K}(Y_2)(t)\|_{L^p(\Omega,\mathbb{H})}$$

$$\leq C \sup_{t \in [0,T]} \left[\frac{1 - e^{-\beta t}}{\beta} + \left(\frac{1 - e^{-2\beta t}}{2\beta}\right)^{1/2}\right] \||Y_1 - Y_2\||_{p,\beta}$$

$$= C\left[\frac{1 - e^{-\beta T}}{\beta} + \left(\frac{1 - e^{-2\beta T}}{2\beta}\right)^{1/2}\right] \||Y_1 - Y_2\||_{p,\beta}$$

for all $Y_1, Y_2 \in \mathscr{H}_{p,\beta}$. Moreover, note that

$$\lim_{\beta \to +\infty} \left[\frac{1 - e^{-\beta T}}{\beta} + \left(\frac{1 - e^{-2\beta T}}{2\beta}\right)^{1/2}\right] = 0.$$

Combining the above estimates, we have shown that $\mathscr{K}$ is a contraction mapping from $\mathscr{H}_{p,\beta}$ to $\mathscr{H}_{p,\beta}$ when β is sufficiently large. Thus, there exists a unique solution of (2.8) which fulfills (2.14).

Step 2: Proof of (2.15). By using Proposition D.5 and the linear growth properties of F and B, we have

$$\mathbb{E}\left[\sup_{t \in [0,T]} \|u(t)\|_{\mathbb{H}}^p\right] \leq C\mathbb{E}\left[\sup_{t \in [0,T]} \|S(t)u_0\|_{\mathbb{H}}^p\right]$$

$$+ C\mathbb{E}\left[\int_0^T \|S(t - s)F(s, u(s))\|_{\mathbb{H}}^p ds\right]$$

$$+ C\mathbb{E}\left[\sup_{t \in [0,T]} \left\|\int_0^t S(t - s)B(s, u(s))dW(s)\right\|_{\mathbb{H}}^p\right]$$

$$\leq C\mathbb{E}\big[\|u_0\|_{\mathbb{H}}^p\big] + C\int_0^T \left(1 + \mathbb{E}\big[\|u(s)\|_{\mathbb{H}}^p\big]\right)ds$$

$$+ C\mathbb{E}\left[\left(\int_0^T \|B(s, u(s))\|_{HS(U_0,\mathbb{H})}^2 ds\right)^{p/2}\right]$$

$$\leq C\mathbb{E}\big[\|u_0\|_{\mathbb{H}}^p\big] + C\int_0^T \Big(1 + \mathbb{E}\big[\|u(s)\|_{\mathbb{H}}^p\big]\Big)ds$$

$$\leq C\mathbb{E}\big[\|u_0\|_{\mathbb{H}}^p\big] + C\int_0^T \Big(1 + \mathbb{E}\big[\sup_{r\in[0,s]}\|u(r)\|_{\mathbb{H}}^p\big]\Big)ds.$$

By the Grönwall inequality, there exists a positive constant C such that

$$\mathbb{E}\Big[\sup_{t\in[0,T]}\|u(t)\|_{\mathbb{H}}^p\Big] \leq C\Big(1 + \mathbb{E}\big[\|u_0\|_{\mathbb{H}}^p\big]\Big).$$

Thus the proof of Theorem 2.2 is finished.   □

Remark 2.4 Consider the stochastic Maxwell equations in the Stratonovich sense

$$\begin{cases} du(t) = \Big[Mu(t) + F(t, u(t))\Big]dt + B(t, u(t)) \circ dW(t), & t \in (0, T], \\ u(0) = u_0. \end{cases}$$

$$(2.16)$$

It is well-known that this system is equivalent to the following system in the Itô sense

$$du(t) = \Big[Mu(t) + F(t, u(t)) - \frac{1}{2}B_u(t, u(t))B(t, u(t))F_Q\Big]dt + B(t, u(t))dW(t),$$

where $F_Q(\mathbf{x}) = \sum_{j\in\mathbb{N}}\big(Q^{\frac{1}{2}}e_j(\mathbf{x})\big)^2$. If the modified coefficient $\tilde{F}(t, u(t)) := F(t, u(t)) - \frac{1}{2}B_u(t, u(t))B(t, u(t))F_Q$ satisfies Assumption 2.2 and the diffusion term B satisfies Assumption 2.3, then by Theorem 2.2, there is a unique mild solution of (2.16).

2.2.2   Non-globally Lipschitz Continuous Case

In this subsection we restrict our attention to (2.8) with Kerr-type nonlinearity, i.e.,

$$F(u(t))(\mathbf{x}) = -|u(t, \mathbf{x})|^q u(t, \mathbf{x}) \qquad (2.17)$$

for $q > 0$. In addition, we assume that $\varepsilon = \mu \equiv 1$. Here and after, we denote by $\langle \cdot, \cdot \rangle$ the dualization between a Banach space V and its dual space V^*. For example, if $V = L^p(D)$ for some $p \in [2, \infty)$, then $V^* = L^{\frac{p}{p-1}}(D)$ and

$$\langle f, g \rangle = \int_D f(\mathbf{x})g(\mathbf{x})d\mathbf{x} \qquad \forall f \in V, g \in V^*.$$

We first introduce some properties of the nonlinear term F in (2.17) as a mapping from $L^{q+2}(D)^6$ to $L^{\frac{q+2}{q+1}}(D)^6$.

Lemma 2.1 *For the term F, it holds*

(i)   there exists a constant $\gamma_0 > 0$ such that

$$\langle F(u) - F(v), u - v \rangle \leq -\gamma_0 \|u - v\|_{L^{q+2}(D)^6}^{q+2} \tag{2.18}$$

   for all $u, v \in L^{q+2}(D)^6$;

(ii)   there exists a constant $C > 0$ such that

$$\|F(u) - F(v)\|_{L^{\frac{q+2}{q+1}}(D)^6} \leq C\Big(\|u\|_{L^{q+2}(D)^6}^q + \|v\|_{L^{q+2}(D)^6}^q\Big)\|u - v\|_{L^{q+2}(D)^6} \tag{2.19}$$

   for all $u, v \in L^{q+2}(D)^6$.

Proof

(i) Note that $\|F(u)\|_{L^{\frac{q+2}{q+1}}(D)^6} = \|u\|_{L^{q+2}(D)^6}$, thus F is a mapping from $L^{q+2}(D)^6$ to $L^{\frac{q+2}{q+1}}(D)^6$. It follows from Proposition A.7 that

$$\langle F(u) - F(v), u - v \rangle = -\langle |u|^q u - |v|^q v, u - v \rangle \leq -\gamma_0 \|u - v\|_{L^{q+2}(D)^6}^{q+2},$$

where the positive constant γ_0 depends on q.

(ii) For any $u, w \in L^{q+2}(D)^6$, F is Fréchet differentiable (see [107, Corollary 9.3]) and its Fréchet derivative is given by

$$F_u(u)w = -q|w|^{q-2}u^\top w - |u|^q w,$$

which implies $\langle F_u(u)w, w \rangle \leq 0$ and $|F_u(u)w(\mathbf{x})| \leq C|u(\mathbf{x})|^q|w(\mathbf{x})|$ for all $\mathbf{x} \in D$. Thus the mean value theorem yields the conclusion. $\square$

Remark 2.5 If $q > 1$, then F is twice continuously Fréchet differentiable with

$$F_{uu}(u)(v, v)(\mathbf{x}) \leq C|u(\mathbf{x})|^{q-1}|v(\mathbf{x})|^2$$

for all $u, v \in L^{q+2}(D)^6$ and $\mathbf{x} \in D$. We refer to [105, Lemma 2.6] for more details of this property.

We are in a position to give the well-posedness of the stochastic Maxwell equations (2.8) with the nonlinear drift term (2.17).

Theorem 2.3 *Let $q > 0$. Suppose that $u_0 \in L^2(\Omega, \mathbb{H})$, and B satisfies Assumption 2.3. Then there exists a unique weak solution u of (2.8), namely, there is an adapted process $u \in L^2(\Omega, C([0, T], \mathbb{H})) \cap L^{q+2}(\Omega \times [0, T] \times D)^6$ satisfying*

$$\langle u(t) - u_0, \phi \rangle_{\mathbb{H}} = \int_0^t \Big(- \langle u(s), M\phi \rangle_{\mathbb{H}} - \langle |u(s)|^q u(s), \phi \rangle_{\mathbb{H}} \Big) ds$$

$$+ \int_0^t \langle \phi, B(s, u(s)) dW(s) \rangle_{\mathbb{H}}, \quad \mathbb{P}\text{-}a.s.$$

for all $t \in [0, T]$ and $\phi \in \mathscr{D}(M) \cap L^{q+2}(D)^6$.

The proof of Theorem 2.3 is mainly based on the Galerkin approximation. Before giving the Galerkin approximation, we first introduce some operators, which will be used to obtain a truncation of (2.8). Define operators $A^{(1)} = A^{(2)} := \nabla \times \nabla \times -\nabla(\nabla \cdot)$ with domains

$$\mathscr{D}(A^{(1)}) := \Big\{ u \in H_0(\text{curl}, D) \cap H(\text{div}, D) : \nabla \times \nabla \times u \in L^2(D)^3,$$

$$\nabla \cdot u \in H_0^1(D) \Big\},$$

$$\mathscr{D}(A^{(2)}) := \Big\{ u \in H(\text{curl}, D) \cap H_0(\text{div}, D) : \nabla \times \nabla \times u \in L^2(D)^3,$$

$$\nabla \cdot u \in H^1(D), \mathbf{n} \times (\nabla \times u)|_{\partial D} = 0 \Big\},$$

respectively. One can check that for $i = 1, 2$, the operator $Id + A^{(i)}$ is strictly positive and self-adjoint on $L^2(D)^3$, and the embedding $\mathscr{D}(A^{(i)}) \hookrightarrow L^2(D)^3$ is compact.

Define the Hodge–Laplacian operator by

$$\Delta_H u := \begin{bmatrix} -A^{(1)} u_1 \\ -A^{(2)} u_2 \end{bmatrix}$$

for all $u = (u_1^\top, u_2^\top)^\top \in \mathscr{D}(\Delta_H) := \mathscr{D}(A^{(1)}) \times \mathscr{D}(A^{(2)})$. Hence, $Id - \Delta_H$ is a densely defined, self-adjoint, and positive definite operator with a compact inverse. Then, there exists an orthonormal basis of eigenvectors $\{g_k\}_{k \in \mathbb{N}}$ to the positive eigenvalues $\{\lambda_k\}_{k \in \mathbb{N}}$ of $Id - \Delta_H$ with $\lambda_k \to \infty$ as $k \to \infty$.

Define the orthogonal projection operator $P_n : \mathbb{H} \to \mathbb{H}$ by

$$P_n u := \sum_{\{k:\, \lambda_k \leq 2^n\}} \langle u, g_k \rangle_{\mathbb{H}} g_k, \quad u \in \mathbb{H}.$$

for any $n \in \mathbb{N}$. Note that $\bigcup_{n \in \mathbb{N}} \mathrm{ran}(P_n)$ is dense in $\mathscr{D}(M)$ and in $L^p(D)^6$ for $p > 1$ (see [105, Corollary 3.6]). Then for each $n \in \mathbb{N}$, we consider the following truncated equation in $\mathrm{ran}(P_n)$:

$$
\begin{cases}
du^{(n)}(t) = \left[P_n M u^{(n)}(t) + P_n F(u^{(n)}(t)) \right] dt + P_n B(t, u^{(n)}(t)) dW(t), \ t \in (0, T], \\
u^{(n)}(0) = P_n u_0.
\end{cases}
$$

$$(2.20)$$

It follows from (2.19) that this is a finite-dimensional stochastic differential equation with a locally Lipschitz continuous drift term. Using the fact $u^\top F(u) = -|u|^{q+2}$, we obtain by [133, Theorem 3.5] that there exists a unique solution $u^{(n)} : \Omega \times [0, T] \to \mathrm{ran}(P_n)$ with continuous paths that solves (2.20).

In order to construct the solution of (2.8), we need some *a priori* estimates for the solution $u^{(n)}$.

Proposition 2.1 *There exists a positive constant C such that*

$$
\sup_{n \in \mathbb{N}} \mathbb{E} \left[\sup_{t \in [0,T]} \|u^{(n)}(t)\|_{\mathbb{H}}^2 \right] + 2 \sup_{n \in \mathbb{N}} \mathbb{E} \left[\int_0^T \|u^{(n)}(s)\|_{L^{q+2}(D)^6}^{q+2} ds \right]
$$
$$
\leq C \left(T + \mathbb{E}\left[\|u_0\|_{\mathbb{H}}^2 \right] \right).
$$

$$(2.21)$$

Proof Applying the Itô formula to $\|u^{(n)}(t)\|_{\mathbb{H}}^2$ yields

$$
\|u^{(n)}(t)\|_{\mathbb{H}}^2 = \|u^{(n)}(0)\|_{\mathbb{H}}^2 + 2 \int_0^t \langle u^{(n)}(s), -|u^{(n)}(s)|^q u^{(n)}(s) \rangle_{\mathbb{H}} ds
$$
$$
+ \int_0^t \sum_{j \in \mathbb{N}} \| P_n B(s, u^{(n)}(s)) Q^{\frac{1}{2}} e_j \|_{\mathbb{H}}^2 ds + 2 M^{(n)}(t), \quad \mathbb{P}\text{-}a.s.
$$

for all $t \in [0, T]$, where

$$
M^{(n)}(t) := \int_0^t \langle u^{(n)}(s), P_n B(s, u^{(n)}(s)) dW(s) \rangle_{\mathbb{H}}.
$$

Hence, for all $t \in [0, T]$,

$$
\|u^{(n)}(t)\|_{\mathbb{H}}^2 \leq \|u_0\|_{\mathbb{H}}^2 - 2 \int_0^t \|u^{(n)}(s)\|_{L^{q+2}(D)^6}^{q+2} ds
$$
$$
+ C \int_0^t \|u^{(n)}(s)\|_{\mathbb{H}}^2 ds + CT + 2 M^{(n)}(t), \quad \mathbb{P}\text{-}a.s.
$$

$$(2.22)$$

due to the linear growth of B.

It follows from Proposition D.4 (ii), (2.11), and the Young inequality that

$$
\mathbb{E}\Big[\sup_{t\in[0,T]} |M^{(n)}(t)| \Big]
$$

$$
\leq C\mathbb{E}\Big[\int_0^t \|u^{(n)}(s)\|_{\mathbb{H}}^2 \|B(s,u^{(n)}(s))\|_{HS(U_0,\mathbb{H})}^2 \mathrm{d}s \Big]^{1/2}
$$

$$
\leq C\mathbb{E}\Big[\Big(\sup_{t\in[0,T]} \|u^{(n)}(t)\|_{\mathbb{H}}^2 \Big)^{1/2} \Big(\int_0^t \big(1+\|u^{(n)}(s)\|_{\mathbb{H}}^2\big)\mathrm{d}s \Big)^{1/2} \Big]
$$

$$
\leq \frac{1}{4}\mathbb{E}\Big[\sup_{t\in[0,T]} \|u^{(n)}(t)\|_{\mathbb{H}}^2 \Big] + C\mathbb{E}\Big[\int_0^T \sup_{r\in[0,s]} \|u^{(n)}(r)\|_{\mathbb{H}}^2 \mathrm{d}s \Big] + CT.
$$

(2.23)

Then by (2.22) and (2.23), and the Grönwall inequality, we have that for any $n \in \mathbb{N}$,

$$
\mathbb{E}\Big[\sup_{t\in[0,T]} \|u^{(n)}(t)\|_{\mathbb{H}}^2 \Big] + 2\mathbb{E}\Big[\int_0^T \|u^{(n)}(s)\|_{L^{q+2}(D)^6}^{q+2} \mathrm{d}s \Big] \leq C\Big(T + \mathbb{E}\big[\|u_0\|_{\mathbb{H}}^2\big] \Big),
$$

where C is a positive constant independent of n.   □

We will need the following lemma; see e.g., [105, Lemma 4.2] and [129, Theorem 4.2.5] for the proof.

Lemma 2.2 *Let $X_0 \in L^2(\Omega, \mathbb{H})$, $Y \in L^{\frac{q+2}{q+1}}(\Omega \times [0,T] \times D)^6$ with $q > 0$, and $Z \in L^2(\Omega \times [0,T], HS(U_0, \mathbb{H}))$ be $\{\mathscr{F}_t\}_{0\leq t\leq T}$-adapted. Define the process X satisfying*

$$
\langle X(t), \phi \rangle_{\mathbb{H}} = \langle X_0, \phi \rangle_{\mathbb{H}} - \int_0^t \Big(\langle X(s), M\phi \rangle_{\mathbb{H}} - \langle Y(s), \phi \rangle_{\mathbb{H}} \Big) \mathrm{d}s
$$

$$
+ \int_0^t \langle \phi, Z(s)\mathrm{d}W(s) \rangle_{\mathbb{H}}, \quad \mathbb{P}\text{-}a.s.
$$

(2.24)

for all $t \in [0,T]$ and $\phi \in \mathscr{D}(M) \cap L^{q+2}(D)^6$. If $X \in L^{q+2}(\Omega \times [0,T] \times D)^6$, then X is an $\mathbb{H}$-valued continuous $\{\mathscr{F}_t\}_{0\leq t\leq T}$-adapted process and satisfies

$$
\|X(t_2)\|_{\mathbb{H}}^2 = \|X(t_1)\|_{\mathbb{H}}^2 + \int_{t_1}^{t_2} \Big(2\langle X(s), Y(s) \rangle + \|Z(s)\|_{HS(U_0,\mathbb{H})}^2 \Big)\mathrm{d}s
$$

$$
+ 2\int_{t_1}^{t_2} \langle X(s), Z(s)\mathrm{d}W(s) \rangle_{\mathbb{H}}, \quad \mathbb{P}\text{-}a.s.
$$

(2.25)

for all $0 \leq t_1 \leq t_2 \leq T$.

Taking into account the truncated equation (2.20) and the estimate of $u^{(n)}$ in Proposition 2.1, we are now in the position to give the proof of the existence and uniqueness of the weak solution for (2.8) with the nonlinear drift term (2.17).

Proof of Theorem 2.3.
Step 1. Existence. Recall that for each $n \in \mathbb{N}$, the process $\{u^{(n)}(t), t \in [0, T]\}$ is the solution of (2.20). By Proposition 2.1, it follows that

$$\left\| F(u^{(n)}) \right\|_{L^{\frac{q+2}{q+1}}(\Omega \times [0,T] \times D)^6}^{\frac{q+2}{q+1}} = \mathbb{E}\left[\int_0^T \| u^{(n)}(s) \|_{L^{q+2}(D)^6}^{q+2} ds \right] \tag{2.26}$$

$$\leq C\left(T + \mathbb{E}[\|u_0\|_{\mathbb{H}}^2] \right).$$

Analogously, by the linear growth of B and Proposition 2.1, we have

$$\left\| B(\cdot, u^{(n)}) \right\|_{L^2(\Omega \times [0,T], HS(U_0, \mathbb{H}))}^2 \leq CT\left(1 + \mathbb{E}[\|u_0\|_{\mathbb{H}}^2] \right). \tag{2.27}$$

Therefore, by the reflexivity of $L^{q+2}(\Omega \times [0, T] \times D)^6$ and $L^2(\Omega \times [0, T], HS(U_0, \mathbb{H}))$, (2.26), (2.27), and Proposition 2.1, one obtains that there exist $u \in L^{q+2}(\Omega \times [0, T] \times D)^6$, $Y \in L^{\frac{q+2}{q+1}}(\Omega \times [0, T] \times D)^6$, $Z \in L^2(\Omega \times [0, T], HS(U_0, \mathbb{H}))$, and a subsequence $\{n_k\}$ such that for $n_k \to \infty$,

(i) $u^{(n_k)} \to u$ weakly in $L^{q+2}(\Omega \times [0, T] \times D)^6$;

(ii) $u^{(n_k)} \to u$ weakly* in $L^2(\Omega, L^\infty([0, T], \mathbb{H}))$;

(iii) $F(u^{(n_k)}) \to Y$ weakly in $L^{\frac{q+2}{q+1}}(\Omega \times [0, T] \times D)^6$;

(iv) $B(\cdot, u^{(n_k)}) \to Z$ weakly in $L^2(\Omega \times [0, T], HS(U_0, \mathbb{H}))$ and

$$\int_0^t P_{n_k} B(s, u^{(n_k)}(s)) dW(s) \to \int_0^t Z(s) dW(s) \quad \forall t \in [0, T]$$

weakly* in $L^\infty([0, T], L^2(\Omega, \mathbb{H}))$.

For any $\rho \in L^{q+2}(\Omega \times [0, T])$ and $\phi \in \bigcup_{n \in \mathbb{N}} \mathrm{ran}(P_n)$, by the symmetry of P_n and the skew-adjointness of M, we obtain from (2.20) that

$$\mathbb{E}\left[\int_0^T \langle u^{(n_k)}(t) - u_0^{(n_k)}, \phi \rangle_{\mathbb{H}} \rho(t) dt \right]$$

$$= \mathbb{E}\left[\int_0^T \rho(t) \int_0^t \left(-\langle u^{(n_k)}(s), M P_{n_k} \phi \rangle_{\mathbb{H}} + \langle F(u^{(n_k)}(s)), P_{n_k} \phi \rangle_{\mathbb{H}} \right) ds dt \right]$$

$$+ \mathbb{E}\left[\int_0^T \rho(t) \int_0^t \langle P_{n_k} \phi, B(s, u^{(n_k)}(s)) dW(s) \rangle_{\mathbb{H}} dt \right].$$

By the weak convergence results (i)–(iv) and the stochastic Fubini theorem, we obtain

$$
\mathbb{E}\Big[\int_0^T \langle u(t) - u_0, \phi\rangle_{\mathbb{H}}\,\rho(t)\mathrm{d}t\Big]
$$

$$
= \mathbb{E}\Big[\int_0^T \rho(t)\int_0^t \Big(-\langle u(s), M\phi\rangle_{\mathbb{H}} + \langle Y(s), \phi\rangle_{\mathbb{H}}\Big)\mathrm{d}s\mathrm{d}t\Big]
$$

$$
+ \mathbb{E}\Big[\int_0^T \rho(t)\int_0^t \langle \phi, Z(s)\mathrm{d}W(s)\rangle_{\mathbb{H}}\mathrm{d}t\Big].
$$

Therefore, by the arbitrariness of ρ, we have

$$
\langle u(t) - u_0, \phi\rangle_{\mathbb{H}} = \int_0^t \Big(-\langle u(s), M\phi\rangle_{\mathbb{H}} + \langle Y(s), \phi\rangle_{\mathbb{H}}\Big)\mathrm{d}s
$$
$$
+ \int_0^t \langle \phi, Z(s)\mathrm{d}W(s)\rangle_{\mathbb{H}}, \quad \mathrm{d}t \otimes \mathbb{P}\text{-}a.e. \tag{2.28}
$$

Using the fact that $\bigcup_{n\in\mathbb{N}} \mathrm{ran}(P_n)$ is dense in $\mathscr{D}(M)\cap L^p(D)^6$ for any $p > 1$, one obtains $\phi \in \mathscr{D}(M)\cap L^{q+2}(D)^6$.

Thus, it remains to verify

$$
B(\cdot, u) = Z, \quad F(u) = Y, \quad \mathrm{d}t \otimes \mathbb{P}\text{-}a.e.
$$

To this end, we first note that for any nonnegative function $\psi \in L^\infty([0, T], \mathbb{R})$, it follows from (i) that

$$
\mathbb{E}\Big[\int_0^T \psi(s)\|u(s)\|_{\mathbb{H}}^2\mathrm{d}s\Big] = \lim_{k\to\infty}\mathbb{E}\Big[\int_0^T \langle \psi(s)u(s), u^{(n_k)}(s)\rangle_{\mathbb{H}}\mathrm{d}s\Big]
$$

$$
\leq \Big(\mathbb{E}\Big[\int_0^T \psi(s)\|u(s)\|_{\mathbb{H}}^2\mathrm{d}s\Big]\Big)^{1/2}\liminf_{k\to\infty}\Big(\mathbb{E}\Big[\int_0^T \psi(s)\|u^{(n_k)}(s)\|_{\mathbb{H}}^2\mathrm{d}s\Big]\Big)^{1/2},
$$

which implies

$$
\mathbb{E}\Big[\int_0^T \psi(s)\|u(s)\|_{\mathbb{H}}^2\mathrm{d}s\Big] \leq \liminf_{k\to\infty}\mathbb{E}\Big[\int_0^T \psi(s)\|u^{(n_k)}(s)\|_{\mathbb{H}}^2\mathrm{d}s\Big]. \tag{2.29}
$$

By using (2.28), Lemma 2.2, and the product rule, we obtain that

$$
\mathbb{E}\big[e^{-ct}\|u(t)\|_{\mathbb{H}}^2\big] - \mathbb{E}\big[\|u_0\|_{\mathbb{H}}^2\big]
$$
$$
= \mathbb{E}\Big[\int_0^t e^{-cs}\Big(2\langle u(s), Y(s)\rangle_{\mathbb{H}} + \|Z(s)\|_{HS(U_0,\mathbb{H})}^2 - c\|u(s)\|_{\mathbb{H}}^2\Big)\mathrm{d}s\Big], \tag{2.30}
$$

where c is a positive constant. Furthermore, for any $\phi \in L^{q+2}(\Omega \times [0, T] \times D)^6$ we have

$$\mathbb{E}\big[e^{-ct}\|u^{(n_k)}(t)\|_{\mathbb{H}}^2\big] - \mathbb{E}\big[\|u_0^{(n_k)}\|_{\mathbb{H}}^2\big]$$

$$= \mathbb{E}\bigg[\int_0^t e^{-cs}\bigg(2\langle u^{(n_k)}(s), F(u^{(n_k)}(s))\rangle_{\mathbb{H}}$$

$$+ \sum_{j \in \mathbb{N}} \|P_{n_k}B(s, u^{(n_k)}(s))Q^{\frac{1}{2}}e_j\|_{\mathbb{H}}^2 - c\|u^{(n_k)}(s)\|_{\mathbb{H}}^2\bigg)ds\bigg]$$

$$\leq \mathbb{E}\bigg[\int_0^t e^{-cs}\bigg(2\langle u^{(n_k)}(s), F(u^{(n_k)}(s))\rangle_{\mathbb{H}}$$

$$+ \|B(s, u^{(n_k)}(s))\|_{HS(U_0,\mathbb{H})}^2 - c\|u^{(n_k)}(s)\|_{\mathbb{H}}^2\bigg)ds\bigg]$$

$$= \mathbb{E}\bigg[\int_0^t e^{-cs}\bigg(2\langle u^{(n_k)}(s) - \phi(s), F(u^{(n_k)}(s)) - F(\phi(s))\rangle_{\mathbb{H}}$$

$$+ \|B(s, u^{(n_k)}(s)) - B(s, \phi(s))\|_{HS(U_0,\mathbb{H})}^2 - c\|u^{(n_k)}(s) - \phi(s)\|_{\mathbb{H}}^2\bigg)ds\bigg]$$

$$+ \mathbb{E}\bigg[\int_0^t e^{-cs}\bigg(2\langle \phi(s), F(u^{(n_k)}(s)) - F(\phi(s))\rangle_{\mathbb{H}}$$

$$+ 2\langle u^{(n_k)}(s), F(\phi(s))\rangle_{\mathbb{H}} + 2\langle B(s, u^{(n_k)}(s)), B(s, \phi(s))\rangle_{HS(U_0,\mathbb{H})}$$

$$- \|B(s, \phi(s))\|_{HS(U_0,\mathbb{H})}^2 - 2c\langle u^{(n_k)}(s), \phi(s)\rangle_{\mathbb{H}} + c\|\phi(s)\|_{\mathbb{H}}^2\bigg)ds\bigg].$$

Note that if we choose c large enough, by the Lipschitz continuity of B and (2.18), we have

$$\mathbb{E}\bigg[\int_0^t e^{-cs}\bigg(2\langle u^{(n_k)}(s) - \phi(s), F(u^{(n_k)}(s)) - F(\phi(s))\rangle_{\mathbb{H}}$$

$$+ \|B(s, u^{(n_k)}(s)) - B(s, \phi(s))\|_{HS(U_0,\mathbb{H})}^2 - c\|u^{(n_k)}(s) - \phi(s)\|_{\mathbb{H}}^2\bigg)ds\bigg] < 0.$$

Hence, by letting $k \to \infty$ we conclude from (i)–(iv) and (2.29) that for every nonnegative function $\psi \in L^\infty([0, T], \mathbb{R})$,

$$\mathbb{E}\bigg[\int_0^T \psi(t)\Big(e^{-ct}\|u(t)\|_{\mathbb{H}}^2 - \|u_0\|_{\mathbb{H}}^2\Big)dt\bigg]$$

$$\leq \mathbb{E}\bigg[\int_0^T \psi(t)\Big(\int_0^t e^{-cs}\Big(2\langle \phi(s), Y(s) - F(\phi(s))\rangle_{\mathbb{H}} \tag{2.31}$$

$$+ 2\langle u(s), F(\phi(s))\rangle_{\mathbb{H}} + 2\langle Z(s), B(s, \phi(s))\rangle_{HS(U_0,\mathbb{H})}$$

$$- \|B(s, \phi(s))\|^2_{HS(U_0,\mathbb{H})} - 2c\langle u(s), \phi(s)\rangle_{\mathbb{H}} + c\|\phi(s)\|^2_{\mathbb{H}}\Big)ds\Big)dt\Big].$$

Inserting (2.30) into the left-hand side of (2.31) yields

$$\mathbb{E}\Big[\int_0^T \psi(t)\Big(\int_0^t e^{-cs}\Big(2\langle\phi(s) - u(s), Y(s) - F(\phi(s))\rangle_{\mathbb{H}} \tag{2.32}$$

$$+ \|Z(s) - B(s, \phi(s))\|^2_{HS(U_0,\mathbb{H})} - c\|u(s) - \phi(s)\|^2_{\mathbb{H}}\Big)ds\Big)dt\Big] \le 0.$$

Then by taking $\phi = u$ in (2.32), we obtain that $B(\cdot, u) = Z$.

Again by taking $\phi = u - \gamma\tilde{\phi}\vartheta$ for $\gamma > 0$, $\vartheta \in \mathbb{H}$, and $\tilde{\phi} \in L^{q+2}([0, T])$ in (2.32), dividing both sides by γ and letting $\gamma \to 0$, it follows from the Lipschitz continuity of B and the dominated convergence theorem that

$$\mathbb{E}\Big[\int_0^T \psi(t)\Big(\int_0^t e^{-cs}\tilde{\phi}(s)\langle\vartheta, Y(s) - F(u(s))\rangle_{\mathbb{H}}ds\Big)dt\Big] \le 0.$$

By the arbitrariness of ψ, $\tilde{\phi}$, and ϑ, we conclude that $F(u) = Y$.

Step 2. Uniqueness. Suppose that u and v are solutions of (2.8) with initial data u_0 and v_0, respectively, i.e., for any $t \in [0, T]$ and $\phi \in \mathscr{D}(M) \cap L^{q+2}(D)^6$, we have

$$\langle u(t) - u_0, \phi\rangle_{\mathbb{H}} = \int_0^t \Big(-\langle u(s), M\phi\rangle_{\mathbb{H}} + \langle F(u(s)), \phi\rangle_{\mathbb{H}}\Big)ds$$

$$+ \int_0^t \langle\phi, B(s, u(s))dW(s)\rangle_{\mathbb{H}}, \quad \mathbb{P}\text{-}a.s.$$

and

$$\langle v(t) - v_0, \phi\rangle_{\mathbb{H}} = \int_0^t \Big(-\langle v(s), M\phi\rangle_{\mathbb{H}} + \langle F(v(s)), \phi\rangle_{\mathbb{H}}\Big)ds$$

$$+ \int_0^t \langle\phi, B(s, v(s))dW(s)\rangle_{\mathbb{H}}, \quad \mathbb{P}\text{-}a.s.$$

Then by Lemma 2.2, for all $t \in [0, T]$, we have

$$\mathbb{E}\big[\|u(t) - v(t)\|^2_{\mathbb{H}}\big] = \mathbb{E}\big[\|u_0 - v_0\|^2_{\mathbb{H}}\big]$$

$$+ \mathbb{E}\Big[\int_0^t \Big(2\langle F(u(s)) - F(v(s)), u(s) - v(s)\rangle_{\mathbb{H}}$$

$$+ \|B(s, u(s)) - B(s, v(s))\|^2_{HS(U_0,\mathbb{H})}\Big)ds\Big]$$

$$\leq \mathbb{E}\big[\|u_0 - v_0\|_{\mathbb{H}}^2\big] + C \int_0^t \mathbb{E}\big[\|u(s) - v(s)\|_{\mathbb{H}}^2\big]\mathrm{d}s,$$

where the last estimate follows from (2.18) and the Lipschitz continuity of B. Using the Grönwall inequality, we obtain

$$\mathbb{E}\big[\|u(t) - v(t)\|_{\mathbb{H}}^2\big] \leq e^{Ct}\mathbb{E}\big[\|u_0 - v_0\|_{\mathbb{H}}^2\big]$$

for all $t \in [0, T]$. Consequently, it follows from $u_0 = v_0$ that for every $t \in [0, T]$,

$$u(t) = v(t), \quad \mathbb{P}\text{-}a.s.$$

The proof of Theorem 2.3 is thus finished. $\qquad\qquad\qquad\qquad\qquad\qquad\square$

2.3 Regularity of the Solution

This section is devoted to the study of the regularity of the solution for the stochastic Maxwell equations (2.8), which plays an important role in the analyses of stochastic structure-preserving algorithms in Chaps. 4 and 5. In Sect. 2.3.1, we present the uniform boundedness of the solution in $L^p(\Omega, \mathscr{D}(M^k))$ ($\mathscr{D}(M^k)$-regularity for short) and the Hölder continuity of the solution in $L^2(\Omega, \mathscr{D}(M^{k-1}))$ for a fixed integer $k \geq 1$. Sect. 2.3.2 gives the uniform boundedness of the solution in $L^2(\Omega, H^k(D)^6)$ (H^k-regularity for short) with $k = 1, 2$.

2.3.1 $\mathscr{D}(M^k)$-Regularity

For $k \geq 1$, define recursively the domain $\mathscr{D}(M^k) := \{u \in \mathscr{D}(M^{k-1}) : M^{k-1}u \in \mathscr{D}(M)\}$ for the k-th power of the Maxwell operator M and $\mathscr{D}(M^0) = \mathbb{H}$. If $\mathscr{D}(M^k)$ is endowed with the norm

$$\|v\|_{\mathscr{D}(M^k)} = \left(\|v\|_{\mathbb{H}}^2 + \|M^k v\|_{\mathbb{H}}^2\right)^{1/2} \quad \forall \, v \in \mathscr{D}(M^k), \ k \geq 1,$$

then $\mathscr{D}(M^k)$ is a Hilbert space. Furthermore, it can be shown that $\|u\|_{\mathscr{D}(M^{k_1})} \leq C\|u\|_{\mathscr{D}(M^{k_2})}$ for all $u \in \mathscr{D}(M^{k_2}), k_1 \leq k_2$.

We fix the integer $k \in \mathbb{N}$ and impose the following assumptions on coefficients F and B of the equations.

Assumption 2.4 *There exists a positive constant C such that, for all $0 \leq \ell \leq k$, we have*

$$\|F(t, u)\|_{\mathscr{D}(M^\ell)} \leq C\big(1 + \|u\|_{\mathscr{D}(M^\ell)}\big), \tag{2.33}$$

$$\|F(t, u) - F(s, v)\|_{\mathscr{D}(M^\ell)} \leq C\big(|t - s| + \|u - v\|_{\mathscr{D}(M^\ell)}\big) \qquad (2.34)$$

for $u, v \in \mathscr{D}(M^\ell)$ and $t, s \in [0, T]$.

Assumption 2.5 *There exists a positive constant C such that, for all $0 \leq \ell \leq k$, we have*

$$\|B(t, u)\|_{HS(U_0, \mathscr{D}(M^\ell))} \leq C\Big(1 + \|u\|_{\mathscr{D}(M^\ell)}^2\Big)^{1/2}, \qquad (2.35)$$

$$\|B(t, u) - B(s, v)\|_{HS(U_0, \mathscr{D}(M^\ell))} \leq C\Big(|t - s| + \|u - v\|_{\mathscr{D}(M^\ell)}\Big) \qquad (2.36)$$

for $u, v \in \mathscr{D}(M^\ell)$ and $t, s \in [0, T]$.

We are now ready to establish the $\mathscr{D}(M^k)$-regularity of the solution of (2.8).

Theorem 2.4 *Suppose that ε, μ satisfy Assumption 2.1 and that F, B satisfy Assumptions 2.4 and 2.5. Assume in addition that u_0 is an $\mathscr{F}_0$-measurable $\mathbb{H}$-valued random variable satisfying $\|u_0\|_{L^p(\Omega, \mathscr{D}(M^k))} < \infty$ for some $p \geq 2$. Then there exists a positive constant $C = C(p, T)$ such that the mild solution of (2.8) satisfies*

$$\mathbb{E}\Big[\sup_{t \in [0,T]} \|u(t)\|_{L^p(\Omega, \mathscr{D}(M^k))}\Big] \leq C\Big(1 + \mathbb{E}[\|u_0\|_{\mathscr{D}(M^k)}]\Big), \quad k \in \mathbb{N}. \qquad (2.37)$$

Proof Using Proposition D.5, (2.33), and (2.35), it holds for the mild solution (2.14) that

$$\mathbb{E}\Big[\sup_{t \in [0,T]} \|u(t)\|_{\mathscr{D}(M^k)}^p\Big]$$

$$\leq C\,\mathbb{E}\Big[\sup_{t \in [0,T]} \|S(t)u_0\|_{\mathscr{D}(M^k)}^p\Big] + C\,\mathbb{E}\Big[\int_0^T \|S(t - s)F(s, u(s))\|_{\mathscr{D}(M^k)}^p \mathrm{d}s\Big]$$

$$+ C\,\mathbb{E}\Big[\sup_{t \in [0,T]} \Big\|\int_0^t S(t - s)B(s, u(s))\mathrm{d}W(s)\Big\|_{\mathscr{D}(M^k)}^p\Big]$$

$$\leq C\,\mathbb{E}[\|u_0\|_{\mathscr{D}(M^k)}^p] + C\int_0^T \Big(1 + \mathbb{E}[\|u(s)\|_{\mathscr{D}(M^k)}^p]\Big)\mathrm{d}s$$

$$+ C\,\mathbb{E}\Big[\Big(\int_0^T \|B(s, u(s))\|_{HS(U_0, \mathscr{D}(M^k))}^2 \mathrm{d}s\Big)^{p/2}\Big]$$

$$\leq C\,\mathbb{E}[\|u_0\|_{\mathscr{D}(M^k)}^p] + C\int_0^T \Big(1 + \mathbb{E}\Big[\sup_{r \in [0,s]} \|u(r)\|_{\mathscr{D}(M^k)}^p\Big]\Big)\mathrm{d}s,$$

which together with the Grönwall inequality leads to the assertion.   □

Now, we turn to the Hölder continuity of the solution of (2.8).

Theorem 2.5 *Let conditions in Theorem 2.4 hold. For any $p \geq 2$, there exists a positive constant $C = C(p, T, u_0)$ such that*

$$\mathbb{E}\left[\|u(t) - u(s)\|_{\mathscr{D}(M^{k-1})}^{p}\right] \leq C(t - s)^{p/2}, \tag{2.38}$$

$$\left\|\mathbb{E}[u(t) - u(s)]\right\|_{\mathscr{D}(M^{k-1})} \leq C(t - s) \tag{2.39}$$

for all $0 \leq s \leq t \leq T$ and $k \geq 1$.

Proof From the mild solution (2.14), we have that for all $0 \leq s \leq t \leq T$,

$$u(t) - u(s) = \left(S(t - s) - Id\right)u(s) + \int_s^t S(t - r)F(r, u(r))dr$$

$$+ \int_s^t S(t - r)B(r, u(r))dW(r), \quad \mathbb{P}\text{-}a.s. \tag{2.40}$$

Therefore,

$$\mathbb{E}\left[\|u(t) - u(s)\|_{\mathscr{D}(M^{k-1})}^{p}\right]$$

$$\leq C\mathbb{E}\left[\|(S(t - s) - Id)u(s)\|_{\mathscr{D}(M^{k-1})}^{p}\right]$$

$$+ C\mathbb{E}\left[\left\|\int_s^t S(t - r)F(r, u(r))dr\right\|_{\mathscr{D}(M^{k-1})}^{p}\right]$$

$$+ C\mathbb{E}\left[\left\|\int_s^t S(t - r)B(r, u(r))dW(r)\right\|_{\mathscr{D}(M^{k-1})}^{p}\right]$$

$$=: I_1 + I_2 + I_3.$$

For the first term I_1, we have

$$I_1 \leq \left\|S(t - s) - Id\right\|_{\mathscr{L}(\mathscr{D}(M^k), \mathscr{D}(M^{k-1}))}^{p} \mathbb{E}\left[\|u(s)\|_{\mathscr{D}(M^k)}^{p}\right]$$

$$\leq C(t - s)^p \mathbb{E}\left[\|u(s)\|_{\mathscr{D}(M^k)}^{p}\right] \tag{2.41}$$

$$\leq C\left(1 + \mathbb{E}\left[\|u_0\|_{\mathscr{D}(M^k)}^{p}\right]\right)(t - s)^p,$$

where Lemma C.1 and Theorem 2.4 are used. For the second term I_2, Assumption 2.4 and Theorem 2.4 imply that

$$
\begin{aligned}
I_2 &= \mathbb{E}\Big[\Big\| \int_s^t S(t-r)F(r, u(r))\mathrm{d}r \Big\|_{\mathscr{D}(M^{k-1})}^p \Big] \\
&\leq C(t-s)^{p-1} \int_s^t \mathbb{E}\Big[\Big\| S(t-r)F(r, u(r)) \Big\|_{\mathscr{D}(M^{k-1})}^p \Big]\mathrm{d}r \\
&\leq C(t-s)^{p-1} \int_s^t \mathbb{E}\Big[\| F(r, u(r)) \|_{\mathscr{D}(M^{k-1})}^p \Big]\mathrm{d}r \\
&\leq C(t-s)^{p-1} \int_s^t \mathbb{E}\Big[1 + \| u(r) \|_{\mathscr{D}(M^{k-1})}^p \Big]\mathrm{d}r \\
&\leq C\Big(1 + \mathbb{E}\big[\| u_0 \|_{\mathscr{D}(M^{k-1})}^p \big]\Big)(t-s)^p.
\end{aligned}
\tag{2.42}
$$

It follows from Proposition D.4 (ii) that

$$
\begin{aligned}
I_3 &\leq C\mathbb{E}\Big[\Big(\int_s^t \Big\| S(t-r)B(r, u(r)) \Big\|_{HS(U_0, \mathscr{D}(M^{k-1}))}^2 \mathrm{d}r \Big)^{p/2} \Big] \\
&\leq C\mathbb{E}\Big[\Big(\int_s^t \| S(t-r) \|_{\mathscr{L}(\mathscr{D}(M^{k-1}))}^2 \| B(r, u(r)) \|_{HS(U_0, \mathscr{D}(M^{k-1}))}^2 \mathrm{d}r \Big)^{p/2} \Big] \\
&\leq C\mathbb{E}\Big[\Big(\int_s^t \big(1 + \| u(r) \|_{\mathscr{D}(M^{k-1})}^2 \big)\mathrm{d}r \Big)^{p/2} \Big] \\
&\leq C\Big(1 + \mathbb{E}\big[\| u_0 \|_{\mathscr{D}(M^{k-1})}^p \big]\Big)(t-s)^{p/2}.
\end{aligned}
\tag{2.43}
$$

Combining (2.41)–(2.43) and the assumption $u_0 \in L^p(\Omega, \mathscr{D}(M^k))$, we obtain (2.38).

To derive (2.39), we take the expectation on both sides of (2.40) and obtain

$$
\mathbb{E}\big[u(t) - u(s)\big] = \mathbb{E}\big[(S(t-s) - Id)u(s)\big] + \mathbb{E}\Big[\int_s^t S(t-r)F(r, u(r))\mathrm{d}r\Big]
$$

for all $0 \leq s \leq t \leq T$. By similar arguments as those in (2.41) and (2.42), it can be shown that

$$
\begin{aligned}
&\big\| \mathbb{E}\big[u(t) - u(s)\big] \big\|_{\mathscr{D}(M^{k-1})} \\
&\leq \mathbb{E}\Big[\big\| (S(t-s) - Id)u(s) \big\|_{\mathscr{D}(M^{k-1})}\Big] + \mathbb{E}\Big[\int_s^t \| S(t-r)F(r, u(r)) \|_{\mathscr{D}(M^{k-1})}\mathrm{d}r\Big] \\
&\leq C(t-s).
\end{aligned}
$$

The proof of Theorem 2.5 is finished. □

2.3.2 H^k-Regularity

In this subsection, we focus on the H^k-regularity ($k = 1, 2$) of the solution for the stochastic Maxwell equations with additive noise

$$\begin{cases} d\mathbf{E}(t) = \nabla \times \mathbf{H}(t)dt + \lambda_1 dW(t), & t \in (0, T], \\ d\mathbf{H}(t) = -\nabla \times \mathbf{E}(t)dt + \lambda_2 dW(t), & t \in (0, T], \\ \mathbf{E}(0) = \mathbf{E}_0, \quad \mathbf{H}(t) = \mathbf{H}_0 \end{cases} \tag{2.44}$$

on a cuboid $D = (a_1^-, a_1^+) \times (a_2^-, a_2^+) \times (a_3^-, a_3^+) \subset \mathbb{R}^3$ with a Lipschitz boundary $\Gamma := \partial D$. Here $\lambda_i \in \mathbb{R}^3$, $i = 1, 2$, describe the scale of the noise, and W is a U-valued Q-Wiener process. To simplify the presentation, we restrict ourselves to the constant parameters case with $\varepsilon = \mu \equiv 1$. Moreover, it is not difficult to prove that the conclusions in this part still hold if we assume $\varepsilon, \mu \in W^{1,\infty}(D) \cap W^{2,3}(D)$ with $\varepsilon, \mu \geq \delta > 0$ for a constant δ.

To derive the H^k-regularity of the solution of (2.44), we impose the following PEC boundary conditions

$$\mathbf{n} \times \mathbf{E} = 0, \quad \mathbf{n} \cdot \mathbf{H} = 0, \quad \text{on } [0, T] \times \Gamma. \tag{2.45}$$

By defining $u = (\mathbf{E}^\top, \mathbf{H}^\top)^\top$, $u_0 = (\mathbf{E}_0^\top, \mathbf{H}_0^\top)^\top$, and $\lambda = (\lambda_1^\top, \lambda_2^\top)^\top$, (2.44) can be rewritten as a stochastic evolution equation

$$\begin{cases} du(t) = Mu(t)dt + \lambda dW(t), & t \in (0, T], \\ u(0) = u_0, \end{cases} \tag{2.46}$$

and the mild solution is given by

$$u(t) = S(t)u_0 + \int_0^t S(t - s)\lambda dW(s), \quad \mathbb{P}\text{-a.s.} \tag{2.47}$$

for each $t \in [0, T]$.

We first investigate the H^1-regularity of the mild solution (2.47).

Theorem 2.6 *Let conditions with $k = 1$ in Theorem 2.4 hold. Assume that $Q^{\frac{1}{2}} \in HS(U, H^1(D))$ and $u_0 \in L^2(\Omega, H^1(D)^6)$. Then the mild solution $u \in L^2(\Omega, H^1(D)^6)$ satisfies*

$$\sup_{t \in [0,T]} \mathbb{E}\big[\|u(t)\|_{H^1(D)^6}^2\big] \leq C\Big(1 + \mathbb{E}\big[\|u_0\|_{H^1(D)^6}^2\big]\Big),$$

where the positive constant C depends on T, $|\lambda|$, and $\|Q^{\frac{1}{2}}\|_{HS(U,H^1(D))}$.

Proof Taking $\nabla\cdot$ on both sides of (2.44) yields

$$\nabla \cdot \mathbf{E}(t) = \nabla \cdot \mathbf{E}_0 + \lambda_1 \cdot (\nabla W(t)), \quad \nabla \cdot \mathbf{H}(t) = \nabla \cdot \mathbf{H}_0 + \lambda_2 \cdot (\nabla W(t)) \tag{2.48}$$

for all $t \in [0, T]$. Hence, there exists a positive constant C such that

$$\mathbb{E}\Big[\|\nabla \cdot (\mathbf{E}(t))\|_U^2\Big] = \mathbb{E}\Big[\|\nabla \cdot (\mathbf{E}_0)\|_U^2\Big] + \mathbb{E}\Big[\|\lambda_1 \cdot (\nabla W(t))\|_U^2\Big]$$

$$\leq C\Big(1 + \mathbb{E}[\|\mathbf{E}_0\|_{H^1(D)^3}^2]\Big) \quad \forall\, t \in [0, T].$$

Similarly, we have

$$\mathbb{E}\Big[\|\nabla \cdot \mathbf{H}(t)\|_U^2\Big] \leq C\Big(1 + \mathbb{E}[\|\mathbf{H}_0\|_{H^1(D)^3}^2]\Big) \quad \forall\, t \in [0, T].$$

Combining these two estimates, Lemma B.2, and Theorem 2.4, we obtain

$$\mathbb{E}\big[\|u(t)\|_{H^1(D)^6}^2\big] \leq C\Big(\mathbb{E}\big[\|\mathbf{E}(t)\|_{L^2(D)^3}^2\big] + \mathbb{E}\big[\|\mathbf{H}(t)\|_{L^2(D)^3}^2\big]$$

$$+\mathbb{E}\big[\|\nabla \times \mathbf{E}(t)\|_{L^2(D)^3}^2\big]$$

$$+\mathbb{E}\big[\|\nabla \times \mathbf{H}(t)\|_{L^2(D)^3}^2\big] + \mathbb{E}\big[\|\nabla \cdot \mathbf{E}(t)\|_U^2\big] + \mathbb{E}\big[\|\nabla \cdot \mathbf{H}(t)\|_U^2\big]\Big)$$

$$\leq C\Big(\mathbb{E}\big[\|u(t)\|_{\mathscr{D}(M)}^2\big] + \mathbb{E}\big[\|\nabla \cdot \mathbf{E}(t)\|_U^2\big] + \mathbb{E}\big[\|\nabla \cdot \mathbf{H}(t)\|_U^2\big]\Big)$$

$$\leq C\Big(1 + \mathbb{E}\big[\|u_0\|_{H^1(D)^6}^2\big]\Big)$$

for all $t \in [0, T]$. $\square$

We now pass to a higher regularity estimation. For simplicity, denote

$$\Gamma_1^\pm = \{\mathbf{x} \in \overline{D} : x = a_1^\pm\}, \quad \Gamma_2^\pm = \{\mathbf{x} \in \overline{D} : y = a_2^\pm\}, \quad \Gamma_3^\pm = \{\mathbf{x} \in \overline{D} : z = a_3^\pm\},$$

and $\Gamma_j := \Gamma_j^- \cup \Gamma_j^+$ for $j = 1, 2, 3$. For a union $\Gamma^* \subseteq \Gamma$ of some faces of D, we set

$$H_{\Gamma^*}^1(D) = \{u \in H^1(D) : u = 0 \text{ on } \Gamma^*\}.$$

By introducing a subspace $H_{00}^1(D)$ of $H^1(D)$:

$$H_{00}^1(D) := \{ f \in H^1(D) : f|_{\Gamma'} \in H_0^{1/2}(\Gamma') \text{ for all faces } \Gamma' \text{ of } D \},$$

we can obtain the H^2-regularity of the mild solution u as stated below.

Theorem 2.7 *Let conditions with $k = 2$ in Theorem 2.4 hold. In addition, assume that $Q^{\frac{1}{2}} \in HS(U, H^2(D))$, $\nabla Q^{\frac{1}{2}} : U \to H_{00}^1(D)^3$, $u_0 \in L^2(\Omega, H^2(D)^6)$, and $\nabla \cdot E_0 \in L^2(\Omega, H_{00}^1(D))$. Then the mild solution $u \in L^2(\Omega, H^2(D)^6)$ satisfies*

$$\sup_{t \in [0,T]} \mathbb{E}\big[\|u(t)\|_{H^2(D)^6}^2\big] \leq C\Big(1 + \mathbb{E}\big[\|u_0\|_{H^2(D)^6}^2\big]\Big),$$

where the positive constant C depends on T, $|\lambda|$, and $\|Q^{\frac{1}{2}}\|_{HS(U,H^2(D))}$. Moreover, the field $(E(t)^\top, H(t)^\top)^\top$ has traces

$$E_2 = E_3 = 0, \quad \partial_y E_2 = \partial_z E_2 = \partial_y E_3 = \partial_z E_3 = 0, \quad on \quad \Gamma_1,$$

$$E_1 = E_3 = 0, \quad \partial_x E_1 = \partial_z E_1 = \partial_x E_3 = \partial_z E_3 = 0, \quad on \quad \Gamma_2,$$

$$E_1 = E_2 = 0, \quad \partial_x E_1 = \partial_y E_1 = \partial_x E_2 = \partial_y E_2 = 0, \quad on \quad \Gamma_3,$$

$$H_1 = 0, \qquad\qquad\qquad \partial_y H_1 = \partial_z H_1 = 0, \quad on \quad \Gamma_1,$$

$$H_2 = 0, \qquad\qquad\qquad \partial_x H_2 = \partial_z H_2 = 0, \quad on \quad \Gamma_2,$$

$$H_3 = 0, \qquad\qquad\qquad \partial_x H_3 = \partial_y H_3 = 0, \quad on \quad \Gamma_3.$$

Proof The PEC boundary conditions (2.45) imply the asserted zero-order traces for E and H. Then, using Lemma B.6 and the H^2-regularity of the solution u, we obtain the first-order traces result.

The proof of the H^2-regularity of the solution of (2.46) consists of three steps. In the first step we prove the mild solution $u \in L^2(\Omega, H_{loc}^2(D)^6)$. In the second and third steps, we show the H^2-regularity for E and H, respectively.

Step 1. It follows from the divergence evolution laws (2.48) that

$$\mathbb{E}\Big[\|\nabla \cdot E(t)\|_{H^1(D)}^2 + \|\nabla \cdot H(t)\|_{H^1(D)}^2\Big] \leq C\Big(1 + \mathbb{E}\big[\|u_0\|_{H^2(D)^6}^2\big]\Big),$$

where the positive constant C depends on T, $|\lambda|$, and $\|Q^{\frac{1}{2}}\|_{HS(U,H^2(D))}$. Moreover, from assumptions of the initial datum and the noise, we have that $\nabla \cdot E(t) \in L^2(\Omega, H_{00}^1(D))$ for all $t \in [0, T]$.

Note that

$$M^2 u = \begin{bmatrix} -\nabla \times (\nabla \times \mathbf{E}) \\ -\nabla \times (\nabla \times \mathbf{H}) \end{bmatrix}$$

and $\Delta \mathbf{E} = \nabla(\nabla \cdot \mathbf{E}) - \nabla \times (\nabla \times \mathbf{E})$. Thus, it follows from Theorem 2.4 with $k = 2$ that $\Delta \mathbf{E} \in L^2(\Omega, L^2(D)^3)$ and

$$\mathbb{E}\big[\|\Delta \mathbf{E}(t)\|^2_{L^2(D)^3}\big] \leq C\Big(\mathbb{E}\big[\|\nabla \cdot \mathbf{E}(t)\|^2_{H^1(D)}\big] + \mathbb{E}\big[\|u(t)\|^2_{\mathscr{D}(M^2)}\big]\Big) \quad \forall t \in [0, T].$$

The field $\mathbf{H}(t)$ can be estimated similarly. Standard interior elliptic regularity result then leads to $\mathbf{E}(t)$, $\mathbf{H}(t) \in L^2(\Omega, H^2_{\text{loc}}(D)^3)$ for all $t \in [0, T]$.

Step 2. To obtain the H^2-regularity of the electric field $\mathbf{E}$, we first consider the H^2-regularity of the first component E_1 of $\mathbf{E}$. Set $\Gamma^* := \Gamma_2 \cup \Gamma_3$. The result in *Step 1* implies that $f := (Id - \Delta)E_1 \in U$. For a given $\varphi \in H^1_{\Gamma^*}(D)$, by employing the cut-off and the mollification in y, z-directions, one can approximate it by a smooth function ψ with support in $[a_1^-, a_1^+] \times [a_2^- + \eta, a_2^+ - \eta] \times [a_3^- + \eta, a_3^+ - \eta]$ for some small constant $\eta := \eta(\psi) > 0$. For each $\kappa \in (0, \eta)$, define $D_\kappa := (a_1^- + \kappa, a_1^- - \kappa) \times (a_2^- + \kappa, a_2^+ - \kappa) \times (a_3^- + \kappa, a_3^+ - \kappa)$ and denote by $\Gamma_1^\pm(\kappa)$ those open faces of D_κ that contain points of the form $(a_1^\mp \pm \kappa, y, z)$. The integration by parts formula and the support of ψ yield that

$$\int_D E_1 \psi \, d\mathbf{x} + \int_D \nabla E_1 \cdot \nabla \psi \, d\mathbf{x} = \lim_{\kappa \to 0} \int_{D_\kappa} E_1 \psi \, d\mathbf{x} + \int_{D_\kappa} \nabla E_1 \cdot \nabla \psi \, d\mathbf{x}$$

$$= \lim_{\kappa \to 0} \left[\int_{D_\kappa} \psi (Id - \Delta) E_1 \, d\mathbf{x} + \int_{\partial D_\kappa} \psi \nabla E_1 \cdot \mathbf{n} \, d\sigma \right]$$

$$= \int_D \psi f \, d\mathbf{x} + \lim_{\kappa \to 0} \int_{\Gamma_1^+(\kappa)} \psi \partial_x E_1 \, d\sigma - \lim_{\kappa \to 0} \int_{\Gamma_1^-(\kappa)} \psi \partial_x E_1 \, d\sigma$$

$$= \int_D \psi f \, d\mathbf{x} + \lim_{\kappa \to 0} \int_{\Gamma_1^+(\kappa)} \psi (\nabla \cdot \mathbf{E} - \partial_y E_2 - \partial_z E_3) \, d\sigma$$

$$\quad - \lim_{\kappa \to 0} \int_{\Gamma_1^-(\kappa)} \psi (\nabla \cdot \mathbf{E} - \partial_y E_2 - \partial_z E_3) \, d\sigma$$

$$= \int_D \psi f \, d\mathbf{x} + \int_{\Gamma_1^+} \psi \nabla \cdot \mathbf{E} \, d\sigma - \int_{\Gamma_1^-} \psi \nabla \cdot \mathbf{E} \, d\sigma,$$

where we used the facts that ψ vanishes on the boundary of Γ^*, and that E_2 and E_3 vanish on Γ_1. Lemma B.5 and $\nabla \cdot \mathbf{E} \in L^2(\Omega, H^1_{00}(D))$ (see *Step 1*) lead to $E_1 \in$

$L^2(\Omega, H^2(D))$. In the same manner, one observes that $E_2, E_3 \in L^2(\Omega, H^2(D))$. Furthermore,

$$\mathbb{E}\big[\|E_j\|^2_{H^2(D)}\big] \le C\Big(\mathbb{E}\big[\|E_j\|^2_U\big] + \mathbb{E}\big[\|\Delta E_j\|^2_U\big] + \mathbb{E}\big[\|\nabla \cdot \mathbf{E}\|^2_{H_0^{1/2}(\Gamma_j)}\big]\Big)$$

$$\le C\Big(\mathbb{E}\big[\|\nabla \cdot \mathbf{E}\|^2_{H^1(D)}\big] + \mathbb{E}\big[\|u\|^2_{\mathscr{D}(M^2)}\big]\Big)$$

for all $j = 1, 2, 3$.

Step 3. Now we consider the first component H_1 of $\mathbf{H}$, and set $\Gamma^* := \Gamma_1$ and $\widetilde{f} := (Id - \Delta)H_1 \in U$. As in *Step 2*, we take a smooth function ψ with support in $[a_1^- + \eta, a_1^+ - \eta] \times [a_2^-, a_2^+] \times [a_3^-, a_3^+]$ for some small constant $\eta := \eta(\psi) > 0$. Choose $\kappa \in (0, \eta)$ so that ψ vanishes on $\Gamma_1^\pm(\kappa)$, then

$$\int_D H_1 \psi \, dx + \int_D \nabla H_1 \cdot \nabla \psi \, dx = \lim_{\kappa \to 0}\Big[\int_{D_\kappa} H_1 \psi \, dx + \int_{D_\kappa} \nabla H_1 \cdot \nabla \psi \, dx\Big]$$

$$= \lim_{\kappa \to 0}\Big[\int_{D_\kappa} \psi(Id - \Delta)H_1 dx + \int_{\partial D_\kappa} \psi \nabla H_1 \cdot \mathbf{n} d\sigma\Big]$$

$$= \int_D \psi \widetilde{f} dx + \lim_{\kappa \to 0}\int_{\partial D_\kappa}\Big[\psi \nabla H_1 \cdot \mathbf{n} - \big((\nabla \times \mathbf{H}) \times \mathbf{n}\big)\cdot(\psi, 0, 0)^\top\Big]d\sigma$$

$$= \int_D \psi \widetilde{f} dx + \lim_{\kappa \to 0}\int_{\partial D_\kappa} \psi \partial_x \mathbf{H} \cdot \mathbf{n} d\sigma$$

$$= \int_D \psi \widetilde{f} dx + \lim_{\kappa \to 0}\Big[\int_{\Gamma_2^+(\kappa)} \psi \partial_x H_2 d\sigma + \int_{\Gamma_3^+(\kappa)} \psi \partial_x H_3 d\sigma\Big]$$

$$\quad - \lim_{\kappa \to 0}\Big[\int_{\Gamma_2^-(\kappa)} \psi \partial_x H_2 d\sigma + \int_{\Gamma_3^-(\kappa)} \psi \partial_x H_3 d\sigma\Big]$$

$$= \int_D \psi \widetilde{f} dx \tag{2.49}$$

due to the integration by parts formula and the fact that H_2 and H_3 vanish on Γ_2 and Γ_3, respectively. Hence, Lemma B.4 leads to

$$\mathbb{E}\big[\|H_1\|^2_{H^2(D)}\big] \le C\Big(\mathbb{E}\big[\|H_1\|^2_U\big] + \mathbb{E}\big[\|\Delta H_1\|^2_U\big]\Big)$$

$$\le C\Big(\mathbb{E}\big[\|\nabla \cdot \mathbf{H}\|^2_{H^1(D)}\big] + \mathbb{E}\big[\|u\|^2_{\mathscr{D}(M^2)}\big]\Big).$$

Components H_2 and H_3 can be treated similarly.

Combining *Steps 1–3* and Theorem 2.4, we have

$$\mathbb{E}\big[\|u(t)\|_{H^2(D)^6}^2\big] \leq C\Big(\mathbb{E}\big[\|u(t)\|_{\mathcal{D}(M^2)}^2\big] + \mathbb{E}\big[\|\nabla \cdot \mathbf{E}(t)\|_{H^1(D)}^2\big]$$

$$+ \mathbb{E}\big[\|\nabla \cdot \mathbf{H}(t)\|_{H^1(D)}^2\big]\Big)$$

$$\leq C\Big(1 + \mathbb{E}\big[\|u_0\|_{H^2(D)^6}^2\big]\Big).$$

The proof of Theorem 2.7 is finished.   □

Remark 2.6 Set $F_Q(\mathbf{x}) := \sum_{j \in \mathbb{N}} \big(Q^{\frac{1}{2}} e_j(\mathbf{x})\big)^2$. Assume $\varepsilon, \mu \in W^{1,\infty}(D) \cap W^{2,3}(D)$ with $\varepsilon, \mu \geq \delta > 0$, $F_Q \in W^{1,\infty}(D)$, $Q^{\frac{1}{2}} \in HS\big(U, H^2(D) \cap H_0^1(D)\big)$. Then conclusions in Theorems 2.6 and 2.7 still hold for the following stochastic Maxwell equations with multiplicative noise in the Stratonovich sense

$$\begin{cases} \varepsilon d\mathbf{E}(t) = \nabla \times \mathbf{H}(t)dt - \lambda_1 \mathbf{H}(t) \circ dW(t), & t \in (0, T], \\ \mu d\mathbf{H}(t) = -\nabla \times \mathbf{E}(t)dt + \lambda_2 \mathbf{E}(t) \circ dW(t), & t \in (0, T], \\ \mathbf{E}(0) = \mathbf{E}_0, \quad \mathbf{H}(t) = \mathbf{H}_0 \end{cases} \tag{2.50}$$

with $\lambda_1, \lambda_2 \in \mathbb{R}$. In fact, by using the equivalent Itô form of the above equations

$$\varepsilon d\mathbf{E}(t) = \Big[\nabla \times \mathbf{H}(t) - \frac{1}{2}\mu^{-1}\lambda_1\lambda_2 F_Q \mathbf{E}(t)\Big]dt - \lambda_1 \mathbf{H}(t)dW(t),$$

$$\mu d\mathbf{H}(t) = \Big[-\nabla \times \mathbf{E}(t) - \frac{1}{2}\varepsilon^{-1}\lambda_1\lambda_2 F_Q \mathbf{H}(t)\Big]dt + \lambda_2 \mathbf{E}(t)dW(t),$$

and the divergence evolution laws given in Proposition 3.2, similar to proofs of the additive noise case, we can obtain the H^k-regularity of the solution of (2.50) for $k = 1, 2$.

2.4 Differentiability with Respect to the Initial Datum

Let $u(t; u_0)$ be the exact solution of (2.8) with the initial datum u_0. First, we show that the solution mapping $u_0 \mapsto u(t; u_0)$ is Lipschitz continuous.

Proposition 2.2 *Suppose that Assumptions 2.1, 2.2, and 2.3 hold. For $p \geq 2$, there exists a constant $C = C(p, T, F, B) > 0$ such that*

$$\mathbb{E}\Big[\|u(t; u_0) - u(t; \tilde{u}_0)\|_{\mathbb{H}}^p\Big] \leq C\|u_0 - \tilde{u}_0\|_{\mathbb{H}}^p, \quad t \in [0, T], \ u_0, \tilde{u}_0 \in \mathbb{H}. \tag{2.51}$$

In addition,

$$\mathbb{E}\Big[\sup_{t\in[0,T]}\|u(t;u_0) - u(t;\tilde{u}_0)\|_{\mathbb{H}}^p\Big] \le C\|u_0 - \tilde{u}_0\|_{\mathbb{H}}^p, \quad u_0, \tilde{u}_0 \in \mathbb{H}. \tag{2.52}$$

Proof It follows from the mild solution (2.14) that

$$\mathbb{E}\Big[\|u(t;u_0) - u(t;\tilde{u}_0)\|_{\mathbb{H}}^p\Big]$$

$$\le C\|S(t)(u_0 - \tilde{u}_0)\|_{\mathbb{H}}^p$$

$$+ C\,\mathbb{E}\Big[\Big\|\int_0^t S(t-r)\Big(F(r,u(r;u_0)) - F(r,u(r;\tilde{u}_0))\Big)dr\Big\|_{\mathbb{H}}^p\Big]$$

$$+ C\,\mathbb{E}\Big[\Big\|\int_0^t S(t-r)\Big(B(r,u(r;u_0)) - B(r,u(r;\tilde{u}_0))\Big)dW(r)\Big\|_{\mathbb{H}}^p\Big]$$

$$\le C\|u_0 - \tilde{u}_0\|_{\mathbb{H}}^p + C\,\mathbb{E}\Big[\int_0^t \|F(r,u(r;u_0)) - F(r,u(r;\tilde{u}_0))\|_{\mathbb{H}}^p dr\Big]$$

$$+ C\,\mathbb{E}\Big[\Big(\int_0^t \|B(r,u(r;u_0)) - B(r,u(r;\tilde{u}_0))\|_{HS(U_0,\mathbb{H})}^2 dr\Big)^{\frac{p}{2}}\Big]$$

$$\le C\|u_0 - \tilde{u}_0\|_{\mathbb{H}}^p + C\int_0^t \mathbb{E}\Big[\|u(r;u_0) - u(r;\tilde{u}_0)\|_{\mathbb{H}}^p\Big]dr,$$

where we used Proposition D.4 (ii) and the Lipschitz continuious properties of F and B. Then, the Grönwall inequality leads to (2.51).

For the derivation of (2.52), it suffices to estimate the term

$$\mathbb{E}\Big[\sup_{t\in[0,T]}\Big\|\int_0^t S(t-r)\Big(B(r,u(r;u_0)) - B(r,u(r;\tilde{u}_0))\Big)dW(r)\Big\|_{\mathbb{H}}^p\Big].$$

In fact, by Proposition D.5, we have

$$\mathbb{E}\Big[\sup_{t\in[0,T]}\Big\|\int_0^t S(t-r)\Big(B(r,u(r;u_0)) - B(r,u(r;\tilde{u}_0))\Big)dW(r)\Big\|_{\mathbb{H}}^p\Big]$$

$$\le C\mathbb{E}\Big[\Big(\int_0^T \|B(r,u(r;u_0)) - B(r,u(r;\tilde{u}_0))\|_{HS(U_0,\mathbb{H})}^2 dr\Big)^{\frac{p}{2}}\Big]$$

$$\le C\mathbb{E}\Big[\int_0^T \|u(r;u_0) - u(r;\tilde{u}_0)\|_{\mathbb{H}}^p dr\Big]$$

$$\le C\|u_0 - \tilde{u}_0\|_{\mathbb{H}}^p,$$

where in the last step we used (2.51). $\qquad\square$

For any $p \geq 2$, denote by $\mathscr{H}_{p,0}$ the Banach space of all $\mathbb{H}$-valued predictable stochastic processes Y defined on the time interval $[0, T]$ such that

$$\|Y\|_{p,0} := \sup_{t \in [0,T]} \left(\mathbb{E}\big[\|Y(t)\|_{\mathbb{H}}^p\big] \right)^{\frac{1}{p}} < \infty.$$

It is known from [62, Theorem 9.8] that for any $t \in [0, T]$, the solution mapping $u_0 \mapsto u(t; u_0)$ is Gâteaux differentiable provided that coefficients F and B are Gâteaux differentiable.

Proposition 2.3 *Let conditions in Proposition 2.2 hold. In addition assume that for $t \in [0, T]$, $F(t, \cdot)$ and $B(t, \cdot)$ are Gâteaux differentiable on $\mathbb{H}$ with derivatives $D_u F(t, u)(v)$ and $D_u B(t, u)(v)$ being continuous in $u, v \in \mathbb{H}$ and satisfying*

$$\|D_u F(t, u)(v)\|_{\mathbb{H}} + \|D_u B(t, u)(v)\|_{HS(U_0, \mathbb{H})} \leq C\|v\|_{\mathbb{H}}$$

with a positive constant C. Then the solution u is Gâteaux differentiable from $\mathbb{H}$ into $\mathscr{H}_{p,0}$ for any $p \geq 2$. Moreover, for any $u_0, \tilde{u}_0 \in \mathbb{H}$, the stochastic process

$$\zeta^{\tilde{u}_0}(t) := D_u u(t; u_0)(\tilde{u}_0), \quad t \in [0, T]$$

is the mild solution of the following equation:

$$\begin{cases} d\zeta^{\tilde{u}_0}(t) = \big[M\zeta^{\tilde{u}_0}(t) + D_u F(t, u(t; u_0))(\zeta^{\tilde{u}_0}(t)) \big] dt \\ \qquad\quad + D_u B(t, u(t; u_0))(\zeta^{\tilde{u}_0}(t)) dW(t), \quad t \in (0, T], \\ \zeta^{\tilde{u}_0}(0) = \tilde{u}_0. \end{cases} \qquad (2.53)$$

Note that (2.53) is a linear stochastic evolution equation. Thus its mild solution reads as

$$\zeta^{\tilde{u}_0}(t) = S(t)\tilde{u}_0 + \int_0^t S(t - r) D_u F(r, u(r; u_0))(\zeta^{\tilde{u}_0}(r)) dr$$

$$+ \int_0^t S(t - r) D_u B(r, u(r; u_0))(\zeta^{\tilde{u}_0}(r)) dW(r), \quad \mathbb{P}\text{-}a.s. \qquad (2.54)$$

for all $t \in [0, T]$. It can be shown that for any $p \geq 2$, there exists a positive constant C such that

$$\mathbb{E}\left[\sup_{t \in [0,T]} \|\zeta^{\tilde{u}_0}(t)\|_{\mathbb{H}}^p \right] \leq C\|\tilde{u}_0\|_{\mathbb{H}}^p. \qquad (2.55)$$

Below we present the Fréchet differentiability of the solution of (2.8).

Theorem 2.8 *Let conditions in Proposition 2.2 hold. Assume that $F(t, \cdot)$ and $B(t, \cdot)$ are Fréchet differentiable on $\mathbb{H}$ with bounded derivatives, i.e.,*

$$\|F_u(t, u)\|_{\mathscr{L}(\mathbb{H})} + \|B_u(t, u)\|_{\mathscr{L}(\mathbb{H}, HS(U_0, \mathbb{H}))} \leq C$$

for all $t \in [0, T]$ and $u \in \mathbb{H}$. Then the solution u of (2.8) is Fréchet differentiable from $\mathbb{H}$ into $\mathscr{H}_{p,0}$ for any $p \geq 2$.

Proof Below we only give the proof for $p = 2$, since the case of $p > 2$ can be proved by the same procedure.

Based on Lemma B.7, it suffices to show that

$$\lim_{\rho \to 0} \frac{u(t; u_0 + \rho \tilde{u}_0) - u(t; u_0) - \rho \zeta^{\tilde{u}_0}(t)}{\rho} = 0 \quad \text{in } \mathscr{H}_{2,0} \tag{2.56}$$

uniformly with respect to $\tilde{u}_0$ in an arbitrary bounded set of $\mathbb{H}$.

Set $u^\rho(t) := u(t; u_0 + \rho \tilde{u}_0)$. According to the mild solutions (2.14) and (2.54), one has

$$\mathbb{E}\left[\left\|\frac{u^\rho(t) - u(t) - \rho \zeta^{\tilde{u}_0}(t)}{\rho}\right\|_{\mathbb{H}}^2\right] \tag{2.57}$$

$$\leq 2\mathbb{E}\left[\left\|\int_0^t S(t - r)\frac{1}{\rho}\Big(F(r, u^\rho(r)) - F(r, u(r)) - \rho F_u(r, u(r))(\zeta^{\tilde{u}_0}(r))\Big)dr\right\|_{\mathbb{H}}^2\right]$$

$$+ 2\mathbb{E}\left[\left\|\int_0^t S(t - r)\frac{1}{\rho}\Big(B(r, u^\rho(r)) - B(r, u(r))\right.\right.$$

$$\left.\left. - \rho B_u(r, u(r))(\zeta^{\tilde{u}_0}(r))\Big)dW(r)\right\|_{\mathbb{H}}^2\right]$$

$$\leq C\int_0^t \mathbb{E}\left[\left\|\frac{1}{\rho}\Big(F(r, u^\rho(r)) - F(r, u(r)) - \rho F_u(r, u(r))(\zeta^{\tilde{u}_0}(r))\Big)\right\|_{\mathbb{H}}^2\right]dr$$

$$+ C\int_0^t \mathbb{E}\left[\left\|\frac{1}{\rho}\Big(B(r, u^\rho(r)) - B(r, u(r)) - \rho B_u(r, u(r))(\zeta^{\tilde{u}_0}(r))\Big)\right\|_{HS(U_0, \mathbb{H})}^2\right]dr.$$

For the first term on the right-hand side of (2.57), we have

$$\left\|\frac{1}{\rho}\Big(F(r, u^\rho(r)) - F(r, u(r)) - \rho F_u(r, u(r))(\zeta^{\tilde{u}_0}(r))\Big)\right\|_{\mathbb{H}}$$

$$\leq \left\|\frac{1}{\rho}\Big(F(r, u^\rho(r)) - F(r, u(r) + \rho \zeta^{\tilde{u}_0}(r))\Big)\right\|_{\mathbb{H}}$$

$$+ \left\|\frac{1}{\rho}\Big(F(r, u(r) + \rho \zeta^{\tilde{u}_0}(r)) - F(r, u(r)) - \rho F_u(r, u(r))(\zeta^{\tilde{u}_0}(r))\Big)\right\|_{\mathbb{H}}$$

$$\leq C \left\| \frac{u^\rho(r) - u(r) - \rho \zeta^{\tilde{u}_0}(r)}{\rho} \right\|_{\mathbb{H}}$$

$$+ \left\| \frac{1}{\rho} \Big(F(r, u(r) + \rho \zeta^{\tilde{u}_0}(r)) - F(r, u(r)) - \rho F_u(r, u(r))(\zeta^{\tilde{u}_0}(r)) \Big) \right\|_{\mathbb{H}},$$

where the Lipschitz continuity of F is utilized. A similar estimate holds for the second term on the right-hand side of (2.57).

Using the Grönwall inequality, we obtain

$$\sup_{t \in [0,T]} \mathbb{E}\left[\left\| \frac{u^\rho(t) - u(t) - \rho \zeta^{\tilde{u}_0}(t)}{\rho} \right\|_{\mathbb{H}}^2 \right]$$

$$\leq C \int_0^T \mathbb{E}\left[\left\| \frac{1}{\rho} \Big(F(r, u(r) + \rho \zeta^{\tilde{u}_0}(r)) - F(r, u(r)) - \rho F_u(r, u(r))(\zeta^{\tilde{u}_0}(r)) \Big) \right\|_{\mathbb{H}}^2 \right] dr$$

$$+ C \int_0^T \mathbb{E}\left[\left\| \frac{1}{\rho} \Big(B(r, u(r) + \rho \zeta^{\tilde{u}_0}(r)) - B(r, u(r)) \right. \right.$$

$$\left. \left. - \rho B_u(r, u(r))(\zeta^{\tilde{u}_0}(r)) \Big) \right\|_{HS(U_0, \mathbb{H})}^2 \right] dr$$

$$=: C\left(I_1^\rho(\zeta^{\tilde{u}_0}) + I_2^\rho(\zeta^{\tilde{u}_0}) \right).$$

It suffices to show that $I_1^\rho(\zeta^{\tilde{u}_0})$ and $I_2^\rho(\zeta^{\tilde{u}_0})$ converge to zero uniformly with respect to $\tilde{u}_0$ as $\rho \to 0$. Namely, we need to show that for any positive constants R and θ, there is $\rho_0 = \rho_0(R, \theta)$ such that for all ρ with $|\rho| < \rho_0$, it holds that $I_j^\rho(\zeta^{\tilde{u}_0}) < \theta$ for all $\|\tilde{u}_0\|_{\mathbb{H}} \leq R$ and $j = 1, 2$.

For any measurable set $A \subset \Omega$, one has

$$I_1(\zeta^{\tilde{u}_0})$$

$$= \int_0^T \mathbb{E}\left[\left\| \frac{1}{\rho} \Big(F(r, u(r) + \rho \zeta^{\tilde{u}_0}(r)) - F(r, u(r)) - \rho F_u(r, u(r))(\zeta^{\tilde{u}_0}(r)) \Big) \right\|_{\mathbb{H}}^2 \right] dr$$

$$= \int_0^T \mathbb{E}\left[\mathbf{1}_A \left\| \frac{1}{\rho} \Big(F(r, u(r) + \rho \zeta^{\tilde{u}_0}(r)) - F(r, u(r)) - \rho F_u(r, u(r))(\zeta^{\tilde{u}_0}(r)) \Big) \right\|_{\mathbb{H}}^2 \right] dr$$

$$+ \int_0^T \mathbb{E}\left[\mathbf{1}_{A^c} \left\| \frac{1}{\rho} \Big(F(r, u(r) + \rho \zeta^{\tilde{u}_0}(r)) - F(r, u(r)) \right. \right.$$

$$\left. \left. - \rho F_u(r, u(r))(\zeta^{\tilde{u}_0}(r)) \Big) \right\|_{\mathbb{H}}^2 \right] dr$$

$$=: I_1^A(\zeta^{\tilde{u}_0}) + I_1^{A^c}(\zeta^{\tilde{u}_0}),$$

where A^c is the complementary set of A. By the Lipschitz continuity of F and the boundedness of $F_u(t, u)$, for any $q > 2$, there exists a positive constant $C =$

$C(q, T)$ such that

$$I_1^A(\zeta^{\tilde{u}_0}) \leq C \int_0^T \mathbb{E}\Big[\mathbf{1}_A \|\zeta^{\tilde{u}_0}(r)\|_{\mathbb{H}}^2\Big] dr \leq C\mathbb{E}\Big[\mathbf{1}_A \sup_{t\in[0,T]} \|\zeta^{\tilde{u}_0}(t)\|_{\mathbb{H}}^2\Big]$$

$$\leq C\big(\mathbb{P}(A)\big)^{\frac{q-2}{q}} \Big(\mathbb{E}\Big[\sup_{t\in[0,T]} \|\zeta^{\tilde{u}_0}\|_{\mathbb{H}}^q\Big]\Big)^{\frac{2}{q}}.$$

By (2.55), we have

$$I_1^A(\zeta^{\tilde{u}_0}) \leq CR^2\big(\mathbb{P}(A)\big)^{\frac{q-2}{q}}$$

provided $\|\tilde{u}_0\|_{\mathbb{H}} \leq R$.

Below we take

$$A_n := \Big\{ \sup_{t\in[0,T]} \|\zeta^{\tilde{u}_0}(t)\|_{\mathbb{H}} > n \Big\}, \quad n \in \mathbb{N}.$$

The Chebyshev inequality yields

$$\mathbb{P}(A_n) \leq \mathbb{P}\Big(\sup_{t\in[0,T]} \|\zeta^{\tilde{u}_0}(t)\|_{\mathbb{H}} \geq n \Big) \leq \frac{\mathbb{E}\big[\sup_{t\in[0,T]} \|\zeta^{\tilde{u}_0}(t)\|_{\mathbb{H}}^2\big]}{n^2} \leq \frac{CR^2}{n^2}.$$

Then for arbitrary $\theta > 0$, there exists a positive constant $n = n(T, R, \theta)$ such that $I_1^{A_n}(\zeta^{\tilde{u}_0}) < \frac{1}{2}\theta$ for all $\|\tilde{u}_0\|_{\mathbb{H}} \leq R$.

For the term $I_1^{A_n^c}(\zeta^{\tilde{u}_0})$, the Fréchet differentiability of F leads to

$$\lim_{\rho\to 0} \sup_{\|v\|_{\mathbb{H}}\leq n} \Big\| \frac{1}{\rho}\big(F(t, u(t, \omega) + \rho v) - F(t, u(t, \omega))\big) - F_u(t, u(t, \omega))(v) \Big\| = 0$$

for almost every $(t, \omega) \in [0, T] \times A_n^c$, where we used the Lipschitz continuity of F

$$\sup_{\|v\|_{\mathbb{H}}\leq n} \Big\| \frac{1}{\rho}\big(F(t, u(t, \omega) + \rho v) - F(t, u(t, \omega))\big) - F_u(t, u(t, \omega))(v) \Big\| \leq 2Cn$$

for almost every $(t, \omega) \in [0, T] \times A_n^c$. Therefore, by the dominated convergence theorem,

$$\lim_{\rho\to 0} \int_0^T \mathbb{E}\Big[\mathbf{1}_{A_n^c} \sup_{\|v\|_{\mathbb{H}}\leq n} \Big\| \frac{1}{\rho}\big(F(r, u(r) + \rho v) - F(r, u(r)) - \rho F_u(r, u(r))(v)\big) \Big\|_{\mathbb{H}}^2\Big] dr$$

$$= 0.$$

Thus for any $\theta > 0$ there exists a positive constant $\rho_1(\theta, n)$ such that

$$\int_0^T \mathbb{E}\Big[\mathbf{1}_{A_n^c} \sup_{\|v\|_{\mathbb{H}} \leq n} \Big\|\frac{1}{\rho}\Big(F(r, u(r) + \rho v) - F(r, u(r))$$

$$- \rho F_u(r, u(r))(v)\Big)\Big\|_{\mathbb{H}}^2\Big]dr < \frac{1}{2}\theta$$

for all $|\rho| < \rho_1(\theta, n)$. We note that

$$\Big\|\frac{1}{\rho}\Big(F(r, u(r) + \rho \zeta^{\tilde{u}_0}(r)) - F(r, u(r)) - \rho F_u(r, u(r))(\zeta^{\tilde{u}_0}(r))\Big)\Big\|_{\mathbb{H}}$$

$$\leq \sup_{\|v\|_{\mathbb{H}} \leq n} \Big\|\frac{1}{\rho}\Big(F(r, u(r) + \rho v) - F(r, u(r)) - \rho F_u(r, u(r))(v)\Big)\Big\|_{\mathbb{H}}$$

for almost every $(t, \omega) \times [0, T] \times A_n^c$, which yields that $I_1^{A_n^c}(\zeta^{\tilde{u}_0}) < \frac{1}{2}\theta$ for all $|\rho| \leq \rho_1(\theta, n)$. Since n depends on R, we conclude that there exists a positive constant $\rho_1 = \rho_1(\theta, R)$ such that $I_1(\zeta^{\tilde{u}_0}) < \theta$ for all $|\rho| < \rho_1$.

The estimate of $I_2(\zeta^{\tilde{u}_0})$ is similar to that of $I_1(\zeta^{\tilde{u}_0})$, which is omitted here. The proof of Theorem 2.8 is finished. $\square$

Summary and Outlook

In this chapter, we investigate the well-posedness of the stochastic Maxwell equations with the drift term being either globally Lipschitz or non-globally Lipschitz continuous. The analyses of the solution on its regularity and the differentiability with respect to the initial datum are given when the drift term is globally Lipschitz continuous. These results are vital in the analyses of structure-preserving algorithms in Chaps. 4 and 5. There are still some problems associated with these topics that are challenging and far from being well understood:

- How to establish the regularities in $L^p(\Omega, \mathscr{D}(M)^k)$ and $L^p(\Omega, H^k(D)^6)$ of the solution of the stochastic Maxwell equations, if the drift term is non-globally Lipschitz continuous?
- How to investigate the differentiability of the solution of the stochastic Maxwell equations with respect to the initial datum, when the drift term is non-globally Lipschitz continuous?
- How to analyze the solution theory of the stochastic Maxwell equations, if the considered domain D is more complicated, or even with certain randomness?

Chapter 3
Intrinsic Properties of Stochastic Maxwell Equations

Finding the invariants of a dynamical system is the prerequisite to constructing structure-preserving algorithms. In this chapter, we study some intrinsic properties of the stochastic Maxwell equations.

We first discuss the stochastic Hamiltonian structure of the stochastic Maxwell equations. In Sect. 3.1, by introducing the integrability lemma, we propose the infinite-dimensional stochastic Hamiltonian structure for the stochastic Maxwell equations and show that the phase flow of the considered model preserves the stochastic symplectic structure. Then we study the stochastic multi-symplectic conservation law for the linear stochastic Maxwell equations in Sect. 3.2. In Sect. 3.3, we investigate the energy and averaged divergence evolution properties for the generalized stochastic Maxwell equations. Sect. 3.4 is devoted to the asymptotic property of the solution of the stochastic Maxwell equations with respect to the scale of the noise. We show that the law of the solution satisfies the large deviations principle with an explicit formulation of a good rate function. Finally, the intrinsic properties of the stochastic Maxwell equations with a damped term are studied in Sect. 3.5. Particularly, we establish the uniform boundedness of the solution with respect to time, which is vital in investigating the ergodicity of the stochastic Maxwell equations.

3.1 Infinite-Dimensional Stochastic Symplectic Structure

We investigate the infinite-dimensional stochastic symplectic structure of the stochastic Maxwell equations in this section. First, we present the general formulation of the infinite-dimensional stochastic Hamiltonian system via a

© The Author(s), under exclusive license to Springer Nature Singapore Pte Ltd. 2023
C. Chen et al., *Numerical Approximations of Stochastic Maxwell Equations*,
Lecture Notes in Mathematics 2341, https://doi.org/10.1007/978-981-99-6686-8_3

generalized variational principle with a stochastic forcing; see [167] for the case of nonconservative systems, and [174, Chap. 4.1] for the case of stochastic ordinary differential equations.

Given two functionals L and $\mathscr{H}_2$, we define the generalized action functional by

$$K(p, q) := \int_{t_0}^{t_1} \left(L(p, q, \dot{p}, \dot{q}) - \mathscr{H}_2(p, q) \circ \gamma \right) dt, \tag{3.1}$$

where $p, q : [t_0, t_1] \times \mathbb{R}^d \to \mathbb{R}$, and $\gamma = \frac{dW(t)}{dt}$ is some space-time noise (see Sect. 1.2.1). Then, the generalized stochastic Hamilton principle reads as

$$\delta K(p, q) \equiv 0 \tag{3.2}$$

under the condition of fixed endpoints, i.e., $\delta p(t_0, \mathbf{x}) = \delta p(t_1, \mathbf{x}) \equiv 0$, $\delta q(t_0, \mathbf{x}) = \delta q(t_1, \mathbf{x}) \equiv 0$. Plugging (3.1) into (3.2) yields

$$\delta K(p, q) = \int_{\mathbb{R}^d} \int_{t_0}^{t_1} \left[\left(\frac{\delta L}{\delta p} - \frac{d}{dt}\left(\frac{\delta L}{\delta \dot{p}} \right) - \frac{\delta \mathscr{H}_2}{\delta p} \circ \gamma \right) \delta p \right.$$

$$\left. + \left(\frac{\delta L}{\delta q} - \frac{d}{dt}\left(\frac{\delta L}{\delta \dot{q}} \right) - \frac{\delta \mathscr{H}_2}{\delta q} \circ \gamma \right) \delta q \right] dt\, d\mathbf{x}$$

$$\equiv 0,$$

from which we have

$$\frac{d}{dt}\left(\frac{\delta L}{\delta \dot{p}} \right) = \frac{\delta L}{\delta p} - \frac{\delta \mathscr{H}_2}{\delta p} \circ \gamma, \qquad \frac{d}{dt}\left(\frac{\delta L}{\delta \dot{q}} \right) = \frac{\delta L}{\delta q} - \frac{\delta \mathscr{H}_2}{\delta q} \circ \gamma.$$

This, combining a variational principle, one can obtain the following infinite-dimensional stochastic Hamiltonian system

$$\begin{cases} dp = -\frac{\delta \mathscr{H}_1}{\delta q} dt - \frac{\delta \mathscr{H}_2}{\delta q} \circ dW(t), & p(t_0) = p_0, \\ dq = \frac{\delta \mathscr{H}_1}{\delta p} dt + \frac{\delta \mathscr{H}_2}{\delta p} \circ dW(t), & q(t_0) = q_0. \end{cases} \tag{3.3}$$

We refer to [31] for more details.

Let $D \subset \mathbb{R}^3$ be an open, bounded, and Lipschitz domain with boundary ∂D. Below we introduce the integrability lemma in order to investigate the geometric structure of the stochastic Maxwell equations.

Lemma 3.1 *Let $G : L^2(D)^6 \to L^2(D)^6$ be Lipschitz continuous and Fréchet differentiable with the derivative $G_u(u) \in \mathscr{L}(L^2(D)^6)$ being a symmetric operator, i.e.,*

$$\langle G_u(u)\phi, \psi \rangle_{L^2(D)^6} = \langle \phi, G_u(u)\psi \rangle_{L^2(D)^6} \quad \forall \phi, \psi \in L^2(D)^6.$$

Then there exists a functional $\widetilde{\mathscr{H}} : L^2(D)^6 \to \mathbb{R}$ such that $G(u) = \frac{\delta\widetilde{\mathscr{H}}(u)}{\delta u}$.

Proof The functional $\widetilde{\mathscr{H}}(u)$ can be defined as

$$\widetilde{\mathscr{H}}(u) = \int_0^1 \langle u, G(\zeta u)\rangle_{L^2(D)^6}\mathrm{d}\zeta + c \tag{3.4}$$

with c being a constant. For any $\phi \in L^2(D)^6$,

$$\left\langle \frac{\delta\widetilde{\mathscr{H}}(u)}{\delta u}, \phi \right\rangle_{L^2(D)^6}$$

$$= \lim_{\rho \to 0} \frac{1}{\rho}\left[\widetilde{\mathscr{H}}(u + \rho\phi) - \widetilde{\mathscr{H}}(u)\right]$$

$$= \lim_{\rho \to 0} \frac{1}{\rho}\left[\int_0^1 \left(\langle u + \rho\phi, G(\zeta u + \rho\zeta\phi)\rangle_{L^2(D)^6} - \langle u, G(\zeta u)\rangle_{L^2(D)^6}\right)\mathrm{d}\zeta\right]$$

$$= \int_0^1 \left\langle u, \lim_{\rho \to 0}\frac{1}{\rho}\left[G(\zeta u + \rho\zeta\phi) - G(\zeta u)\right]\right\rangle_{L^2(D)^6}\mathrm{d}\zeta$$

$$+ \lim_{\rho \to 0}\int_0^1 \left\langle \phi, G(\zeta u + \rho\zeta\phi)\right\rangle_{L^2(D)^6}\mathrm{d}\zeta,$$

where in the last step we used the dominated convergence theorem and the Lipschitz continuity of G. Therefore, we have

$$\left\langle \frac{\delta\widetilde{\mathscr{H}}(u)}{\delta u}, \phi \right\rangle_{L^2(D)^6} = \int_0^1 \zeta\langle u, G_u(\zeta u)\phi\rangle_{L^2(D)^6}\mathrm{d}\zeta + \int_0^1 \langle \phi, G(\zeta u)\rangle_{L^2(D)^6}\mathrm{d}\zeta$$

$$= \left\langle \int_0^1 \left(\zeta G_u(\zeta u)u + G(\zeta u)\right)\mathrm{d}\zeta, \phi \right\rangle_{L^2(D)^6}$$

due to the symmetry of $G_u(u)$. Thus,

$$\frac{\delta \widetilde{\mathscr{H}}(u)}{\delta u} = \int_0^1 \left(\zeta G_u(\zeta u)u + G(\zeta u) \right) d\zeta = \int_0^1 \frac{d}{d\zeta} \left(\zeta G(\zeta u) \right) d\zeta = G(u),$$

which finishes the proof.   □

Now let us turn to the infinite-dimensional stochastic symplectic structure for the stochastic Maxwell equations in the Stratonovich sense:

$$\varepsilon d\mathbf{E} = \left[\nabla \times \mathbf{H} - \mathbf{J}_e(t, \mathbf{x}, \mathbf{E}, \mathbf{H}) \right] dt - \mathbf{J}_e^r(t, \mathbf{x}, \mathbf{E}, \mathbf{H}) \circ dW(t),$$

$$\mu d\mathbf{H} = [-\nabla \times \mathbf{E} - \mathbf{J}_m(t, \mathbf{x}, \mathbf{E}, \mathbf{H})] dt - \mathbf{J}_m^r(t, \mathbf{x}, \mathbf{E}, \mathbf{H}) \circ dW(t). \tag{3.5}$$

For arbitrary $t \in [0, T]$ and $\mathbf{x} \in D$, we set $u = (\mathbf{E}^\top, \mathbf{H}^\top)^\top$ and

$$G_1(t, u(t))(\mathbf{x}) := \begin{bmatrix} -\mu^{-1} \mathbf{J}_m(t, \mathbf{x}, \mathbf{E}(t, \mathbf{x}), \mathbf{H}(t, \mathbf{x})) \\ \varepsilon^{-1} \mathbf{J}_e(t, \mathbf{x}, \mathbf{E}(t, \mathbf{x}), \mathbf{H}(t, \mathbf{x})) \end{bmatrix},$$

$$G_2(t, u(t))(\mathbf{x}) := \begin{bmatrix} -\mu^{-1} \mathbf{J}_m^r(t, \mathbf{x}, \mathbf{E}(t, \mathbf{x}), \mathbf{H}(t, \mathbf{x})) \\ \varepsilon^{-1} \mathbf{J}_e^r(t, \mathbf{x}, \mathbf{E}(t, \mathbf{x}), \mathbf{H}(t, \mathbf{x})) \end{bmatrix}.$$

If G_1 and G_2 satisfy conditions in Lemma 3.1, then there exist functionals $\widetilde{\mathscr{H}_1}$ and $\mathscr{H}_2$ such that

$$\frac{\delta \widetilde{\mathscr{H}_1}}{\delta \mathbf{E}} = -\mu^{-1} \mathbf{J}_m, \qquad \frac{\delta \widetilde{\mathscr{H}_1}}{\delta \mathbf{H}} = \varepsilon^{-1} \mathbf{J}_e,$$

$$\frac{\delta \mathscr{H}_2}{\delta \mathbf{E}} = -\mu^{-1} \mathbf{J}_m^r, \qquad \frac{\delta \mathscr{H}_2}{\delta \mathbf{H}} = \varepsilon^{-1} \mathbf{J}_e^r.$$

One can see that if, in addition, ε and μ are positive constants, then (3.5) is an infinite-dimensional stochastic Hamiltonian system admitting the following formulation:

$$d \begin{bmatrix} \mathbf{E} \\ \mathbf{H} \end{bmatrix} = \begin{bmatrix} 0 & -Id \\ Id & 0 \end{bmatrix} \begin{bmatrix} -\mu^{-1} \nabla \times \mathbf{E} - \mu^{-1} \mathbf{J}_m \\ -\varepsilon^{-1} \nabla \times \mathbf{H} + \varepsilon^{-1} \mathbf{J}_e \end{bmatrix} dt$$

$$+ \begin{bmatrix} 0 & -Id \\ Id & 0 \end{bmatrix} \begin{bmatrix} -\mu^{-1} \mathbf{J}_m^r \\ \varepsilon^{-1} \mathbf{J}_e^r \end{bmatrix} \circ dW(t)$$

$$=: \mathbb{J}^{-1} \begin{bmatrix} \frac{\delta \widetilde{\mathscr{H}_1}}{\delta \mathbf{E}} \\ \frac{\delta \widetilde{\mathscr{H}_1}}{\delta \mathbf{H}} \end{bmatrix} dt + \mathbb{J}^{-1} \begin{bmatrix} \frac{\delta \mathscr{H}_2}{\delta \mathbf{E}} \\ \frac{\delta \mathscr{H}_2}{\delta \mathbf{H}} \end{bmatrix} \circ dW(t), \tag{3.6}$$

where $\mathbb{J}$ is the standard symplectic matrix and the Hamiltonian $\mathscr{H}_1$ is given by

$$\mathscr{H}_1(\mathbf{E}, \mathbf{H}) = - \int_D \frac{1}{2} \left(\mu^{-1} \mathbf{E} \cdot \nabla \times \mathbf{E} + \varepsilon^{-1} \mathbf{H} \cdot \nabla \times \mathbf{H} \right) dx + \widetilde{\mathscr{H}_1}(\mathbf{E}, \mathbf{H}).$$

Particularly, if (3.5) is driven by additive noise, i.e., $\mathbf{J}_e^r$ and $\mathbf{J}_m^r$ are independent of $\mathbf{E}$ and $\mathbf{H}$, then we have the explicit expression of $\mathscr{H}_2$:

$$\mathscr{H}_2(\mathbf{E}, \mathbf{H}) = \int_D \left(\varepsilon^{-1} \mathbf{J}_e^r \cdot \mathbf{H} - \mu^{-1} \mathbf{J}_m^r \cdot \mathbf{E} \right) dx.$$

If (3.5) is driven by linear multiplicative noise, that is $\mathbf{J}_e^r = \lambda_1 \mathbf{H}$, $\mathbf{J}_m^r = -\lambda_2 \mathbf{E}$ with constants $\lambda_1, \lambda_2 \in \mathbb{R}$, then

$$\mathscr{H}_2(\mathbf{E}, \mathbf{H}) = \frac{1}{2} \int_D \left(\lambda_1 \varepsilon^{-1} |\mathbf{H}|^2 + \lambda_2 \mu^{-1} |\mathbf{E}|^2 \right) dx.$$

Theorem 3.1 *Suppose that ε and μ are two positive constants. Let conditions in Theorem 2.8 and Lemma 3.1 hold. Then the phase flow of (3.6) preserves the stochastic symplectic structure under the homogeneous boundary condition, i.e., for $t \in (0, T]$,*

$$\varpi(t) := \int_D d\mathbf{E}(t, \mathbf{x}) \wedge d\mathbf{H}(t, \mathbf{x}) dx = \int_D d\mathbf{E}_0(\mathbf{x}) \wedge d\mathbf{H}_0(\mathbf{x}) dx = \varpi(0), \quad \mathbb{P}\text{-}a.s.$$

Proof For simplicity in notations, we denote $\mathbf{E}_0$ and $\mathbf{H}_0$ by $\mathbf{e}$ and $\mathbf{h}$, respectively. We use notations $\mathbf{E_e} = \frac{\partial \mathbf{E}}{\partial \mathbf{e}}$, $\mathbf{E_h} = \frac{\partial \mathbf{E}}{\partial \mathbf{h}}$, $\mathbf{H_e} = \frac{\partial \mathbf{H}}{\partial \mathbf{e}}$ and $\mathbf{H_h} = \frac{\partial \mathbf{H}}{\partial \mathbf{h}}$ in the sequel.

The chain rule for differentials leads to

$$\varpi(t) = \int_D \left[d\mathbf{e} \wedge (\mathbf{E_e})^\top \mathbf{H_e} d\mathbf{e} \right] dx + \int_D \left[d\mathbf{h} \wedge (\mathbf{E_h})^\top \mathbf{H_h} d\mathbf{h} \right] dx$$

$$+ \int_D \left[d\mathbf{e} \wedge \left((\mathbf{E_e})^\top \mathbf{H_h} - (\mathbf{H_e})^\top \mathbf{E_h} \right) d\mathbf{h} \right] dx, \quad t \in [0, T],$$

from which we obtain

$$d\varpi(t) = \int_D \left[d\mathbf{e} \wedge d\left(\mathbf{E_e}^\top \mathbf{H_e} \right) d\mathbf{e} + d\mathbf{h} \wedge d\left(\mathbf{E_h}^\top \mathbf{H_h} \right) d\mathbf{h} \right] dx$$

$$+ \int_D \left[d\mathbf{e} \wedge d\left(\mathbf{E_e}^\top \mathbf{H_h} - \mathbf{H_e}^\top \mathbf{E_h} \right) d\mathbf{h} \right] dx, \quad t \in [0, T].$$

$$(3.7)$$

Theorem 2.8 implies the Fréchet differentiability of the solution of (3.5). This yields that for $t \in [0, T]$,

$$
\begin{aligned}
d\mathbf{E_e} =& \left(\varepsilon^{-1} \nabla \times \mathbf{H_e} - \frac{\delta^2 \widetilde{\mathscr{H}_1}}{\delta \mathbf{E} \delta \mathbf{H}} \mathbf{E_e} - \frac{\delta^2 \widetilde{\mathscr{H}_1}}{\delta \mathbf{H}^2} \mathbf{H_e} \right) dt \\
& - \left(\frac{\delta^2 \mathscr{H}_2}{\delta \mathbf{E} \delta \mathbf{H}} \mathbf{E_e} + \frac{\delta^2 \mathscr{H}_2}{\delta \mathbf{H}^2} \mathbf{H_e} \right) \circ dW(t), \\
d\mathbf{H_e} =& \left(-\mu^{-1} \nabla \times \mathbf{E_e} + \frac{\delta^2 \widetilde{\mathscr{H}_1}}{\delta \mathbf{E}^2} \mathbf{E_e} + \frac{\delta^2 \widetilde{\mathscr{H}_1}}{\delta \mathbf{E} \delta \mathbf{H}} \mathbf{H_e} \right) dt \\
& + \left(\frac{\delta^2 \mathscr{H}_2}{\delta \mathbf{E}^2} \mathbf{E_e} + \frac{\delta^2 \mathscr{H}_2}{\delta \mathbf{E} \delta \mathbf{H}} \mathbf{H_e} \right) \circ dW(t), \\
d\mathbf{E_h} =& \left(\varepsilon^{-1} \nabla \times \mathbf{H_h} - \frac{\delta^2 \widetilde{\mathscr{H}_1}}{\delta \mathbf{E} \delta \mathbf{H}} \mathbf{E_h} - \frac{\delta^2 \widetilde{\mathscr{H}_1}}{\delta \mathbf{H}^2} \mathbf{H_h} \right) dt \\
& - \left(\frac{\delta^2 \mathscr{H}_2}{\delta \mathbf{E} \delta \mathbf{H}} \mathbf{E_h} + \frac{\delta^2 \mathscr{H}_2}{\delta \mathbf{H}^2} \mathbf{H_h} \right) \circ dW(t), \\
d\mathbf{H_h} =& \left(-\mu^{-1} \nabla \times \mathbf{E_h} + \frac{\delta^2 \widetilde{\mathscr{H}_1}}{\delta \mathbf{E}^2} \mathbf{E_h} + \frac{\delta^2 \widetilde{\mathscr{H}_1}}{\delta \mathbf{E} \delta \mathbf{H}} \mathbf{H_h} \right) dt \\
& + \left(\frac{\delta^2 \mathscr{H}_2}{\delta \mathbf{E}^2} \mathbf{E_h} + \frac{\delta^2 \mathscr{H}_2}{\delta \mathbf{E} \delta \mathbf{H}} \mathbf{H_h} \right) \circ dW(t)
\end{aligned}
\tag{3.8}
$$

with $\mathbf{E_e}(0) = Id$, $\mathbf{H_e}(0) = 0$, $\mathbf{E_h}(0) = 0$, and $\mathbf{H_h}(0) = Id$.

Plugging (3.8) into (3.7) yields

$$
\begin{aligned}
d\varpi(t) =& \int_D \left[d\mathbf{e} \wedge \left(\varepsilon^{-1} (\nabla \times \mathbf{H_e})^\top \mathbf{H_e} - \mu^{-1} \mathbf{E_e}^\top \nabla \times \mathbf{E_e} \right) d\mathbf{e} \right] d\mathbf{x} dt \\
& + \int_D \left[d\mathbf{h} \wedge \left(\varepsilon^{-1} (\nabla \times \mathbf{H_h})^\top \mathbf{H_h} - \mu^{-1} \mathbf{E_h}^\top \nabla \times \mathbf{E_h} \right) d\mathbf{h} \right] d\mathbf{x} dt \\
& + \int_D \left[d\mathbf{e} \wedge \left(\varepsilon^{-1} (\nabla \times \mathbf{H_e})^\top \mathbf{H_h} - \mu^{-1} \mathbf{E_e}^\top \nabla \times \mathbf{E_h} \right) d\mathbf{h} \right] d\mathbf{x} dt \\
& + \int_D \left[d\mathbf{e} \wedge \left(\mu^{-1} (\nabla \times \mathbf{E_e})^\top \mathbf{E_h} - \varepsilon^{-1} \mathbf{H_e}^\top \nabla \times \mathbf{H_h} \right) d\mathbf{h} \right] d\mathbf{x} dt \\
=& \int_D \varepsilon^{-1} \left[d\mathbf{e} \wedge (\nabla \times \mathbf{H_e})^\top \mathbf{H_e} d\mathbf{e} + d\mathbf{h} \wedge (\nabla \times \mathbf{H_h})^\top \mathbf{H_h} d\mathbf{h} \right. \\
& \left. + d\mathbf{e} \wedge (\nabla \times \mathbf{H_e})^\top \mathbf{H_h} d\mathbf{h} - d\mathbf{e} \wedge \mathbf{H_e}^\top \nabla \times \mathbf{H_h} d\mathbf{h} \right] d\mathbf{x} dt \\
& + \int_D \mu^{-1} \left[d\mathbf{e} \wedge (\nabla \times \mathbf{E_e})^\top \mathbf{E_e} d\mathbf{e} + d\mathbf{h} \wedge (\nabla \times \mathbf{E_h})^\top \mathbf{E_h} d\mathbf{h} \right. \\
& \left. + d\mathbf{e} \wedge (\nabla \times \mathbf{E_e})^\top \mathbf{E_h} d\mathbf{h} - d\mathbf{e} \wedge \mathbf{E_e}^\top \nabla \times \mathbf{E_h} d\mathbf{h} \right] d\mathbf{x} dt.
\end{aligned}
$$

Using the fact that $df \wedge Adg = A^\top df \wedge dg$ for differential 1-forms df and dg, it holds that

$$
\begin{aligned}
d\varpi &= \int_D \left[\varepsilon^{-1} d(\nabla \times \mathbf{H}) \wedge d\mathbf{H} + \mu^{-1} d(\nabla \times \mathbf{E}) \wedge d\mathbf{E} \right] dxdt \\
&= \int_D \varepsilon^{-1} \left[\partial_x (dH_2 \wedge dH_3) + \partial_y (dH_3 \wedge dH_1) + \partial_z (dH_1 \wedge dH_2) \right] dxdt \\
&\quad + \int_D \mu^{-1} \left[\partial_x (dE_2 \wedge dE_3) + \partial_y (dE_3 \wedge dE_1) + \partial_z (dE_1 \wedge dE_2) \right] dxdt \\
&= 0,
\end{aligned}
$$

(3.9)

where the homogeneous boundary condition is utilized in the last step. This finishes the proof of Theorem 3.1. □

3.2 Stochastic Multi-Symplectic Structure

When (3.5) is regarded as an infinite-dimensional stochastic evolution equation, we have shown that the stochastic Maxwell equations possess the stochastic symplectic structure under the assumption that ε and μ are constant in Sect. 3.1. When the space variable is of interest for a stochastic Hamiltonian partial differential equation, the stochastic multi-symplectic structure is usually involved, which is first investigated in [112] for the stochastic nonlinear Schrödinger equation.

Definition 3.1 A stochastic partial differential equation is called a stochastic Hamiltonian partial differential equation if it can be written in the form

$$
\mathbb{F}du + \sum_{i=1}^{d} \mathbb{K}_i \partial_{x_i} u dt = \nabla_u S_0(u) dt + \nabla_u S_1(u) \circ dW(t),
$$

where $u : [0, T] \times \mathbb{R}^d \to \mathbb{R}^n, n \geq 3, \mathbb{F}$ and $\mathbb{K}_i$ $(i = 1, 2, \ldots, d)$ are skew-symmetric matrices in $\mathbb{R}^{n \times n}$, and S_0 and S_1 are smooth functionals of the state variable u.

To simplify notations, we let $\varepsilon = \mu \equiv 1$ and consider the following stochastic Maxwell equations with multiplicative noise

$$
\begin{cases}
d\mathbf{E}(t) = \nabla \times \mathbf{H}(t) dt - \lambda_1 \mathbf{H} \circ dW(t), & t \in (0, T], \\
d\mathbf{H}(t) = -\nabla \times \mathbf{E}(t) dt + \lambda_2 \mathbf{E} \circ dW(t), & t \in (0, T], \\
\mathbf{E}(0) = \mathbf{E}_0, \quad \mathbf{H}(0) = \mathbf{H}_0,
\end{cases}
$$

(3.10)

where $\lambda_1, \lambda_2 \in \mathbb{R}$ are two constants.

By denoting $u = (\mathbf{E}^\top, \mathbf{H}^\top)^\top$ and $S(u) = \frac{\lambda_2}{2}|\mathbf{E}|^2 + \frac{\lambda_1}{2}|\mathbf{H}|^2$, (3.10) can be reformulated in the form of a stochastic Hamiltonian partial differential equation:

$$\mathbb{F}du + \mathbb{K}_1\partial_x u dt + \mathbb{K}_2\partial_y u dt + \mathbb{K}_3\partial_z u dt = \nabla_u S(u) \circ dW(t), \tag{3.11}$$

where the skew-symmetric matrices $\mathbb{F}$ and $\mathbb{K}_p$ $(p = 1, 2, 3)$ are defined in (1.23), i.e.,

$$\mathbb{F} = \begin{bmatrix} 0 & Id \\ -Id & 0 \end{bmatrix}, \quad \mathbb{K}_p = \begin{bmatrix} \mathbb{D}_p & 0 \\ 0 & \mathbb{D}_p \end{bmatrix} \tag{3.12}$$

and

$$\mathbb{D}_1 = \begin{bmatrix} 0 & 0 & 0 \\ 0 & 0 & -1 \\ 0 & 1 & 0 \end{bmatrix}, \quad \mathbb{D}_2 = \begin{bmatrix} 0 & 0 & 1 \\ 0 & 0 & 0 \\ -1 & 0 & 0 \end{bmatrix}, \quad \mathbb{D}_3 = \begin{bmatrix} 0 & -1 & 0 \\ 1 & 0 & 0 \\ 0 & 0 & 0 \end{bmatrix}.$$

Theorem 3.2 *The equation* (3.10) *preserves the stochastic multi-symplectic conservation law*

$$d\varpi + \partial_x \kappa_1 dt + \partial_y \kappa_2 dt + \partial_z \kappa_3 dt = 0, \quad \mathbb{P}\text{-}a.s., \tag{3.13}$$

which means

$$\int_{z_0}^{z_1} \int_{y_0}^{y_1} \int_{x_0}^{x_1} \varpi(t_1, x, y, z) dx dy dz + \int_{z_0}^{z_1} \int_{y_0}^{y_1} \int_{t_0}^{t_1} \kappa_1(t, x_1, y, z) dt dy dz$$

$$+ \int_{z_0}^{z_1} \int_{x_0}^{x_1} \int_{t_0}^{t_1} \kappa_2(t, x, y_1, z) dt dx dz + \int_{y_0}^{y_1} \int_{x_0}^{x_1} \int_{t_0}^{t_1} \kappa_3(t, x, y, z_1) dt dx dy$$

$$= \int_{z_0}^{z_1} \int_{y_0}^{y_1} \int_{x_0}^{x_1} \varpi(t_0, x, y, z) dx dy dz + \int_{z_0}^{z_1} \int_{y_0}^{y_1} \int_{t_0}^{t_1} \kappa_1(t, x_0, y, z) dt dy dz$$

$$+ \int_{z_0}^{z_1} \int_{x_0}^{x_1} \int_{t_0}^{t_1} \kappa_2(t, x, y_0, z) dt dx dz$$

$$+ \int_{y_0}^{y_1} \int_{x_0}^{x_1} \int_{t_0}^{t_1} \kappa_3(t, x, y, z_0) dt dx dy, \quad \mathbb{P}\text{-}a.s.,$$

where $\varpi = \frac{1}{2}du \wedge \mathbb{F}du$ *and* $\kappa_p = \frac{1}{2}du \wedge \mathbb{K}_p du$, $p = 1, 2, 3$ *are differential 2-forms associated with the skew-symmetric matrices* $\mathbb{F}$ *and* $\mathbb{K}_p$, *respectively, and* $(t_0, t_1) \times (x_0, x_1) \times (y_0, y_1) \times (z_0, z_1) \subset [0, T] \times D$ *is the local domain of* u.

Proof Denote by $du_{1,x,y,z,i}$ (resp. $du_{0,x,y,z,i}$) the i-th component of the differential 1-form $du(t_1, x, y, z)$ (resp. $du(t_0, x, y, z)$), and by $du_{0,i}$ the i-th component of the differential 1-form $du(t_0, x_0, y_0, z_0)$. The same definitions hold for $du_{t,1,y,z,i}$, $du_{t,0,y,z,i}$, $du_{t,x,1,z,i}$, $du_{t,x,0,z,i}$, $du_{t,x,y,1,i}$, and $du_{t,x,y,0,i}$.

Rewriting $\mathbb{F}$ and $\mathbb{K}_p$ by their components, i.e., $\mathbb{F} = (\mathbb{F}_{ij})_{i,j=1}^6$, $\mathbb{K}_p = (\mathbb{K}_{ij}^p)_{i,j=1}^6$, we have

$$
I_1 := \int_{z_0}^{z_1}\int_{y_0}^{y_1}\int_{x_0}^{x_1} \varpi(t_1, x, y, z)\mathrm{d}x\mathrm{d}y\mathrm{d}z - \int_{z_0}^{z_1}\int_{y_0}^{y_1}\int_{x_0}^{x_1} \varpi(t_0, x, y, z)\mathrm{d}x\mathrm{d}y\mathrm{d}z
$$

$$
= \frac{1}{2}\int_{z_0}^{z_1}\int_{y_0}^{y_1}\int_{x_0}^{x_1} \Big[\sum_{i=1}^6 \Big(du_{1,x,y,z,i} \wedge \sum_{j=1}^6 \mathbb{F}_{ij}du_{1,x,y,z,j}\Big)
$$

$$
- \sum_{i=1}^6 \Big(du_{0,x,y,z,i} \wedge \sum_{j=1}^6 \mathbb{F}_{ij}du_{0,x,y,z,j}\Big)\Big]\mathrm{d}x\mathrm{d}y\mathrm{d}z
$$

$$
= \frac{1}{2}\int_{z_0}^{z_1}\int_{y_0}^{y_1}\int_{x_0}^{x_1} \sum_{i,j=1}^6 \mathbb{F}_{ij}\Big(du_{1,x,y,z,i} \wedge du_{1,x,y,z,j}
$$

$$
- du_{0,x,y,z,i} \wedge du_{0,x,y,z,j}\Big)\mathrm{d}x\mathrm{d}y\mathrm{d}z.
$$

It follows from

$$
du_{1,x,y,z,i} = \sum_{l=1}^6 \frac{\partial u_{1,x,y,z,i}}{\partial u_{0,l}}du_{0,l}, \quad du_{0,x,y,z,i} = \sum_{l=1}^6 \frac{\partial u_{0,x,y,z,i}}{\partial u_{0,l}}du_{0,l}
$$

that

$$
I_1 = \frac{1}{2}\sum_{l,k=1}^6 a_{l,k}(t_1, x_1, y_1, z_1)du_{0,l} \wedge du_{0,k},
$$

where

$$
a_{l,k}(t_1, x_1, y_1, z_1)
$$

$$
= \sum_{i,j=1}^6 \mathbb{F}_{ij}\int_{z_0}^{z_1}\int_{y_0}^{y_1}\int_{x_0}^{x_1} \Big(\frac{\partial u_{1,x,y,z,i}}{\partial u_{0,l}}\frac{\partial u_{1,x,y,z,j}}{\partial u_{0,k}} \tag{3.14}
$$

$$
- \frac{\partial u_{0,x,y,z,i}}{\partial u_{0,l}}\frac{\partial u_{0,x,y,z,j}}{\partial u_{0,k}}\Big)\mathrm{d}x\mathrm{d}y\mathrm{d}z.
$$

In a similar manner, we have

$$I_2 := \int_{z_0}^{z_1} \int_{y_0}^{y_1} \int_{t_0}^{t_1} \kappa_1(t, x_1, y, z) dt dy dz - \int_{z_0}^{z_1} \int_{y_0}^{y_1} \int_{t_0}^{t_1} \kappa_1(t, x_0, y, z) dt dy dz$$

$$= \frac{1}{2} \sum_{l,k=1}^{6} b_{l,k}(t_1, x_1, y_1, z_1) du_{0,l} \wedge du_{0,k},$$

$$I_3 := \int_{z_0}^{z_1} \int_{x_0}^{x_1} \int_{t_0}^{t_1} \kappa_2(t, x, y_1, z) dt dx dz - \int_{z_0}^{z_1} \int_{x_0}^{x_1} \int_{t_0}^{t_1} \kappa_2(t, x, y_0, z) dt dx dz$$

$$= \frac{1}{2} \sum_{l,k=1}^{6} c_{l,k}(t_1, x_1, y_1, z_1) du_{0,l} \wedge du_{0,k},$$

$$I_4 := \int_{y_0}^{y_1} \int_{x_0}^{x_1} \int_{t_0}^{t_1} \kappa_3(t, x, y, z_1) dt dx dy - \int_{y_0}^{y_1} \int_{x_0}^{x_1} \int_{t_0}^{t_1} \kappa_3(t, x, y, z_0) dt dx dy$$

$$= \frac{1}{2} \sum_{l,k=1}^{6} d_{l,k}(t_1, x_1, y_1, z_1) du_{0,l} \wedge du_{0,k},$$

where

$$b_{l,k}(t_1, x_1, y_1, z_1)$$

$$= \sum_{i,j=1}^{6} \mathbb{K}_{ij}^1 \int_{z_0}^{z_1} \int_{y_0}^{y_1} \int_{t_0}^{t_1} \left(\frac{\partial u_{t,1,y,z,i}}{\partial u_{0,l}} \frac{\partial u_{t,1,y,z,j}}{\partial u_{0,k}} - \frac{\partial u_{t,0,y,z,i}}{\partial u_{0,l}} \frac{\partial u_{t,0,y,z,j}}{\partial u_{0,k}} \right) dt dy dz,$$

$$c_{l,k}(t_1, x_1, y_1, z_1)$$

$$= \sum_{i,j=1}^{6} \mathbb{K}_{ij}^2 \int_{z_0}^{z_1} \int_{x_0}^{x_1} \int_{t_0}^{t_1} \left(\frac{\partial u_{t,x,1,z,i}}{\partial u_{0,l}} \frac{\partial u_{t,x,1,z,j}}{\partial u_{0,k}} - \frac{\partial u_{t,x,0,z,i}}{\partial u_{0,l}} \frac{\partial u_{t,x,0,z,j}}{\partial u_{0,k}} \right) dt dx dz,$$

$$d_{l,k}(t_1, x_1, y_1, z_1)$$

$$= \sum_{i,j=1}^{6} \mathbb{K}_{ij}^3 \int_{y_0}^{y_1} \int_{x_0}^{x_1} \int_{t_0}^{t_1} \left(\frac{\partial u_{t,x,y,1,i}}{\partial u_{0,l}} \frac{\partial u_{t,x,y,1,j}}{\partial u_{0,k}} - \frac{\partial u_{t,x,y,0,i}}{\partial u_{0,l}} \frac{\partial u_{t,x,y,0,j}}{\partial u_{0,k}} \right) dt dx dy.$$

$$(3.15)$$

Combining the above identities for I_i, $i = 1, 2, 3, 4$, we conclude that

$$I_1 + I_2 + I_3 + I_4 = \frac{1}{2} \sum_{l,k=1}^{6} \left[a_{l,k} + b_{l,k} + c_{l,k} + d_{l,k} \right] (t_1, x_1, y_1, z_1) du_{0,l} \wedge du_{0,k}.$$

Then, (3.13) is fulfilled if and only if

$$\sum_{l,k=1}^{6} \Big[a_{l,k} + b_{l,k} + c_{l,k} + d_{l,k}\Big](t_1, x_1, y_1, z_1)du_{\mathbf{0},l} \wedge du_{\mathbf{0},k} \equiv 0. \tag{3.16}$$

Based on definitions of $a_{l,k}, b_{l,k}, c_{l,k}, d_{l,k}$ in (3.14) and (3.15), we have

$$a_{l,k}(t_0, x_1, y_1, z_1) = b_{l,k}(t_0, x_1, y_1, z_1) = c_{l,k}(t_0, x_1, y_1, z_1)$$
$$= d_{l,k}(t_0, x_1, y_1, z_1) = 0$$

for all $l, k = 1, 2, \ldots, 6$. Then it suffices to prove that

$$\frac{d}{dt}\Big[a_{l,k} + b_{l,k} + c_{l,k} + d_{l,k}\Big](t, x_1, y_1, z_1) = 0, \quad l, k = 1, 2, \ldots, 6. \tag{3.17}$$

Consider the i-th component equation of (3.11)

$$\sum_{j=1}^{6} \Big(\mathbb{F}_{ij} du_{\cdot,x,y,z,j} + \mathbb{K}_{ij}^1 \partial_x u_{\cdot,x,y,z,j} dt + \mathbb{K}_{ij}^2 \partial_y u_{\cdot,x,y,z,j} dt + \mathbb{K}_{ij}^3 \partial_z u_{\cdot,x,y,z,j} dt \Big)$$

$$= \frac{\partial S(u)}{\partial u_{\cdot,x,y,z,i}} \circ dW(t).$$

Taking partial derivatives with respect to $u_{\mathbf{0},k}, k = 1, 2, \ldots, 6$, on both sides of the above equation yields

$$\sum_{j=1}^{6} \mathbb{F}_{ij} d\Big(\frac{\partial u_{\cdot,x,y,z,j}}{\partial u_{\mathbf{0},k}}\Big)$$

$$= -\sum_{j=1}^{6} \mathbb{K}_{ij}^1 \partial_x \Big(\frac{\partial u_{\cdot,x,y,z,j}}{\partial u_{\mathbf{0},k}}\Big) dt - \sum_{j=1}^{6} \mathbb{K}_{ij}^2 \partial_y \Big(\frac{\partial u_{\cdot,x,y,z,j}}{\partial u_{\mathbf{0},k}}\Big) dt$$

$$- \sum_{j=1}^{6} \mathbb{K}_{ij}^3 \partial_z \Big(\frac{\partial u_{\cdot,x,y,z,j}}{\partial u_{\mathbf{0},k}}\Big) dt + \sum_{j=1}^{6} \frac{\partial^2 S(u)}{\partial u_{\cdot,x,y,z,i} \partial u_{\cdot,x,y,z,j}} \Big(\frac{\partial u_{\cdot,x,y,z,j}}{\partial u_{\mathbf{0},k}}\Big) \circ dW(t).$$

$$\tag{3.18}$$

Similarly, we take the partial derivatives with respect to $u_{0,l}$, $l = 1, 2 \ldots, 6$, use the skew-symmetry of F and permute i, j to obtain

$$
\sum_{i=1}^{6} \mathbb{F}_{ij} d\left(\frac{\partial u_{\cdot,x,y,z,i}}{\partial u_{0,l}}\right)
$$

$$
= \sum_{i=1}^{6} \mathbb{K}_{ji}^{1} \partial_x \left(\frac{\partial u_{\cdot,x,y,z,i}}{\partial u_{0,l}}\right) dt + \sum_{i=1}^{6} \mathbb{K}_{ji}^{2} \partial_y \left(\frac{\partial u_{\cdot,x,y,z,i}}{\partial u_{0,l}}\right) dt
$$

$$
+ \sum_{i=1}^{6} \mathbb{K}_{ji}^{3} \partial_z \left(\frac{\partial u_{\cdot,x,y,z,i}}{\partial u_{0,l}}\right) dt - \sum_{i=1}^{6} \frac{\partial^2 S(u)}{\partial u_{\cdot,x,y,z,j} \partial u_{1,x,y,z,i}} \left(\frac{\partial u_{\cdot,x,y,z,i}}{\partial u_{0,l}}\right) \circ dW(t).
$$

$$(3.19)$$

Due to (3.14), we have

$$
da_{l,k} = \sum_{i=1}^{6} \int_{z_0}^{z_1} \int_{y_0}^{y_1} \int_{x_0}^{x_1} \frac{\partial u_{\cdot,x,y,z,i}}{\partial u_{0,l}} \sum_{j=1}^{d} \mathbb{F}_{ij} d\left(\frac{\partial u_{\cdot,x,y,z,j}}{\partial u_{0,k}}\right) dxdydz
$$

$$(3.20)$$

$$
+ \sum_{j=1}^{6} \int_{z_0}^{z_1} \int_{y_0}^{y_1} \int_{x_0}^{x_1} \frac{\partial u_{\cdot,x,y,z,j}}{\partial u_{0,k}} \sum_{i=1}^{d} \mathbb{F}_{ij} d\left(\frac{\partial u_{\cdot,x,y,z,i}}{\partial u_{0,l}}\right) dxdydz.
$$

Plugging (3.18) and (3.19) into (3.20), and using

$$
\frac{\partial^2 S(u)}{\partial u_{\cdot,x,y,z,j} \partial u_{\cdot,x,y,z,i}} = \frac{\partial^2 S(u)}{\partial u_{\cdot,x,y,z,i} \partial u_{\cdot,x,y,z,j}},
$$

one obtains

$$
\frac{da_{l,k}}{dt} = - \sum_{i,j=1}^{6} \int_{y_0}^{y_1} \int_{z_0}^{z_1} \mathbb{K}_{ij}^{1} \left[\frac{\partial u_{\cdot,1,y,z,i}}{\partial u_{0,l}} \frac{\partial u_{\cdot,1,y,z,j}}{\partial u_{0,k}} - \frac{\partial u_{\cdot,0,y,z,i}}{\partial u_{0,l}} \frac{\partial u_{\cdot,0,y,z,j}}{\partial u_{0,k}}\right] dzdy
$$

$$
- \sum_{i,j=1}^{6} \int_{x_0}^{x_1} \int_{z_0}^{z_1} \mathbb{K}_{ij}^{2} \left[\frac{\partial u_{\cdot,x,1,z,i}}{\partial u_{0,l}} \frac{\partial u_{\cdot,x,1,z,j}}{\partial u_{0,k}} - \frac{\partial u_{\cdot,x,0,z,i}}{\partial u_{0,l}} \frac{\partial u_{\cdot,x,0,z,j}}{\partial u_{0,k}}\right] dzdx
$$

$$
- \sum_{i,j=1}^{6} \int_{y_0}^{y_1} \int_{x_0}^{x_1} \mathbb{K}_{ij}^{3} \left[\frac{\partial u_{\cdot,x,y,1,i}}{\partial u_{0,l}} \frac{\partial u_{\cdot,x,y,1,j}}{\partial u_{0,k}} - \frac{\partial u_{\cdot,x,y,0,i}}{\partial u_{0,l}} \frac{\partial u_{\cdot,x,y,0,j}}{\partial u_{0,k}}\right] dxdy.
$$

Similarly, according to (3.15), we derive

$$\frac{db_{l,k}}{dt} = \sum_{i,j=1}^{6} \int_{y_0}^{y_1} \int_{z_0}^{z_1} \mathbb{K}_{ij}^1 \left[\frac{\partial u_{\cdot,1,y,z,i}}{\partial u_{0,l}} \frac{\partial u_{\cdot,1,y,z,j}}{\partial u_{0,k}} - \frac{\partial u_{\cdot,0,y,z,i}}{\partial u_{0,l}} \frac{\partial u_{\cdot,0,y,z,j}}{\partial u_{0,k}} \right] dzdy,$$

$$\frac{dc_{l,k}}{dt} = \sum_{i,j=1}^{6} \int_{x_0}^{x_1} \int_{z_0}^{z_1} \mathbb{K}_{ij}^2 \left[\frac{\partial u_{\cdot,x,1,z,i}}{\partial u_{0,l}} \frac{\partial u_{\cdot,x,1,z,j}}{\partial u_{0,k}} - \frac{\partial u_{\cdot,x,0,z,i}}{\partial u_{0,l}} \frac{\partial u_{\cdot,x,0,z,j}}{\partial u_{0,k}} \right] dzdx,$$

$$\frac{dd_{l,k}}{dt} = \sum_{i,j=1}^{6} \int_{y_0}^{y_1} \int_{x_0}^{x_1} \mathbb{K}_{ij}^3 \left[\frac{\partial u_{\cdot,x,y,1,i}}{\partial u_{0,l}} \frac{\partial u_{\cdot,x,y,1,j}}{\partial u_{0,k}} - \frac{\partial u_{\cdot,x,y,0,i}}{\partial u_{0,l}} \frac{\partial u_{\cdot,x,y,0,j}}{\partial u_{0,k}} \right] dxdy.$$

Then, the identity (3.17) results from adding up the above equations. Thus, the proof of Theorem 3.2 is finished.   □

Remark 3.1 When the stochastic Maxwell equations are driven by additive noise, e.g.,

$$\begin{cases} d\mathbf{E}(t) = \nabla \times \mathbf{H}(t)dt + \lambda_1 dW(t), & t \in (0, T], \\ d\mathbf{H}(t) = -\nabla \times \mathbf{E}(t)dt + \lambda_2 dW(t), & t \in (0, T] \end{cases} \tag{3.21}$$

with $\lambda_1, \lambda_2 \in \mathbb{R}^3$, the functional $S(u)$ in the form (3.11) of the stochastic Hamiltonian partial differential equation is given by $S(u) = \lambda_2 \cdot \mathbf{E} - \lambda_1 \cdot \mathbf{H}$. For more details of the stochastic multi-symplectic structure of the stochastic Maxwell equations with additive noise, we refer readers to [38, 102].

3.3 Physical Properties

Recall that the deterministic Maxwell equations (1.13) in the lossless medium have both the energy conservation law and the divergence conservation law. In this section, we aim to investigate the corresponding physical properties of the stochastic Maxwell equations (2.8), i.e.,

$$\begin{cases} du(t) = \left[Mu(t) + F(t, u(t)) \right]dt + B(t, u(t))dW(t), & t \in (0, T], \\ u(0) = u_0. \end{cases}$$
$$\tag{3.22}$$

The energy of the solution for (3.22) is defined as

$$\mathscr{E}(u(t)) := \|u(t)\|_{\mathbb{H}}^2 = \int_D \Big(\varepsilon|\mathbf{E}(t,\mathbf{x})|^2 + \mu|\mathbf{H}(t,\mathbf{x})|^2\Big)d\mathbf{x}, \quad t \in [0,T].$$

Proposition 3.1 *Under conditions in Theorem 2.2, the energy of the solution of (3.22) satisfies that for $t \in [0,T]$,*

$$\mathscr{E}(u(t)) = \mathscr{E}(u_0) + \int_0^t \Big(2\langle u(s), F(s, u(s))\rangle_{\mathbb{H}} + \|B(s, u(s))\|_{HS(U_0, \mathbb{H})}^2\Big)ds$$

$$+ 2\int_0^t \Big\langle u(s), B(s, u(s))dW(s)\Big\rangle_{\mathbb{H}}, \quad \mathbb{P}\text{-}a.s.$$

$$(3.23)$$

Proof Since $\mathscr{E}(u)$ is Fréchet differentiable, the first and second order derivatives of $\mathscr{E}(u)$ are

$$\mathscr{E}_u(u)(\phi) = 2\langle u, \phi\rangle_{\mathbb{H}}, \quad \mathscr{E}_{uu}(u)(\phi, \varphi) = 2\langle \varphi, \phi\rangle_{\mathbb{H}} \quad \forall \phi, \varphi \in \mathbb{H}, \quad (3.24)$$

respectively. By the Itô formula given in Theorem D.2, we obtain that for all $t \in [0,T]$,

$$\mathscr{E}(u(t)) = \mathscr{E}(u_0) + \int_0^t \mathscr{E}_u(u(s))\Big(B(s, u(s))dW(s)\Big)$$

$$+ \int_0^t \mathscr{E}_u(u(s))\Big(Mu(s) + F(s, u(s))\Big)ds$$

$$+ \frac{1}{2}\int_0^t \text{Tr}\Big[\mathscr{E}_{uu}(u(s))(B(s, u(s))Q^{\frac{1}{2}})(B(s, u(s))Q^{\frac{1}{2}})^*\Big]ds, \quad \mathbb{P}\text{-}a.s.$$

$$(3.25)$$

Plugging (3.24) into (3.25) leads to

$$\mathscr{E}(u(t)) = \mathscr{E}(u_0) + 2\int_0^t \langle u(s), Mu(s) + F(s, u(s))\rangle_{\mathbb{H}}ds$$

$$+ 2\int_0^t \langle u(s), B(s, u(s))dW(s)\rangle_{\mathbb{H}} + \int_0^t \|B(s, u(s))\|_{HS(U_0, \mathbb{H})}^2 ds, \quad \mathbb{P}\text{-}a.s.$$

for all $t \in [0,T]$. Then, the assertion follows immediately from the skew-adjointness of the Maxwell operator M. $\qquad\square$

Remark 3.2 Two special cases of (3.23) are given as follows.

(a) If

$$\varepsilon = \mu \equiv 1, \quad F(t, u(t)) \equiv 0, \quad B(t, u(t)) \equiv (\lambda_1^\top, \lambda_2^\top)^\top,$$

then the averaged energy $\mathbb{E}\big[\mathscr{E}(u(t))\big]$ satisfies the following linear growth law

$$\mathbb{E}\big[\mathscr{E}(u(t))\big] = \mathbb{E}\big[\mathscr{E}(u_0)\big] + \varkappa t, \quad t \in [0, T],$$

where $\varkappa = \Big(|\lambda_1|^2 + |\lambda_2|^2\Big)\mathrm{Tr}(Q)$ represents the growth rate.

(b) If

$$\varepsilon = \mu \equiv 1, \quad F(t, u(t)) = -\frac{1}{2}\lambda^2 u F_Q, \quad B(t, u(t)) = \lambda(-\mathbf{H}(t)^\top, \mathbf{E}(t)^\top)^\top,$$

then the energy is conserved, i.e., for $t \in [0, T]$,

$$\mathscr{E}(u(t)) = \mathscr{E}(u_0), \quad \mathbb{P}\text{-}a.s.,$$

where $F_Q(\mathbf{x}) := \sum_{j \in \mathbb{N}} \big(Q^{\frac{1}{2}} e_j(\mathbf{x})\big)^2$.

The following proposition states the divergence evolution law for (3.22).

Proposition 3.2 *Suppose that* $\mathbf{J}_e, \mathbf{J}_m \in H(\mathrm{div}, D)$, *and* $\mathbf{J}_e^r, \mathbf{J}_m^r \in HS(U_0, H(\mathrm{div}, D))$. *Then averaged divergences of the solution* $u = (\mathbf{E}^\top, \mathbf{H}^\top)^\top$ *of (3.22) satisfy*

$$\mathbb{E}\Big[\nabla \cdot (\varepsilon \mathbf{E}(t))\Big] = \mathbb{E}\Big[\nabla \cdot (\varepsilon \mathbf{E}_0)\Big] - \mathbb{E}\Big[\int_0^t \nabla \cdot \mathbf{J}_e(s)ds\Big],$$

$$\mathbb{E}\Big[\nabla \cdot (\mu \mathbf{H}(t))\Big] = \mathbb{E}\Big[\nabla \cdot (\mu \mathbf{H}_0)\Big] - \mathbb{E}\Big[\int_0^t \nabla \cdot \mathbf{J}_m(s)ds\Big]$$

$$(3.26)$$

for all $t \in [0, T]$.

Proof Let $\Psi(\mathbf{E}) := \nabla \cdot (\varepsilon \mathbf{E})$. Since Ψ is Fréchet differentiable, the first and second order derivatives of Ψ are

$$\Psi_{\mathbf{E}}(\mathbf{E})(\phi) = \nabla \cdot (\varepsilon \phi), \quad \Psi_{\mathbf{EE}}(\mathbf{E})(\phi, \varphi) = 0 \quad \forall \phi, \varphi \in L^2(D)^3. \qquad (3.27)$$

By applying the Itô formula to $\Psi(\mathbf{E})$, we obtain

$$
\begin{aligned}
\Psi(\mathbf{E}(t)) &= \Psi(\mathbf{E}_0) + \int_0^t \Psi_{\mathbf{E}}(\mathbf{E}(s))\Big[\varepsilon^{-1}\nabla \times \mathbf{H}(s) - \varepsilon^{-1}\mathbf{J}_e\Big]ds \\
&\quad + \int_0^t \Psi_{\mathbf{E}}(\mathbf{E}(s))\Big(-\varepsilon^{-1}\mathbf{J}_e^r dW(s)\Big) \\
&= \Psi(\mathbf{E}_0) + \int_0^t \Big[\nabla \cdot (\nabla \times \mathbf{H}(s)) - \nabla \cdot \mathbf{J}_e\Big]ds - \int_0^t \nabla \cdot \Big(\mathbf{J}_e^r dW(s)\Big) \\
&= \Psi(\mathbf{E}_0) - \int_0^t \nabla \cdot \mathbf{J}_e ds - \int_0^t \nabla \cdot \Big(\mathbf{J}_e^r dW(s)\Big), \quad \mathbb{P}\text{-}a.s.
\end{aligned}
$$

$$(3.28)$$

for all $t \in [0, T]$, where in the last step, we used the fact $\nabla \cdot (\nabla \times \boldsymbol{\psi}) = 0$ for vector function $\boldsymbol{\psi} = (\psi_1, \psi_2, \psi_3)^\top$. In a similar manner, applying the Itô formula to the functional $\Psi(\mathbf{H}(t)) = \nabla \cdot (\mu\mathbf{H}(t))$ gives that for all $t \in [0, T]$,

$$
\Psi(\mathbf{H}(t)) = \Psi(\mathbf{H}_0) - \int_0^t \nabla \cdot \mathbf{J}_m ds - \int_0^t \nabla \cdot \Big(\mathbf{J}_m^r dW(s)\Big), \quad \mathbb{P}\text{-}a.s. \qquad (3.29)
$$

Then, the assertions follow from taking the expectation on both sides of (3.28) and (3.29). $\square$

Remark 3.3 If functions $\mathbf{J}_e$ and $\mathbf{J}_m$ are divergence-free, i.e., $\nabla \cdot \mathbf{J}_e = 0$ and $\nabla \cdot \mathbf{J}_m = 0$, then

$$
\mathbb{E}\Big[\nabla \cdot (\varepsilon\mathbf{E}(t))\Big] = \mathbb{E}\Big[\nabla \cdot (\varepsilon\mathbf{E}_0)\Big], \qquad \mathbb{E}\Big[\nabla \cdot (\mu\mathbf{H}(t))\Big] = \mathbb{E}\Big[\nabla \cdot (\mu\mathbf{H}_0)\Big]
$$

for all $t \in [0, T]$.

3.4 Small Noise Asymptotics

In this part, we are devoted to studying the asymptotic property of the solution of the stochastic Maxwell equations with small noise:

$$
\begin{cases}
du(t) = Mu(t)dt - \sqrt{\lambda}dW(t), & t \in (0, T], \\
u(0) = u_0,
\end{cases}
\qquad (3.30)
$$

where $\lambda \in \mathbb{R}^+$, and $W(t) = (\varepsilon^{-1} W_1(t)^\top, \mu^{-1} W_2(t)^\top)^\top$ is an $\mathbb{H}$-valued Q-Wiener process with

$$Q = \begin{bmatrix} \varepsilon^{-1} Q_1 & 0 \\ 0 & \mu^{-1} Q_2 \end{bmatrix}$$

and ε, μ satisfying Assumption 2.1. Here, W_i are $L^2(D)^3$-valued Q_i-Wiener processes with Q_i, $i = 1, 2$ being nonnegative and symmetric operators with finite traces. Assume that W_1 and W_2 are independent.

Remark 3.4 Note that for any $u = (u_1^\top, u_2^\top)^\top$, $v = (v_1^\top, v_2^\top)^\top \in \mathbb{H}$, we have

$$\mathbb{E}\big[\langle W(t), u \rangle_{\mathbb{H}} \langle W(t), v \rangle_{\mathbb{H}} \big]$$

$$= \mathbb{E}\big[\big(\langle W_1(t), u_1 \rangle_{L^2(D)^3} + \langle W_2(t), u_2 \rangle_{L^2(D)^3} \big) \big(\langle W_1(t), v_1 \rangle_{L^2(D)^3} + \langle W_2(t), v_2 \rangle_{L^2(D)^3} \big) \big]$$

$$= t \langle Q_1 u_1, v_1 \rangle_{L^2(D)^3} + t \langle Q_2 u_2, v_2 \rangle_{L^2(D)^3} = t \langle Q u, v \rangle_{\mathbb{H}}.$$

Using Assumption 2.1 and the fact

$$\mathbb{E}\big[\| W(t) \|_{\mathbb{H}}^2 \big] = t \Big(\| \varepsilon^{-\frac{1}{2}} Q_1^{\frac{1}{2}} \|_{HS(L^2(D)^3, L^2(D)^3)}^2 + \| \mu^{-\frac{1}{2}} Q_2^{\frac{1}{2}} \|_{HS(L^2(D)^3, L^2(D)^3)}^2 \Big),$$

we know that the operator Q is still symmetric and nonnegative on $\mathbb{H}$ with the finite trace

$$\mathrm{Tr}(Q) = \Big(\| \varepsilon^{-\frac{1}{2}} Q_1^{\frac{1}{2}} \|_{HS(L^2(D)^3, L^2(D)^3)}^2 + \| \mu^{-\frac{1}{2}} Q_2^{\frac{1}{2}} \|_{HS(L^2(D)^3, L^2(D)^3)}^2 \Big)$$

$$\leq \delta^{-1} \Big(\mathrm{Tr}(Q_1) + \mathrm{Tr}(Q_2) \Big) < \infty.$$

To emphasize the dependence of the solution u of (3.30) on parameters λ and u_0, below we write it as $u^{u_0, \lambda}$. It is observed that $u^{u_0, \lambda}$ converges in a certain sense to its deterministic counterpart $u^{u_0, 0}$ which is the solution of (3.30) with $\lambda = 0$. In many circumstances, one may be interested in the asymptotics of the probability $\mathbb{P}\big(\| u^{u_0, \lambda}(T) - u^{u_0, 0}(T) \|_{\mathbb{H}} > \iota \big)$ for some $\iota > 0$, which is usually characterized by the large deviations principle of $\{ u^{u_0, \lambda}(T) \}_{\lambda > 0}$.

The large deviations principle is concerned with the exponential decay of probabilities of rare events, which can be regarded as an extension or refinement of the law of large numbers and the central limit theorem. It is usually used to describe the asymptotic behavior of stochastic processes (see e.g., [69]).

We first introduce some concepts on the theory of large deviations (see e.g., [69, 116]).

Definition 3.2 Let $\mathscr{X}$ be a Polish space. $\mathbb{I} : \mathscr{X} \to [0, \infty]$ is called a rate function if it is lower semi-continuous. Particularly, $\mathbb{I}$ is called a good rate function if all level sets $\mathbb{I}^{-1}([0, a])$, $a \in [0, \infty)$ are compact.

Definition 3.3 Let $\mathbb{I}$ be a rate function and $\{\mu_\lambda\}_{\lambda>0}$ be a family of probability measures on $\mathscr{X}$. $\{\mu_\lambda\}_{\lambda>0}$ is said to satisfy a large deviations principle with the rate function $\mathbb{I}$ if

(LDP1) $\displaystyle\liminf_{\lambda\to0} \lambda \log(\mu_\lambda(U)) \geq -\inf \mathbb{I}(U)$ for every open $U \subset \mathscr{X}$,

(LDP2) $\displaystyle\limsup_{\lambda\to0} \lambda \log(\mu_\lambda(G)) \leq -\inf \mathbb{I}(G)$ for every closed $G \subset \mathscr{X}$.

For a family $\{X_\lambda\}_{\lambda>0}$ of random variables from $(\Omega, \mathscr{F}, \mathbb{P})$ to $(\mathscr{X}, \mathscr{B}(\mathscr{X}))$, we say that $\{X_\lambda\}_{\lambda>0}$ satisfies a large deviations principle with the rate function $\mathbb{I}$ if its distribution $\{\mathbb{P} \circ X_\lambda^{-1}\}_{\lambda>0}$ satisfies conditions in Definition 3.3. The following lemma plays an important role in dealing with the large deviations principle for a family of Gaussian random variables (see [62, Proposition 12.10]).

Lemma 3.2 *Assume that X is a Gaussian random variable with distribution $N(0, \widetilde{Q})$ on some arbitrary Hilbert space H. Then the family $\{X_\lambda := \sqrt{\lambda}X\}_{\lambda>0}$ of random variables satisfies a large deviations principle with the good rate function*

$$\mathbb{I}(x) = \begin{cases} \frac{1}{2}\|\widetilde{Q}^{-\frac{1}{2}}x\|_H^2, & \text{if } x \in \widetilde{Q}^{\frac{1}{2}}(H), \\ +\infty, & \text{otherwise}, \end{cases} \tag{3.31}$$

where $\widetilde{Q}^{-\frac{1}{2}}$ is the pseudo-inverse of $\widetilde{Q}^{\frac{1}{2}}$.

Denote the stochastic convolution by $W_M(t) := \int_0^t S(t-r)\mathrm{d}W(r)$, $t \geq 0$. Then for arbitrary $T > 0$, $W_M(T)$ is Gaussian on $\mathbb{H}$ with mean zero and covariance operator

$$Q_T := \mathrm{Cov}\big(W_M(T)\big) = \int_0^T S(r)QS^*(r)\mathrm{d}r.$$

Note that if Q commutes with M, then $Q_T^{\frac{1}{2}}(\mathbb{H}) = Q^{\frac{1}{2}}(\mathbb{H})$. In fact, $Q_T = TQ$ in this case.

Based on Lemma 3.2, we obtain the following asymptotic behavior of the solution of (3.30), which states that the law of the solution satisfies a large deviations principle.

Theorem 3.3 *For any $T > 0$ and $u_0 \in \mathbb{H}$, the family $\{u^{u_0,\lambda}(T)\}_{\lambda>0}$ of random variables satisfies a large deviations principle with the good rate function*

$$
\mathbb{I}_T^{u_0}(v) =
\begin{cases}
\frac{1}{2}\|Q_T^{-\frac{1}{2}}(v - S(T)u_0)\|_{\mathbb{H}}^2, & \text{if } v - S(T)u_0 \in Q_T^{\frac{1}{2}}(\mathbb{H}), \\
+\infty, & \text{otherwise,}
\end{cases}
\tag{3.32}
$$

where $Q_T^{-\frac{1}{2}}$ is the pseudo-inverse of $Q_T^{\frac{1}{2}}$, and $u^{u_0,\lambda}(t)$ is the mild solution of (3.30) given by

$$
u^{u_0,\lambda}(t) = S(t)u_0 - \sqrt{\lambda}W_M(t), \quad t \geq 0.
$$

Proof We define a process $Y^\lambda(t) := u^{u_0,\lambda}(t) - S(t)u_0$, $t \in [0, T]$, which satisfies (3.30) with the initial datum $Y^\lambda(0) = 0$. This means that $Y^\lambda(t) = -\sqrt{\lambda}W_M(t)$. Then it follows from Lemma 3.2 that the good rate function of the large deviations principle of $\{Y^\lambda(T)\}_{\lambda>0}$ is

$$
\mathbb{I}_T^0(v) =
\begin{cases}
\frac{1}{2}\|Q_T^{-\frac{1}{2}}v\|_{\mathbb{H}}^2, & \text{if } v \in Q_T^{\frac{1}{2}}(\mathbb{H}), \\
+\infty, & \text{otherwise.}
\end{cases}
\tag{3.33}
$$

Now, we want to find the rate function of $\{u^{u_0,\lambda}(T)\}_{\lambda>0}$. It suffices to show that (LDP1) and (LDP2) in Definition 3.3 hold with the rate function $\mathbb{I}_T^{u_0}(v)$.

Let $A \in \mathscr{B}(\mathbb{H})$ be closed. Then $A - S(T)u_0 := \{v - S(T)u_0, v \in A\}$ is still closed in $\mathscr{B}(\mathbb{H})$ and hence

$$
\limsup_{\lambda \to 0} \left[\lambda \log \mathbb{P}\big(u^{u_0,\lambda}(T) \in A\big)\right]
$$

$$
= \limsup_{\lambda \to 0} \left[\lambda \log \mathbb{P}\big(Y^\lambda(T) \in A - S(T)u_0\big)\right]
$$

$$
\leq - \inf_{\overline{v} \in A - S(T)u_0} \mathbb{I}_T^0(\overline{v})
$$

$$
= - \inf_{v \in A} \mathbb{I}_T^0(v - S(T)u_0) = - \inf_{v \in A} \mathbb{I}_T^{u_0}(v).
$$

Similarly, we can check that for any open set $B \in \mathscr{B}(\mathbb{H})$,

$$
\liminf_{\lambda \to 0} \left[\lambda \log \mathbb{P}\big(u^{u_0,\lambda}(T) \in B\big)\right] \geq - \inf_{v \in B} \mathbb{I}_T^0(v - S(T)u_0) = - \inf_{v \in B} \mathbb{I}_T^{u_0}(v).
$$

Thus, $\mathbb{I}_T^{u_0}$ is the required good rate function of the large deviations principle of $\{u^{u_0,\lambda}(T)\}_{\lambda>0}$. $\square$

3.5   Intrinsic Properties for the Damped Case

In recent years, the deterministic Maxwell equations with damping have aroused much attention in the literature (see e.g., [19, 76, 177]). Some subjects such as the asymptotic behavior and global existence of the solution of the damped Maxwell equations have been investigated in the aforementioned works. Here we consider a stochastic version and study the corresponding intrinsic properties. Consider the following stochastic Maxwell equations with damping

$$
\begin{cases}
\mathrm{d}u(t) = [Mu(t) - \sigma u(t)]\mathrm{d}t + \lambda \mathbb{J}^{-1}u(t) \circ \mathrm{d}W_1(t) + \boldsymbol{\theta}\mathrm{d}W_2(t), & t > 0, \\
u(0) = u_0,
\end{cases}
$$

$$(3.34)$$

where $u = (\mathbf{E}^\top, \mathbf{H}^\top)^\top$, $u_0 = (\mathbf{E}_0^\top, \mathbf{H}_0^\top)^\top$ and

$$
M = \begin{bmatrix} 0 & \nabla\times \\ -\nabla\times & 0 \end{bmatrix}, \quad \boldsymbol{\theta} = \begin{bmatrix} \boldsymbol{\theta}_1 \\ \boldsymbol{\theta}_2 \end{bmatrix} \in \mathbb{R}^6, \quad \lambda \in \mathbb{R}.
$$

We impose the PEC boundary conditions

$$
\mathbf{n} \times \mathbf{E} = 0, \quad \mathbf{n} \cdot \mathbf{H} = 0, \quad \text{on} \quad [0, T] \times \partial D, \tag{3.35}
$$

and assume that the damped coefficient σ satisfies

$$
\sigma \in W^{1,\infty}(D), \quad \sigma \ge \sigma_0 > 0
$$

with σ_0 being a constant. Here, W_i $(i = 1, 2)$ are two independent U-valued Q_i-Wiener processes which have the following Karhunen–Loève expansions

$$
W_i(t) = \sum_{k \in \mathbb{N}} (Q_i)^{\frac{1}{2}} e_k \beta_k^i(t) = \sum_{k \in \mathbb{N}} \sqrt{\eta_k^i} e_k \beta_k^i(t), \quad \eta_k^i \ge 0 \text{ and } \sum_{k \in \mathbb{N}} \eta_k^i < \infty
$$

with $\{e_k\}_{k \in \mathbb{N}}$ being an orthonormal basis of U, and $\{\beta_k^i\}_{k \in \mathbb{N}}^{i=1,2}$ being a family of independent standard Brownian motions.

Set $F_{Q_1}(\mathbf{x}) := \sum_{k \in \mathbb{N}} (Q_1^{\frac{1}{2}} e_k(\mathbf{x}))^2$. We rewrite (3.34) as its equivalent form in the Itô sense and obtain

$$
\mathrm{d}u(t) = \left[Mu(t) - \sigma u(t) - \frac{1}{2}\lambda^2 F_{Q_1} u(t) \right]\mathrm{d}t + \lambda \mathbb{J}^{-1}u(t)\mathrm{d}W_1(t) + \boldsymbol{\theta}\mathrm{d}W_2(t) \tag{3.36}
$$

for all $t \ge 0$.

Let $u_0 \in L^2(\Omega, \mathbb{H})$ and assume that $Q_i^{\frac{1}{2}} \in HS(U, H^{\gamma_i}(D))$ with some $\gamma_i \geq 0$, $i = 1, 2$. Similar to the one performed in Sect. 2.2.1, (3.34) have a unique mild solution given by

$$u(t) = \hat{S}(t)u_0 - \frac{1}{2}\lambda^2 \int_0^t \hat{S}(t-r)F_{Q_1}u(r)dr + \lambda \int_0^t \hat{S}(t-r)\mathbb{J}^{-1}u(r)dW_1(r)$$

$$+ \int_0^t \hat{S}(t-r)\theta dW_2(r), \quad \mathbb{P}\text{-}a.s.$$

for each $t \geq 0$, where $\{\hat{S}(t) = e^{t(M-\sigma Id)}, \ t \geq 0\}$ is a C_0-semigroup generated by $M - \sigma Id$.

3.5.1 Energy Evolution Law

Now, we give the energy evolution law of the solution u of (3.36).

Theorem 3.4 *Assume $\sigma \geq \sigma_0 > 0$ with a constant σ_0. Then for all $t \geq 0$,*

$$\|u(t)\|_{\mathbb{H}}^2 = \|u_0\|_{\mathbb{H}}^2 - 2 \int_0^t \langle u(s), \sigma u(s)\rangle_{\mathbb{H}}ds + |\theta|^2 \text{Tr}(Q_2)t$$

$$+ 2 \int_0^t \langle u(s), \theta dW_2(s)\rangle_{\mathbb{H}}, \quad \mathbb{P}\text{-}a.s.$$

Moreover, there exists a positive constant $C = C(\sigma_0, \theta, u_0, Q_2)$ such that

$$\mathbb{E}[\|u(t)\|_{\mathbb{H}}^2] \leq e^{-2\sigma_0 t}\mathbb{E}[\|u_0\|_{\mathbb{H}}^2] + |\theta|^2 \text{Tr}(Q_2)\frac{1 - e^{-2\sigma_0 t}}{2\sigma_0} \leq C$$

for all $t \geq 0$.

Proof Applying the Itô formula to $\|u(t)\|_{\mathbb{H}}^2$, we have

$$d\|u(t)\|_{\mathbb{H}}^2 = 2\Big\langle u(t), Mu(t) - \sigma u(t) - \frac{1}{2}\lambda^2 F_{Q_1}u(t)\Big\rangle_{\mathbb{H}}dt$$

$$+ 2\lambda\Big\langle u(t), \mathbb{J}^{-1}u(t)dW_1(t)\Big\rangle_{\mathbb{H}} + 2\langle u(t), \theta dW_2(t)\rangle_{\mathbb{H}}$$

$$+ \lambda^2 \langle |u(t)|^2, F_{Q_1}\rangle_U dt + |\theta|^2 \text{Tr}(Q_2)dt$$

$$= -2\langle u(t), \sigma u(t)\rangle_{\mathbb{H}}dt + 2\langle u(t), \theta dW_2(t)\rangle_{\mathbb{H}} + |\theta|^2 \text{Tr}(Q_2)dt, \quad \mathbb{P}\text{-}a.s.$$

By taking the expectation and using the Grönwall inequality in the differential form given in Proposition A.5, we have

$$\mathbb{E}\big[\|u(t)\|_{\mathbb{H}}^2\big] \le e^{-2\sigma_0 t}\,\mathbb{E}\big[\|u_0\|_{\mathbb{H}}^2\big] + |\boldsymbol{\theta}|^2 \mathrm{Tr}(Q_2)\frac{1 - e^{-2\sigma_0 t}}{2\sigma_0}$$

$$\le \mathbb{E}\big[\|u_0\|_{\mathbb{H}}^2\big] + \frac{|\boldsymbol{\theta}|^2 \mathrm{Tr}(Q_2)}{2\sigma_0},$$

which leads to the uniform boundedness of the solution u with respect to time.   □

From Theorem 3.4, we can observe that if $\sigma \equiv \sigma_0 > 0$ and $\boldsymbol{\theta} \equiv 0$, then for $t \ge 0$,

$$\|u(t)\|_{\mathbb{H}} = e^{-\sigma_0 t}\|u_0\|_{\mathbb{H}}, \quad \mathbb{P}\text{-}a.s.$$

3.5.2 Stochastic Conformal Multi-Symplectic Conservation Law

Due to the existence of the damped term in (3.36), the properties of stochastic Maxwell equations have a huge difference. For instance, the multi-symplectic conservation law is not preserved anymore. Instead, the stochastic Maxwell equations possess the stochastic conformal multi-symplectic conservation law.

Denoting

$$S_1(u) := \frac{\lambda}{2}\Big(|\mathbf{E}|^2 + |\mathbf{H}|^2\Big), \quad S_2(u) := \boldsymbol{\theta}_2 \cdot \mathbf{E} - \boldsymbol{\theta}_1 \cdot \mathbf{H},$$

we reformate (3.34) as a damped stochastic Hamiltonian partial differential equation

$$\mathbb{F}\mathrm{d}u + \mathbb{K}_1\partial_x u\mathrm{d}t + \mathbb{K}_2\partial_y u\mathrm{d}t + \mathbb{K}_3\partial_z u\mathrm{d}t = -\sigma\mathbb{F}u\mathrm{d}t + \sum_{i=1}^{2}\nabla_u S_i(u)\circ\mathrm{d}W_i(t), \quad (3.37)$$

where $\mathbb{F}$, $\mathbb{K}_1$, $\mathbb{K}_2$, and $\mathbb{K}_3$ are defined in (3.12). One can show that the damped stochastic Maxwell equations (3.34) possess the following stochastic conformal multi-symplectic conservation law. The proof is analogous to that of Theorem 3.2 and thus is omitted.

Theorem 3.5 *Equation (3.34) preserves the stochastic conformal multi-symplectic conservation law*

$$\mathrm{d}\varpi + \partial_x\kappa_1\mathrm{d}t + \partial_y\kappa_2\mathrm{d}t + \partial_z\kappa_3\mathrm{d}t = -2\sigma\varpi\mathrm{d}t, \quad \mathbb{P}\text{-}a.s.,$$

which means

$$\int_{z_0}^{z_1} \int_{y_0}^{y_1} \int_{x_0}^{x_1} \varpi(t_1, x, y, z) \mathrm{d}x \mathrm{d}y \mathrm{d}z + \int_{z_0}^{z_1} \int_{y_0}^{y_1} \int_{t_0}^{t_1} \kappa_1(t, x_1, y, z) \mathrm{d}t \mathrm{d}y \mathrm{d}z$$

$$+ \int_{z_0}^{z_1} \int_{x_0}^{x_1} \int_{t_0}^{t_1} \kappa_2(t, x, y_1, z) \mathrm{d}t \mathrm{d}x \mathrm{d}z + \int_{y_0}^{y_1} \int_{x_0}^{x_1} \int_{t_0}^{t_1} \kappa_3(t, x, y, z_1) \mathrm{d}t \mathrm{d}x \mathrm{d}y$$

$$- \int_{z_0}^{z_1} \int_{y_0}^{y_1} \int_{x_0}^{x_1} \varpi(t_0, x, y, z) \mathrm{d}x \mathrm{d}y \mathrm{d}z - \int_{z_0}^{z_1} \int_{y_0}^{y_1} \int_{t_0}^{t_1} \kappa_1(t, x_0, y, z) \mathrm{d}t \mathrm{d}y \mathrm{d}z$$

$$- \int_{z_0}^{z_1} \int_{x_0}^{x_1} \int_{t_0}^{t_1} \kappa_2(t, x, y_0, z) \mathrm{d}t \mathrm{d}x \mathrm{d}z - \int_{y_0}^{y_1} \int_{x_0}^{x_1} \int_{t_0}^{t_1} \kappa_3(t, x, y, z_0) \mathrm{d}t \mathrm{d}x \mathrm{d}y$$

$$= -2 \int_{t_0}^{t_1} \int_{z_0}^{z_1} \int_{y_0}^{y_1} \int_{x_0}^{x_1} \sigma \varpi(t, x, y, z) \mathrm{d}x \mathrm{d}y \mathrm{d}z \mathrm{d}t,$$

where $\varpi = \frac{1}{2} \mathrm{d}u \wedge \mathbb{F} \mathrm{d}u$ and $\kappa_p = \frac{1}{2} \mathrm{d}u \wedge \mathbb{K}_p \mathrm{d}u$, $p = 1, 2, 3$ are differential 2-forms associated with the skew-symmetric matrices $\mathbb{F}$ and $\mathbb{K}_p$, respectively, and $(t_0, t_1) \times (x_0, x_1) \times (y_0, y_1) \times (z_0, z_1) \subset [0, T] \times D$ is the local domain of u.

Remark 3.5 The conclusion of Theorem 3.5 can be generalized to the following damped stochastic Hamiltonian partial differential equation

$$\widetilde{\mathbb{F}} \mathrm{d}v + \sum_{i=1}^{d} \widetilde{\mathbb{K}}_i \partial_{x_i} v \mathrm{d}t = \widetilde{\mathbb{G}} v \mathrm{d}t + \nabla_v S_1(v) \mathrm{d}t + \nabla_v S_2(v) \circ \mathrm{d}W(t), \qquad (3.38)$$

where $v : [0, \infty) \times \mathbb{R}^d \to \mathbb{R}^n$, $\widetilde{\mathbb{F}}$ and $\widetilde{\mathbb{K}}_i$ are skew-symmetric matrices, and $\widetilde{\mathbb{G}} = -\frac{a}{2} \widetilde{\mathbb{F}} - \sum_{i=1}^{d} \frac{b_i}{2} \widetilde{\mathbb{K}}_i$ with $a, b_i \in \mathbb{R}^+$ ($i = 1, 2, \ldots, d$) being constants. It can be shown that (3.38) possesses the following stochastic conformal multi-symplectic conservation law

$$\mathrm{d}\widetilde{\varpi}(t, \mathbf{x}) + \sum_{i=1}^{d} \partial_{x_i} \widetilde{\kappa}_i(t, \mathbf{x}) \mathrm{d}t = \left(-a\widetilde{\omega}(t, \mathbf{x}) - \sum_{i=1}^{d} b_i \widetilde{\kappa}_i(t, \mathbf{x}) \right) \mathrm{d}t, \qquad \mathbb{P}\text{-}a.s.$$

with $\widetilde{\varpi} = \frac{1}{2} \mathrm{d}v \wedge \widetilde{\mathbb{F}} \mathrm{d}v$ and $\widetilde{\kappa}_i = \frac{1}{2} \mathrm{d}v \wedge \widetilde{\mathbb{K}}_i \mathrm{d}v$, $i = 1, 2, \ldots, d$.

In fact, a large class of stochastic partial differential equations can be represented in the form (3.38), e.g., the damped stochastic nonlinear Schrödinger equation (see e.g., [39, 101]) and the damped stochastic wave equation (see e.g., [150, 154]).

3.5.3 Ergodicity

To have a better understanding of the long-time behavior for the solution of (3.36), we investigate the ergodicity of the solution in this part.

We first give a brief introduction to the invariant measure and ergodicity.

Definition 3.4 (see e.g., [61]) A probability measure $\pi \in \mathscr{P}(\mathbb{H})$ is said to be invariant for a Markov semigroup $\{P_t, t \geq 0\}$, if

$$\int_{\mathbb{H}} P_t \varphi \mathrm{d}\pi = \int_{\mathbb{H}} \varphi \mathrm{d}\pi =: \pi(\varphi) \quad \forall \varphi \in \mathbf{B}_b(\mathbb{H}), \ t \geq 0,$$

where $\mathscr{P}(\mathbb{H})$ and $\mathbf{B}_b(\mathbb{H})$ denote the space of all probability measures on $\mathbb{H}$ and the space of all measurable and bounded functions defined on $\mathbb{H}$, respectively.

Definition 3.5 (see e.g., [101]) Let π be an invariant measure of a stochastic process $u : \Omega \times [0, T] \to \mathbb{H}$.

(i) u is said to be ergodic on $\mathbb{H}$ if

$$\lim_{T \to \infty} \frac{1}{T} \int_0^T \mathbb{E}[\varphi(u(t))]\mathrm{d}t = \pi(\varphi) \quad \text{in } L^2(\mathbb{H}, \pi)$$

for all $\varphi \in L^2(\mathbb{H}, \pi)$, where $L^2(\mathbb{H}, \pi)$ denotes the space of all functions defined on $\mathbb{H}$ which are square integrable with respect to measure π.

(ii) u is said to be exponentially mixing on $\mathbb{H}$ if there exists a positive constant ρ and a positive function $K(\cdot)$ such that for any bounded Lipschitz continuous function φ on $\mathbb{H}$, $t > 0$, and $u_0 \in \mathbb{H}$,

$$\left| P_t \varphi(u_0) - \pi(\varphi) \right| \leq K(u_0) L_\varphi e^{-\rho t},$$

where L_φ denotes the Lipschitz constant of φ.

Before we proceed, we need to obtain the following uniform boundedness of the solution u of (3.36) in the $H^1(D)^6$-norm with respect to time.

Lemma 3.3 *Let $u_0 \in L^2(\Omega, H^1(D)^6)$ and $F_{Q_1} \in W^{1,\infty}(D)$. In addition, let $Q_i^{\frac{1}{2}} \in HS(U, H^{\gamma_i}(D))$, $i = 1, 2$ for any $\gamma_1 > 5/2$ and $\gamma_2 \geq 1$, and let $\sigma \in W^{1,\infty}(D)$, $\sigma \geq \sigma_0 > 0$ with a constant σ_0. Then the solution of (3.36) is uniformly bounded in the $H^1(D)^6$-norm and satisfies*

$$\mathbb{E}\left[\|u(t)\|^2_{H^1(D)^6}\right] \leq C\left(1 + e^{-\sigma_0 t} \mathbb{E}\left[\|u_0\|^2_{H^1(D)^6}\right]\right) \quad \forall t \geq 0,$$

where the positive constant C depends on σ_0, λ, $|\boldsymbol{\theta}|$, $\mathrm{Tr}(Q_2)$, $\|\sigma\|_{W^{1,\infty}(D)}$, $\|F_{Q_1}\|_{W^{1,\infty}(D)}$, $\mathbb{E}[\|u_0\|_{\mathbb{H}}^2]$, and $\|Q_i^{\frac{1}{2}}\|_{HS(U,H^{\gamma_i}(D))}$, $i = 1, 2$.

Proof *Step 1. Estimate of* $\mathbb{E}[\|Mu(t)\|_{\mathbb{H}}^2]$. Applying the Itô formula to $\|Mu\|_{\mathbb{H}}^2$ and taking the expectation, we have

$$
\begin{aligned}
d\mathbb{E}\big[\|Mu(t)\|_{\mathbb{H}}^2\big] = & -2\mathbb{E}\Big[\Big\langle Mu(t), M\big((\sigma + \tfrac{1}{2}\lambda_1^2 F_{Q_1})u(t)\big)\Big\rangle_{\mathbb{H}}\Big]dt \\
& + \lambda^2 \mathbb{E}\Big[\sum_{k\in\mathbb{N}} \big\|M(\mathbb{J}^{-1}u(t)Q_1^{\frac{1}{2}}e_k)\big\|_{\mathbb{H}}^2\Big]dt \\
& + \sum_{k\in\mathbb{N}} \big\|M(\boldsymbol{\theta} Q_2^{\frac{1}{2}}e_k)\big\|_{\mathbb{H}}^2 dt.
\end{aligned}
\tag{3.39}
$$

For the first term on the right-hand side of (3.39), we note that

$$
\begin{aligned}
& -2\Big\langle Mu, M\big((\sigma + \tfrac{1}{2}\lambda^2 F_{Q_1})u\big)\Big\rangle_{\mathbb{H}} \\
& = -2\Big\langle Mu, (\sigma + \tfrac{1}{2}\lambda^2 F_{Q_1})Mu\Big\rangle_{\mathbb{H}} - 2\Big\langle Mu, \begin{bmatrix} \nabla(\sigma + \tfrac{1}{2}\lambda^2 F_{Q_1}) \times \mathbf{H} \\ -\nabla(\sigma + \tfrac{1}{2}\lambda^2 F_{Q_1}) \times \mathbf{E} \end{bmatrix}\Big\rangle_{\mathbb{H}},
\end{aligned}
\tag{3.40}
$$

where we used the fact that $\nabla \times (fv) = f\nabla \times v + (\nabla f) \times v$ for scalar function f and vector function v. It follows from the Young inequality that

$$
\begin{aligned}
& -2\mathbb{E}\Big[\Big\langle Mu, \begin{bmatrix} \nabla(\sigma + \tfrac{1}{2}\lambda^2 F_{Q_1}) \times \mathbf{H} \\ -\nabla(\sigma + \tfrac{1}{2}\lambda^2 F_{Q_1}) \times \mathbf{E} \end{bmatrix}\Big\rangle_{\mathbb{H}}\Big] \\
& \leq 4\mathbb{E}\Big[\|Mu\|_{\mathbb{H}}\|u\|_{\mathbb{H}}\|\nabla(\sigma + \tfrac{1}{2}\lambda^2 F_{Q_1})\|_{L^\infty(D)^3}\Big] \\
& \leq \tfrac{1}{2}\sigma_0\mathbb{E}\big[\|Mu\|_{\mathbb{H}}^2\big] + C\mathbb{E}\big[\|u\|_{\mathbb{H}}^2\big],
\end{aligned}
\tag{3.41}
$$

where the positive constant C depends on σ_0, λ, $\|\sigma\|_{W^{1,\infty}(D)}$, and $\|F_{Q_1}\|_{W^{1,\infty}(D)}$. Plugging (3.41) into (3.40), we have

$$
\begin{aligned}
&-2\mathbb{E}\Big[\Big\langle Mu, M\big((\sigma + \tfrac{1}{2}\lambda^2 F_{Q_1})u\big)\Big\rangle_{\mathbb{H}}\Big]\\
&\leq -2\mathbb{E}\Big[\Big\langle Mu, (\sigma + \tfrac{1}{2}\lambda^2 F_{Q_1})Mu\Big\rangle_{\mathbb{H}}\Big] + \tfrac{1}{2}\sigma_0 \mathbb{E}\big[\|Mu\|_{\mathbb{H}}^2\big] + C\mathbb{E}\big[\|u\|_{\mathbb{H}}^2\big]\\
&\leq -\tfrac{3}{2}\sigma_0 \mathbb{E}\big[\|Mu\|_{\mathbb{H}}^2\big] - \lambda^2 \mathbb{E}\Big[\big\langle Mu, F_{Q_1}Mu\big\rangle_{\mathbb{H}}\Big] + C\mathbb{E}\big[\|u\|_{\mathbb{H}}^2\big]\\
&\leq -\tfrac{3}{2}\sigma_0 \mathbb{E}\big[\|Mu\|_{\mathbb{H}}^2\big] - \lambda^2 \mathbb{E}\Big[\big\langle Mu, F_{Q_1}Mu\big\rangle_{\mathbb{H}}\Big] + C
\end{aligned}
\tag{3.42}
$$

due to the assumption $\sigma \geq \sigma_0 > 0$ and Theorem 3.4. Similarly, by using the Sobolev embedding $H^\gamma(D) \hookrightarrow L^\infty(D)$ with $\gamma > 3/2$, the Hölder inequality and the Young inequality, it holds that

$$
\begin{aligned}
&\lambda^2 \mathbb{E}\Big[\sum_{k\in\mathbb{N}} \big\| M(\mathbb{J}^{-1}u Q_1^{\frac{1}{2}}e_k)\big\|_{\mathbb{H}}^2\Big] + \sum_{k\in\mathbb{N}} \big\| M(\theta Q_2^{\frac{1}{2}}e_k)\big\|_{\mathbb{H}}^2\\
&\leq \lambda^2 \mathbb{E}\Big[\big\langle Mu, F_{Q_1}Mu\big\rangle_{\mathbb{H}}\Big] + 4\lambda^2 \mathbb{E}\big[\|u\|_{\mathbb{H}}^2\big] \sum_{k\in\mathbb{N}} \big\|\nabla(Q_1^{\frac{1}{2}}e_k)\big\|_{L^\infty(D)^3}^2\\
&\quad + 4\lambda^2 \mathbb{E}\big[\|Mu\|_{\mathbb{H}}\|u\|_{\mathbb{H}}\big] \sum_{k\in\mathbb{N}} \Big[\big\|Q_1^{\frac{1}{2}}e_k\big\|_{L^\infty(D)} \big\|\nabla(Q_1^{\frac{1}{2}}e_k)\big\|_{L^\infty(D)^3}\Big]\\
&\quad + 2|\theta|^2 \sum_{k\in\mathbb{N}} \big\|\nabla(Q_2^{\frac{1}{2}}e_k)\big\|_{L^2(D)^3}^2\\
&\leq \lambda^2 \mathbb{E}\Big[\big\langle Mu, F_{Q_1}Mu\big\rangle_{\mathbb{H}}\Big] + \tfrac{1}{2}\sigma_0 \mathbb{E}\big[\|Mu\|_{\mathbb{H}}^2\big] + C.
\end{aligned}
\tag{3.43}
$$

Plugging (3.42) and (3.43) into (3.39), we obtain

$$
\mathrm{d}\mathbb{E}\big[\|Mu(t)\|_{\mathbb{H}}^2\big] \leq -\sigma_0 \mathbb{E}\big[\|Mu(t)\|_{\mathbb{H}}^2\big]\mathrm{d}t + C\mathrm{d}t,
\tag{3.44}
$$

where the positive constant C depends on σ_0, λ, $|\theta|$, $\mathrm{Tr}(Q_2)$, $\|\sigma\|_{W^{1,\infty}(D)}$, $\|F_{Q_1}\|_{W^{1,\infty}(D)}$, $\mathbb{E}\big[\|u_0\|_{\mathbb{H}}^2\big]$, and $\|Q_i^{\frac{1}{2}}\|_{HS(U,H^{\gamma_i}(D))}$, $i = 1, 2$. Thus, the Grönwall inequality in differential form yields

$$
\mathbb{E}\big[\|Mu(t)\|_{\mathbb{H}}^2\big] \leq C + e^{-\sigma_0 t}\mathbb{E}\big[\|Mu_0\|_{\mathbb{H}}^2\big].
$$

Step 2. Estimate of $\mathbb{E}\Big[\|\nabla \cdot \mathbf{E}(t)\|_U^2 + \|\nabla \cdot \mathbf{H}(t)\|_U^2\Big]$. Applying the Itô formula to $\|\nabla \cdot \mathbf{E}\|_U^2$ and then taking the expectation, we arrive at

$$
d\mathbb{E}\big[\|\nabla \cdot \mathbf{E}(t)\|_U^2\big] = -2\mathbb{E}\Big[\Big\langle \nabla \cdot \mathbf{E}(t), \nabla \cdot \big((\sigma + \tfrac{1}{2}\lambda^2 F_{Q_1})\mathbf{E}(t)\big)\Big\rangle_U\Big]dt
$$

$$
+ \lambda^2 \mathbb{E}\Big[\sum_{k \in \mathbb{N}} \|\nabla \cdot (\mathbf{H}(t)Q_1^{\frac{1}{2}}e_k)\|_U^2\Big]dt \qquad (3.45)
$$

$$
+ \sum_{k \in \mathbb{N}} \|\nabla \cdot (\boldsymbol{\theta}_1 Q_2^{\frac{1}{2}}e_k)\|_U^2\, dt.
$$

Notice that $\nabla \cdot (fv) = f(\nabla \cdot v) + (\nabla f) \cdot v$ for scalar function f and vector function v. By using the Hölder inequality and the Young inequality, the assumptions $\sigma, F_{Q_1} \in W^{1,\infty}(D)$ and $\sigma \geq \sigma_0 > 0$, and Theorem 3.4, we have

$$
-2\mathbb{E}\Big[\Big\langle \nabla \cdot \mathbf{E}, \nabla \cdot \big((\sigma + \tfrac{1}{2}\lambda^2 F_{Q_1})\mathbf{E}\big)\Big\rangle_U\Big]
$$

$$
\leq 2\mathbb{E}\Big[\|\nabla \cdot \mathbf{E}\|_U \|\mathbf{E}\|_{L^2(D)^3} \big\|\nabla\big(\sigma + \tfrac{1}{2}\lambda^2 F_{Q_1}\big)\big\|_{L^\infty(D)^3}\Big]
$$

$$
-2\sigma_0\mathbb{E}\big[\|\nabla \cdot \mathbf{E}\|_U^2\big] - \lambda^2\mathbb{E}\big[\langle \nabla \cdot \mathbf{E}, F_{Q_1} \nabla \cdot \mathbf{E}\rangle_U\big] \qquad (3.46)
$$

$$
\leq -\tfrac{3}{2}\sigma_0\mathbb{E}\big[\|\nabla \cdot \mathbf{E}\|_U^2\big] - \lambda^2\mathbb{E}\big[\langle \nabla \cdot \mathbf{E}, F_{Q_1} \nabla \cdot \mathbf{E}\rangle_U\big] + C,
$$

where the positive constant C depends on σ_0, $|\boldsymbol{\theta}|$, $\mathrm{Tr}(Q_2)$, $\|\sigma\|_{W^{1,\infty}(D)}$, and $\|F_{Q_1}\|_{W^{1,\infty}(D)}$.

On the other hand, it follows from the Sobolev embedding $H^\gamma(D) \hookrightarrow L^\infty(D)$ with $\gamma > 3/2$, the Hölder inequality, the Young inequality, and Theorem 3.4 that

$$
\lambda^2 \mathbb{E}\Big[\sum_{k \in \mathbb{N}} \|\nabla \cdot (\mathbf{H}Q_1^{\frac{1}{2}}e_k)\|_U^2\Big]
$$

$$
= \lambda^2 \mathbb{E}\big[\langle \nabla \cdot \mathbf{H}, F_{Q_1} \nabla \cdot \mathbf{H}\rangle_U\big] + \lambda^2 \mathbb{E}\Big[\sum_{k \in \mathbb{N}} \|\mathbf{H} \cdot \nabla(Q_1^{\frac{1}{2}}e_k)\|_U^2\Big]
$$

$$
+ 2\lambda^2 \mathbb{E}\Big[\sum_{k \in \mathbb{N}} \langle Q_1^{\frac{1}{2}}e_k \nabla \cdot \mathbf{H}, \mathbf{H} \cdot \nabla(Q_1^{\frac{1}{2}}e_k)\rangle_U\Big]
$$

$$
\leq \lambda^2 \mathbb{E}\big[\langle \nabla \cdot \mathbf{H}, F_{Q_1} \nabla \cdot \mathbf{H}\rangle_U\big] + \lambda^2 \mathbb{E}\big[\|\mathbf{H}\|_{L^2(D)^3}^2\big]\sum_{k \in \mathbb{N}} \|\nabla(Q_1^{\frac{1}{2}}e_k)\|_{L^\infty(D)^3}^2
$$

$$+ 2\lambda^2 \mathbb{E}\big[\|\mathbf{H}\|_{L^2(D)^3}\|\nabla \cdot \mathbf{H}\|_U\big] \sum_{k \in \mathbb{N}} \Big[\|\nabla(Q_1^{\frac{1}{2}} e_k)\|_{L^\infty(D)^3} \|Q_1^{\frac{1}{2}} e_k\|_{L^\infty(D)}\Big]$$

$$\leq \lambda^2 \mathbb{E}\Big[\langle \nabla \cdot \mathbf{H}, F_{Q_1} \nabla \cdot \mathbf{H}\rangle_U\Big] + \frac{1}{2}\sigma_0 \mathbb{E}\big[\|\nabla \cdot \mathbf{H}\|_U^2\big] + C, \tag{3.47}$$

where the positive constant C depends on σ_0, λ, $|\boldsymbol{\theta}|$, $\mathrm{Tr}(Q_2)$, and $\|Q_1^{\frac{1}{2}}\|_{HS(U,H^{\gamma_1}(D))}$.
Note that

$$\sum_{k \in \mathbb{N}} \|\nabla \cdot (\boldsymbol{\theta}_1 Q_2^{\frac{1}{2}} e_k)\|_U^2 \leq C(|\boldsymbol{\theta}_1|, \|Q_2^{\frac{1}{2}}\|_{HS(U,H^{\gamma_2}(D))}). \tag{3.48}$$

Plugging (3.46)–(3.48) into (3.45) yields

$$\begin{aligned}
d\mathbb{E}\big[\|\nabla \cdot \mathbf{E}(t)\|_U^2\big] \leq\ & -\frac{3}{2}\sigma_0 \mathbb{E}\big[\|\nabla \cdot \mathbf{E}(t)\|_U^2\big] dt + \frac{1}{2}\sigma_0 \mathbb{E}\big[\|\nabla \cdot \mathbf{H}(t)\|_U^2\big] dt \\
& - \lambda^2 \mathbb{E}\Big[\big\langle \nabla \cdot \mathbf{E}(t), F_{Q_1} \nabla \cdot \mathbf{E}(t)\big\rangle_U\Big] dt \\
& + \lambda^2 \mathbb{E}\Big[\big\langle \nabla \cdot \mathbf{H}(t), F_{Q_1} \nabla \cdot \mathbf{H}(t)\big\rangle_U\Big] dt + C dt.
\end{aligned}$$

By similar arguments, it can be shown that

$$\begin{aligned}
d\mathbb{E}\big[\|\nabla \cdot \mathbf{H}(t)\|_U^2\big] \leq\ & -\frac{3}{2}\sigma_0 \mathbb{E}\big[\|\nabla \cdot \mathbf{H}(t)\|_U^2\big] dt + \frac{1}{2}\sigma_0 \mathbb{E}\big[\|\nabla \cdot \mathbf{E}(t)\|_U^2\big] dt \\
& - \lambda^2 \mathbb{E}\Big[\big\langle \nabla \cdot \mathbf{H}(t), F_{Q_1} \nabla \cdot \mathbf{H}(t)\big\rangle_U\Big] dt \\
& + \lambda^2 \mathbb{E}\Big[\big\langle \nabla \cdot \mathbf{E}(t), F_{Q_1} \nabla \cdot \mathbf{E}(t)\big\rangle_U\Big] dt + C dt.
\end{aligned}$$

Combining these two estimates, we obtain

$$\begin{aligned}
d\Big(\mathbb{E}\big[\|\nabla \cdot \mathbf{E}(t)\|_U^2 + \|\nabla \cdot \mathbf{H}(t)\|_U^2\big]\Big) \leq\ & -\sigma_0 \mathbb{E}\big[\|\nabla \cdot \mathbf{E}(t)\|_U^2 + \|\nabla \cdot \mathbf{H}(t)\|_U^2\big] dt \\
& + C dt.
\end{aligned}$$

Hence, the Grönwall inequality in the differential form implies

$$\mathbb{E}\big[\|\nabla \cdot \mathbf{E}(t)\|_U^2 + \|\nabla \cdot \mathbf{H}(t)\|_U^2\big] \leq C + e^{-\sigma_0 t}\mathbb{E}\big[\|\nabla \cdot \mathbf{E}_0\|_U^2 + \|\nabla \cdot \mathbf{H}_0\|_U^2\big],$$

where the positive constant C is independent of time.

Step 3. By using Lemma B.2 and combining *Step 1*, *Step 2*, and Theorem 3.4, we conclude that

$$\mathbb{E}\big[\|u(t)\|^2_{H^1(D)^6}\big] \le C\Big(\mathbb{E}\big[\|u(t)\|^2_{\mathbb{H}}\big] + \mathbb{E}\big[\|\nabla \times \mathbf{E}(t)\|^2_{L^2(D)^3}\big] + \mathbb{E}\big[\|\nabla \cdot \mathbf{E}(t)\|^2_U\big]$$

$$+ \mathbb{E}\big[\|\nabla \times \mathbf{H}(t)\|^2_{L^2(D)^3}\big] + \mathbb{E}\big[\|\nabla \cdot \mathbf{H}(t)\|^2_U\big]\Big)$$

$$\le C\Big(1 + e^{-\sigma_0 t}\mathbb{E}\big[\|u_0\|^2_{H^1(D)^6}\big]\Big)$$

for any $t \ge 0$. The proof of Lemma 3.3 is thus finished. $\quad\square$

Based on the above $H^1(D)^6$-regularity of the solution u, we can obtain the ergodicity of the stochastic Maxwell equations. Moreover, the probability distribution $P_t^*\pi$ of $u(t)$ is shown to converge towards the invariant measure π^* of the solution $u(t)$ in the L^2-Wasserstein distance as $t \to \infty$. Here P_t^* is the transpose operator of P_t. To this end, we denote

$$\mathscr{P}_2(\mathbb{H}) = \Big\{\mu \in \mathscr{P}(\mathbb{H}) : \int_{\mathbb{H}} \|w\|^2_{\mathbb{H}}\mu(dw) < \infty\Big\},$$

then the L^2-Wasserstein distance of two measures $\mu, \nu \in \mathscr{P}_2(\mathbb{H})$ is defined by

$$\mathscr{W}_2(\mu, \nu) = \inf_{\Theta \in \Gamma(\mu,\nu)} \Big(\int_{\mathbb{H}\times\mathbb{H}} \|u - w\|^2_{\mathbb{H}}\Theta(du, dw)\Big)^{\frac{1}{2}}, \qquad (3.49)$$

where $\Gamma(\mu, \nu)$ is the set of coupling of the measure (μ, ν) satisfying $\Theta(A \times \mathbb{H}) = \mu(A)$ and $\Theta(\mathbb{H} \times A) = \nu(A)$ for any $\Theta \in \Gamma(\mu, \nu)$ and $A \in \mathscr{B}(\mathbb{H})$. We refer to [171] for more details.

The main result of the present subsection is given below.

Theorem 3.6 *Under conditions in Lemma 3.3, the following statements hold.*

(i) *Let u and $\tilde{u}$ be solutions of (3.36) with initial data u_0 and $\tilde{u}_0$, respectively. Then*

$$\mathbb{E}\big[\|u(t) - \tilde{u}(t)\|^2_{\mathbb{H}}\big] \le e^{-2\sigma_0 t}\mathbb{E}\big[\|u_0 - \tilde{u}_0\|^2_{\mathbb{H}}\big] \quad \forall t \ge 0. \qquad (3.50)$$

(ii) *The solution u of (3.36) possesses a unique invariant measure $\pi^* \in \mathscr{P}_2(\mathbb{H})$. Thus u is ergodic. Moreover, u is exponentially mixing.*

(iii) *For arbitrary distributions $\pi_1, \pi_2 \in \mathscr{P}_2(\mathbb{H})$, and $t \ge 0$,*

$$\mathscr{W}_2(P_t^*\pi_1, P_t^*\pi_2) \le e^{-\sigma_0 t}\mathscr{W}_2(\pi_1, \pi_2), \qquad (3.51)$$

which implies for any distribution $\pi \in \mathscr{P}_2(\mathbb{H})$,

$$\mathscr{W}_2(P_t^*\pi, \pi^*) \leq e^{-\sigma_0 t} \mathscr{W}_2(\pi, \pi^*) \quad \forall\, t \geq 0.$$

Proof

(i) It follows from (3.36) that for $t \geq 0$,

$$d(u - \tilde{u})(t) = \left[M(u - \tilde{u})(t) - \sigma(u - \tilde{u})(t) - \frac{1}{2}\lambda^2 F_{Q_1}(u - \tilde{u})(t) \right]dt$$

$$+ \lambda \mathbb{J}^{-1}(u - \tilde{u})(t)dW_1(t)$$

with $(u - \tilde{u})(0) = u_0 - \tilde{u}_0$. Applying the Itô formula to $\|u - \tilde{u}\|_{\mathbb{H}}^2$ and taking the expectation, we have

$$d\mathbb{E}\big[\|u(t) - \tilde{u}(t)\|_{\mathbb{H}}^2\big] = -2\mathbb{E}\Big[\big\langle u(t) - \tilde{u}(t), \sigma(u(t) - \tilde{u}(t))\big\rangle_{\mathbb{H}}\Big]dt$$

$$- \lambda^2 \mathbb{E}\Big[\big\langle u(t) - \tilde{u}(t), F_{Q_1}(u(t) - \tilde{u}(t))\big\rangle_{\mathbb{H}}\Big]dt$$

$$+ \lambda^2 \mathbb{E}\Big[\sum_{k \in \mathbb{N}} \Big\| \mathbb{J}^{-1}\big(u(t) - \tilde{u}(t)\big)Q_1^{\frac{1}{2}}e_k\Big\|_{\mathbb{H}}^2\Big]dt.$$

Since $\sum_{k \in \mathbb{N}} \|\mathbb{J}^{-1}(u - \tilde{u})Q_1^{\frac{1}{2}}e_k\|_{\mathbb{H}}^2 = \langle u - \tilde{u}, F_{Q_1}(u - \tilde{u})\rangle_{\mathbb{H}}$, we obtain that

$$d\mathbb{E}\big[\|u(t) - \tilde{u}(t)\|_{\mathbb{H}}^2\big] = -2\mathbb{E}\big[\langle u(t) - \tilde{u}(t), \sigma(u(t) - \tilde{u}(t))\rangle_{\mathbb{H}}\big]dt$$

$$\leq -2\sigma_0 \mathbb{E}\big[\|u(t) - \tilde{u}(t)\|_{\mathbb{H}}^2\big]dt,$$

which yields the assertion via the Grönwall inequality.

(ii) The existence and the uniqueness of the invariant measure follow from the Krylov–Bogoliubov theorem (see [61, Proposition 7.10]) and the general Harris theorem (see [91, Theorem 4.8]).

The key ingredient lies in showing that $\|\cdot\|_{H^1(D)^6}$ is a proper Lyapunov functional whose level sets are compact. We note that $H^1(D)^6$ is compactly embedded in $\mathbb{H}$. Therefore, the level sets $K_\alpha := \{v \in \mathbb{H} : \|v\|_{H^1(D)^6} \leq \alpha\}$ are compact for any constant $\alpha > 0$, then the Krylov–Bogoliubov theorem implies the existence of the invariant measure π^* for u. Combining the contraction property derived in (i) and the general Harris theorem, we obtain the uniqueness of the invariant measure π^* for u. Thus, u is ergodic.

Moreover, it follows from the uniform boundedness of the solution u in the $\mathbb{H}$-norm given in Theorem 3.4 that

$$
\begin{aligned}
\int_{\mathbb{H}} \|w\|_{\mathbb{H}}^2 \pi^*(dw) &= \int_{\mathbb{H}} P_t(\|w\|_{\mathbb{H}}^2)\pi^*(dw) \\
&= \int_{\mathbb{H}} \mathbb{E}\big[\|u(t; w)\|_{\mathbb{H}}^2\big]\pi^*(dw) \\
&\leq \int_{\mathbb{H}} \left(e^{-2\sigma_0 t}\|w\|_{\mathbb{H}}^2 + |\theta|^2 \mathrm{Tr}(Q_2)\frac{1 - e^{-2\sigma_0 t}}{2\sigma_0}\right)\pi^*(dw) \\
&= e^{-2\sigma_0 t}\int_{\mathbb{H}} \|w\|_{\mathbb{H}}^2 \pi^*(dw) + |\theta|^2 \mathrm{Tr}(Q_2)\frac{1 - e^{-2\sigma_0 t}}{2\sigma_0},
\end{aligned}
$$

from which we have

$$
\int_{\mathbb{H}} \|w\|_{\mathbb{H}}^2 \pi^*(dw) \leq \frac{|\theta|^2 \mathrm{Tr}(Q_2)}{2\sigma_0} < \infty.
$$

Thus, $\pi^* \in \mathscr{P}_2(\mathbb{H})$.

In addition, for any bounded Lipschitz continuous function φ on $\mathbb{H}$, $t > 0$, and the deterministic initial datum $u_0 \in \mathbb{H}$, it yields

$$
\begin{aligned}
|P_t\varphi(u_0) - \pi^*(\varphi)| &= \left|\mathbb{E}\big[\varphi(u(t; u_0))\big] - \int_{\mathbb{H}} \mathbb{E}\big[\varphi(u(t; w))\big]\pi^*(dw)\right| \\
&= \left|\int_{\mathbb{H}} \mathbb{E}\big[\varphi(u(t; u_0)) - \varphi(u(t; w))\big]\pi^*(dw)\right| \\
&\leq L_\varphi \int_{\mathbb{H}} \mathbb{E}\big[\|u(t; u_0) - u(t; w)\|\big]\pi^*(dw) \\
&\leq L_\varphi \int_{\mathbb{H}} \big(\mathbb{E}\big[\|u(t; u_0) - u(t; w)\|_{\mathbb{H}}^2\big]\big)^{\frac{1}{2}}\pi^*(dw) \\
&\leq L_\varphi e^{-\sigma_0 t} \int_{\mathbb{H}} \|u_0 - w\|_{\mathbb{H}}\pi^*(dw) \\
&\leq (C + \|u_0\|_{\mathbb{H}})L_\varphi e^{-\sigma_0 t} =: C(u_0)L_\varphi e^{-\sigma_0 t}.
\end{aligned}
$$

Hence the exponentially mixing property is proved.

(iii) We fix the deterministic initial datum $(u_0, \tilde{u}_0) \subset \mathbb{H} \times \mathbb{H}$ and prove (3.51) for Dirac measures δ_{u_0} and $\delta_{\tilde{u}_0}$. Let $u(\cdot, u_0)$ and $u(\cdot, \tilde{u}_0)$ be solutions of (3.36) with initial data u_0 and $\tilde{u}_0$, respectively. Denote the joint distribution of $(u(t; u_0), u(t; \tilde{u}_0))$ by $\Pi(P_t^*\delta_{u_0}, P_t^*\delta_{\tilde{u}_0})$ which belongs to $\Gamma(P_t^*\delta_{u_0}, P_t^*\delta_{\tilde{u}_0})$. By the definition of L^2-

Wasserstein distance and (3.50), we have

$$
\begin{aligned}
\mathscr{W}_2(P_t^*\delta_{u_0}, P_t^*\delta_{\tilde{u}_0}) &= \inf_{\Theta \in \Gamma(P_t^*\delta_{u_0}, P_t^*\delta_{\tilde{u}_0})} \left(\int_{\mathbb{H}\times\mathbb{H}} \|v - w\|_{\mathbb{H}}^2 \Theta(\mathrm{d}v, \mathrm{d}w) \right)^{\frac{1}{2}} \\
&\leq \left(\int_{\mathbb{H}\times\mathbb{H}} \|v - w\|_{\mathbb{H}}^2 \Pi(P_t^*\delta_{u_0}, P_t^*\delta_{\tilde{u}_0})(\mathrm{d}v, \mathrm{d}w) \right)^{\frac{1}{2}} \qquad (3.52) \\
&= \left(\int_{\Omega} \|u(t; u_0) - u(t; \tilde{u}_0)\|_{\mathbb{H}}^2 \mathrm{d}\mathbb{P} \right)^{\frac{1}{2}} \\
&\leq e^{-\sigma_0 t} \|u_0 - \tilde{u}_0\|_{\mathbb{H}} = e^{-\sigma_0 t} \mathscr{W}_2(\delta_{u_0}, \delta_{\tilde{u}_0}).
\end{aligned}
$$

By the convexity of the L^2-Wasserstein distance (see e.g., [171, Theorem 4.8]), (3.52), and the Hölder inequality, we obtain that for any coupling γ of (π_1, π_2),

$$
\begin{aligned}
\mathscr{W}_2(P_t^*\pi_1, P_t^*\pi_2) &\leq \int_{\mathbb{H}\times\mathbb{H}} \mathscr{W}_2(P_t^*\delta_{u_0}, P_t^*\delta_{\tilde{u}_0}) \gamma(\mathrm{d}u_0, \mathrm{d}\tilde{u}_0) \\
&\leq e^{-\sigma_0 t} \int_{\mathbb{H}\times\mathbb{H}} \mathscr{W}_2(\delta_{u_0}, \delta_{\tilde{u}_0}) \gamma(\mathrm{d}u_0, \mathrm{d}\tilde{u}_0) \\
&= e^{-\sigma_0 t} \int_{\mathbb{H}\times\mathbb{H}} \|u_0 - \tilde{u}_0\|_{\mathbb{H}} \gamma(\mathrm{d}u_0, \mathrm{d}\tilde{u}_0) \\
&\leq e^{-\sigma_0 t} \left(\int_{\mathbb{H}\times\mathbb{H}} \|u_0 - \tilde{u}_0\|_{\mathbb{H}}^2 \gamma(\mathrm{d}u_0, \mathrm{d}\tilde{u}_0) \right)^{\frac{1}{2}}.
\end{aligned}
$$

Thus, by the arbitrariness of γ, we finish the proof of Theorem 3.6. $\qquad\square$

Summary and Outlook

In this chapter, the geometric structures, physical properties, and asymptotic properties of the stochastic Maxwell equations are presented. The phase flow of the stochastic Maxwell equations preserves the stochastic symplectic structure when the equations are regarded as an infinite-dimensional stochastic Hamiltonian system, while the stochastic Maxwell equations possess the multi-symplectic conservation law when the equations are interpreted as stochastic Hamiltonian partial differential equations. There are lots of works on studying the geometric structure for other kinds of stochastic partial differential equations. We refer to [31, 40, 58, 59, 112] and references therein for the case of the stochastic nonlinear Schrödinger equation, and to [111] for the case of the stochastic Korteweg-de Vries equation.

In the deterministic case, it has been shown that symplectic methods have remarkable superiority compared to non-symplectic ones when applied to Hamiltonian systems, such as long-time behavior, geometric structure-preserving, and physical properties-preserving (see e.g. [80, 90]). For the stochastic case, large quantities of numerical experiments suggest that stochastic symplectic methods possess excellent long-time stability. To theoretically explain the superiority of the stochastic symplectic method, the backward error analysis technique is exploited in [8, 66, 103, 153] and references therein. Recently, several researchers applied the large deviations principle to investigate the probabilistic superiority of the stochastic symplectic method (see e.g., [44, 46]). In Sect. 3.4, we preliminarily studied the asymptotic property of the solution of the linear stochastic Maxwell equations with small noise by investigating the large deviations principle of the solution. Whereas in certain circumstances, the coefficients of the stochastic Maxwell equations may be nonlinear. Hence it is of interest to study the large deviations principle for the general stochastic Maxwell equations and their numerical approximations.

As we all know, ergodicity is an important long-time property of stochastic partial differential equations. There have been fruitful works on studying the ergodicity of the original system and constructing numerical algorithms which can inherit the ergodicity of the considered system. For instance, we refer to [1, 103, 135, 136, 163, 164] for stochastic ordinary differential equations, to [20, 21] for parabolic stochastic partial differential equations, to [24, 71] for stochastic Navier–Stokes equations, and to [39, 58, 67, 101] for stochastic Schrödinger equations. In the last section of this chapter, we showed that the damped stochastic Maxwell equations also possess ergodicity, as well as the stochastic conformal multi-symplectic structure (see also [45] for more details). However, there are still some unsolved problems, such as

- Does the general (damped) stochastic Hamiltonian partial differential equation possess a unique invariant measure?
- What is the large deviations principle related to the invariant measures of the stochastic Maxwell equations?

Chapter 4
Structure-Preserving Algorithms for Stochastic Maxwell Equations

In this chapter, we focus on several structure-preserving algorithms which can inherit the intrinsic properties studied in Chap. 3 for the stochastic Maxwell equations.

In Sect. 4.1, we mainly present some temporally semi-discrete algorithms, including stochastic symplectic Runge–Kutta methods and the exponential-type methods, for the stochastic Maxwell equations with Stratonovich noise. We further give *a priori* estimates of the numerical solutions associated with the temporal semi-discretizations. These results will be used in Chap. 5 to obtain the mean-square convergence order for the proposed structure-preserving algorithms.

In Sect. 4.2, we turn to the construction and analysis of fully discrete structure-preserving algorithms for the stochastic Maxwell equations by exploiting the FDTD method, the wavelet interpolation method, and the dG method. More precisely, we present three types of stochastic multi-symplectic algorithms for the stochastic Maxwell equations driven by additive noise via the FDTD method. After introducing the basic theory of wavelets, we design an energy-conserving stochastic multi-symplectic wavelet algorithm for the stochastic Maxwell equations driven by multiplicative noise. We then study dG algorithms for the stochastic Maxwell equations driven by additive noise.

In Sect. 4.3, we focus on developing the splitting technique for solving the stochastic Maxwell equations efficiently. The three-dimensional stochastic Maxwell equations can be split into three local one-dimensional Hamiltonian systems. If we apply the previous algorithms to these three subsystems, plenty of highly efficient and structure-preserving algorithms for the stochastic Maxwell equations can be obtained.

4.1 Temporally Semi-Discrete Algorithms

This section concentrates on temporal semi-discretizations for the stochastic Maxwell equations studied in Sect. 3.1:

$$
\begin{cases}
\mathrm{d}u(t) = \big[Mu(t) + F(t, u(t))\big]\mathrm{d}t + B(t, u(t)) \circ \mathrm{d}W(t), & t \in (0, T], \\
u(0) = u_0.
\end{cases}
$$

$$(4.1)$$

Assume that there exist Hamiltonians $\widetilde{\mathscr{H}_1}$ and $\widetilde{\mathscr{H}_2}$ such that

$$
F(t, u(t)) = \mathbb{J}^{-1}\frac{\delta\widetilde{\mathscr{H}_1}}{\delta u}, \quad B(t, u(t)) = \mathbb{J}^{-1}\frac{\delta\widetilde{\mathscr{H}_2}}{\delta u},
$$

$$(4.2)$$

where $\mathbb{J}$ is the standard symplectic matrix. It has been shown in Theorem 3.1 that the phase flow of (4.1) preserves the stochastic symplectic structure under the homogeneous boundary condition.

For the semi-discretization of (4.1) in the temporal direction, we introduce the partition $0 = t_0 < t_1 < \cdots < t_N = T$ with the uniform time step size $\tau = T/N$. Let $D \subset \mathbb{R}^3$ be an open, bounded, and Lipschitz domain with boundary ∂D. The definition of the infinite-dimensional stochastic symplectic algorithm is given below. We refer to [100] for the systematic study of stochastic symplectic algorithms of stochastic Hamiltonian systems.

Definition 4.1 The discrete stochastic flow $\phi_\tau : \mathbb{H} \to \mathbb{H}$, $((\mathbf{E}^n)^\top, (\mathbf{H}^n)^\top)^\top \mapsto ((\mathbf{E}^{n+1})^\top, (\mathbf{H}^{n+1})^\top)^\top$ of (4.1) is said to be an infinite-dimensional stochastic symplectic algorithm if it preserves the infinite-dimensional stochastic symplectic structure, i.e., for all $n = 0, 1, \ldots, N - 1$,

$$
\int_D d\mathbf{E}^{n+1}(\mathbf{x}) \wedge d\mathbf{H}^{n+1}(\mathbf{x})\mathrm{d}\mathbf{x} = \int_D d\mathbf{E}^n(\mathbf{x}) \wedge d\mathbf{H}^n(\mathbf{x})\mathrm{d}\mathbf{x}, \quad \mathbb{P}\text{-}a.s.
$$

4.1.1 Stochastic Symplectic Runge–Kutta Methods

Applying the s-stage stochastic Runge–Kutta method, which depends only on the increments of the Wiener process, to (4.1) in the temporal direction, we obtain that for $i = 1, 2, \ldots, s$ and $n = 0, 1, \ldots, N - 1$,

$$
U_i^n = u^n + \tau \sum_{j=1}^s a_{ij}\big(MU_j^n + F^{n,j}(U_j^n)\big) + \sum_{j=1}^s \widetilde{a}_{ij} B^{n,j}(U_j^n)\Delta W^{n+1}, \quad (4.3)
$$

$$
u^{n+1} = u^n + \tau \sum_{i=1}^s b_i\big(MU_i^n + F^{n,i}(U_i^n)\big) + \sum_{i=1}^s \widetilde{b}_i B^{n,i}(U_i^n)\Delta W^{n+1}, \quad (4.4)
$$

where $F^{n,i}(U_i^n) := F(t_n + c_i\tau, U_i^n)$, $B^{n,i}(U_i^n) := B(t_n + \tilde{c}_i\tau, U_i^n)$, $i = 1, 2, \ldots, s$, and $\Delta W^{n+1} := W(t_{n+1}) - W(t_n)$ is the increment of the Wiener process with $t_n = n\tau$ and $n = 0, 1, \ldots, N - 1$.

Denote

$$A = \left(a_{ij}\right)_{i,j=1}^s, \quad b = (b_1, \ldots, b_s)^\top, \quad c_i = \sum_{j=1}^s a_{ij},$$

$$\tilde{A} = \left(\tilde{a}_{ij}\right)_{i,j=1}^s, \quad \tilde{b} = (\tilde{b}_1, \ldots, \tilde{b}_s)^\top, \quad \tilde{c}_i = \sum_{j=1}^s \tilde{a}_{ij},$$

and set $U_n := \left((U_1^n)^\top, \ldots, (U_s^n)^\top\right)^\top$, $F^n(U_n) := \left(F^{n1}(U_1^n)^\top, \ldots, F^{ns}(U_s^n)^\top\right)^\top$, and $B^n(U_n) := \left(B^{n1}(U_1^n)^\top, \ldots, B^{ns}(U_s^n)^\top\right)^\top$. Utilizing the Kronecker product, we can rewrite (4.3)–(4.4) in a compact form,

$$U_n = \mathbf{1}_s \otimes u^n + \tau\left(A \otimes M\right)U_n + \tau\left(A \otimes Id\right)F^n(U_n) + \left(\tilde{A} \otimes Id\right)B^n(U_n)\Delta W^{n+1}, \tag{4.5}$$

$$u^{n+1} = u^n + \tau\left(b^\top \otimes M\right)U_n + \tau\left(b^\top \otimes Id\right)F^n(U_n) + \left(\tilde{b}^\top \otimes Id\right)B^n(U_n)\Delta W^{n+1}, \tag{4.6}$$

where $\mathbf{1}_s = (1, \ldots, 1)^\top \in \mathbb{R}^s$.

The following proposition gives conditions on coefficients A, $\tilde{A}$, b, and $\tilde{b}$ to ensure that the stochastic Runge–Kutta semi-discretization (4.3)–(4.4) inherits the symplectic structure of the original equations.

Proposition 4.1 *Suppose that conditions in Theorem 3.1 hold. In addition, assume that the coefficients of the stochastic Runge–Kutta method (4.3)–(4.4) satisfy*

$$m_{ij} := b_i a_{ij} + b_j a_{ji} - b_i b_j = 0,$$

$$\tilde{m}_{ij} := \tilde{b}_i a_{ij} + b_j \tilde{a}_{ji} - \tilde{b}_i b_j = 0, \tag{4.7}$$

$$\tilde{\tilde{m}}_{ij} := \tilde{b}_i \tilde{a}_{ij} + \tilde{b}_j \tilde{a}_{ji} - \tilde{b}_i \tilde{b}_j = 0$$

for all $i, j = 1, 2, \ldots, s$. Then the stochastic Runge–Kutta method (4.3)–(4.4) is stochastic symplectic under the homogeneous boundary condition, that is, for all $n = 0, 1, \ldots, N - 1$,

$$\varpi^{n+1} := \int_D d\mathbf{E}^{n+1}(\mathbf{x}) \wedge d\mathbf{H}^{n+1}(\mathbf{x})d\mathbf{x} = \int_D d\mathbf{E}^n(\mathbf{x}) \wedge d\mathbf{H}^n(\mathbf{x})d\mathbf{x} = \varpi^n, \quad \mathbb{P}\text{-}a.s.$$

Proof For the stochastic Runge–Kutta method (4.3)–(4.4), it follows from (4.2) that

$$
dU_i^n = du^n + \tau \sum_{j=1}^{s} a_{ij} M dU_j^n + \tau \sum_{j=1}^{s} a_{ij} \mathbb{J}^{-1} \frac{\delta^2 \widetilde{\mathcal{H}_1}}{\delta u^2} dU_j^n
$$
$$
+ \sum_{j=1}^{s} \tilde{a}_{ij} \mathbb{J}^{-1} \frac{\delta^2 \widetilde{\mathcal{H}_2}}{\delta u^2} dU_j^n \Delta W^{n+1}
$$

(4.8)

and

$$
du^{n+1} = du^n + \tau \sum_{i=1}^{s} b_i M dU_i^n + \tau \sum_{i=1}^{s} b_i \mathbb{J}^{-1} \frac{\delta^2 \widetilde{\mathcal{H}_1}}{\delta u^2} dU_i^n
$$
$$
+ \sum_{i=1}^{s} \tilde{b}_i \mathbb{J}^{-1} \frac{\delta^2 \widetilde{\mathcal{H}_2}}{\delta u^2} dU_i^n \Delta W^{n+1}.
$$

(4.9)

It follows from (4.9) that

$$
du^{n+1} \wedge \mathbb{J} du^{n+1} - du^n \wedge \mathbb{J} du^n
$$

$$
= \tau \sum_{i=1}^{s} b_i \left(du^n \wedge \mathbb{J} M dU_i^n + M dU_i^n \wedge \mathbb{J} du^n \right)
$$

$$
+ \tau \sum_{i=1}^{s} b_i \left(du^n \wedge \frac{\delta^2 \widetilde{\mathcal{H}_1}}{\delta u^2} dU_i^n - \frac{\delta^2 \widetilde{\mathcal{H}_1}}{\delta u^2} dU_i^n \wedge du^n \right)
$$

$$
+ \sum_{i=1}^{s} \tilde{b}_i \left(du^n \wedge \frac{\delta^2 \widetilde{\mathcal{H}_2}}{\delta u^2} dU_i^n \Delta W^{n+1} - \frac{\delta^2 \widetilde{\mathcal{H}_2}}{\delta u^2} dU_i^n \Delta W^{n+1} \wedge du^n \right)
$$

$$
+ \tau^2 \sum_{i,j=1}^{s} b_i b_j \left[\left(M dU_i^n + \mathbb{J}^{-1} \frac{\delta^2 \widetilde{\mathcal{H}_1}}{\delta u^2} dU_i^n \right) \wedge \left(\mathbb{J} M dU_j^n + \frac{\delta^2 \widetilde{\mathcal{H}_1}}{\delta u^2} dU_j^n \right) \right]
$$

$$
+ 2\tau \sum_{i,j=1}^{s} b_i \tilde{b}_j \left[\left(M dU_i^n + \mathbb{J}^{-1} \frac{\delta^2 \widetilde{\mathcal{H}_1}}{\delta u^2} dU_i^n \right) \wedge \frac{\delta^2 \widetilde{\mathcal{H}_2}}{\delta u^2} dU_j^n \Delta W^{n+1} \right]
$$

$$
+ \sum_{i,j=1}^{s} \tilde{b}_i \tilde{b}_j \left(\mathbb{J}^{-1} \frac{\delta^2 \widetilde{\mathcal{H}_2}}{\delta u^2} dU_i^n \Delta W^{n+1} \wedge \frac{\delta^2 \widetilde{\mathcal{H}_2}}{\delta u^2} dU_j^n \Delta W^{n+1} \right).
$$

(4.10)

From (4.8), one has

$$du^n = dU_i^n - \tau \sum_{j=1}^{s} a_{ij} M dU_j^n - \tau \sum_{j=1}^{s} a_{ij} \mathbb{J}^{-1} \frac{\delta^2 \widetilde{\mathscr{H}_1}}{\delta u^2} dU_j^n$$

$$- \sum_{j=1}^{s} \tilde{a}_{ij} \mathbb{J}^{-1} \frac{\delta^2 \widetilde{\mathscr{H}_2}}{\delta u^2} dU_j^n \Delta W^{n+1}.$$

Plugging the above equation into the first three terms on the right-hand side of (4.10), it yields

$$du^{n+1} \wedge \mathbb{J} du^{n+1} - du^n \wedge \mathbb{J} du^n$$

$$= \tau \sum_{i=1}^{s} b_i \left(dU_i^n \wedge \mathbb{J} M dU_i^n + M dU_i^n \wedge \mathbb{J} dU_i^n \right)$$

$$+ \tau \sum_{i=1}^{s} b_i \left(dU_i^n \wedge \frac{\delta^2 \widetilde{\mathscr{H}_1}}{\delta u^2} dU_i^n - \frac{\delta^2 \widetilde{\mathscr{H}_1}}{\delta u^2} dU_i^n \wedge dU_i^n \right)$$

$$+ \sum_{i=1}^{s} \tilde{b}_i \left(dU_i^n \wedge \frac{\delta^2 \widetilde{\mathscr{H}_2}}{\delta u^2} dU_i^n \Delta W^{n+1} - \frac{\delta^2 \widetilde{\mathscr{H}_2}}{\delta u^2} dU_i^n \Delta W^{n+1} \wedge dU_i^n \right)$$

$$- \tau^2 \sum_{i,j=1}^{s} m_{ij} \left[\left(M dU_i^n + \mathbb{J}^{-1} \frac{\delta^2 \widetilde{\mathscr{H}_1}}{\delta u^2} dU_i^n \right) \wedge \left(\mathbb{J} M dU_j^n + \frac{\delta^2 \widetilde{\mathscr{H}_1}}{\delta u^2} dU_j^n \right) \right]$$

$$- 2\tau \sum_{i,j=1}^{s} \tilde{m}_{ij} \left[\left(M dU_i^n + \mathbb{J}^{-1} \frac{\delta^2 \widetilde{\mathscr{H}_1}}{\delta u^2} dU_i^n \right) \wedge \frac{\delta^2 \widetilde{\mathscr{H}_2}}{\delta u^2} dU_j^n \Delta W^{n+1} \right]$$

$$- \sum_{i,j=1}^{s} \tilde{\tilde{m}}_{ij} \left[\mathbb{J}^{-1} \frac{\delta^2 \widetilde{\mathscr{H}_2}}{\delta u^2} dU_i^n \Delta W^{n+1} \wedge \frac{\delta^2 \widetilde{\mathscr{H}_2}}{\delta u^2} dU_j^n \Delta W^{n+1} \right]$$

$$= 2\tau \sum_{i=1}^{s} b_i \left(dU_i^n \wedge \mathbb{J} M dU_i^n \right),$$

where the last equality follows from the symmetry of $\frac{\delta^2 \widetilde{\mathscr{H}_p}}{\delta u^2}$, $p = 1, 2$, and the condition (4.7). Recalling $u = (\mathbf{E}^\top, \mathbf{H}^\top)^\top$ and the definition of the Maxwell operator M, it holds that

$$d\mathbf{E}^{n+1} \wedge d\mathbf{H}^{n+1} - d\mathbf{E}^n \wedge d\mathbf{H}^n$$

$$= \frac{1}{2} \left(du^{n+1} \wedge \mathbb{J} du^{n+1} - du^n \wedge \mathbb{J} du^n \right)$$

$$= \tau \sum_{i=1}^{s} b_i \left(dU_i^n \wedge \mathbb{J}M dU_i^n \right)$$

$$= -\tau \sum_{i=1}^{s} b_i \left[\mu^{-1} d\mathbf{E}_{ni} \wedge (\nabla \times d\mathbf{E}_{ni}) + \varepsilon^{-1} d\mathbf{H}_{ni} \wedge (\nabla \times d\mathbf{H}_{ni}) \right].$$

Therefore, using the similar approach in the last two steps of (3.9) leads to

$$\int_D d\mathbf{E}^{n+1}(\mathbf{x}) \wedge d\mathbf{H}^{n+1}(\mathbf{x}) d\mathbf{x} - \int_D d\mathbf{E}^n(\mathbf{x}) \wedge d\mathbf{H}^n(\mathbf{x}) d\mathbf{x} = 0$$

for all $n = 0, 1, \ldots, N - 1$. The proof of Proposition 4.1 is finished.   □

Remark 4.1 A stochastic Runge–Kutta method (4.3)–(4.4) satisfying the symplectic condition (4.7) is called a stochastic symplectic Runge–Kutta method.

Specially, if (4.1) is driven by additive noise (i.e., $B(t, u(t)) = B(t)$), then the symplectic condition (4.7) becomes

$$m_{ij} := b_i a_{ij} + b_j a_{ji} - b_i b_j = 0 \tag{4.11}$$

for all $i, j = 1, 2, \ldots, s$. We refer to [42] for further discussion.

As a typical example of the stochastic symplectic Runge–Kutta method, the stochastic midpoint method for (4.1)

$$U_1^n = u^n + \frac{\tau}{2} \left(M U_1^n + F(t_{n+\frac{1}{2}}, U_1^n) \right) + \frac{1}{2} B(t_{n+\frac{1}{2}}, U_1^n) \Delta W^{n+1},$$

$$u^{n+1} = u^n + \tau \left(M U_1^n + F(t_{n+\frac{1}{2}}, U_1^n) \right) + B(t_{n+\frac{1}{2}}, U_1^n) \Delta W^{n+1}$$

will be studied in the following subsection, where $t_{n+\frac{1}{2}} := t_n + \frac{\tau}{2}$. The above method can be rewritten into a compact form by eliminating intermediate variable U_1^n, namely,

$$u^{n+1} = u^n + \tau M u^{n+\frac{1}{2}} + \tau F^{n+\frac{1}{2}}(u^{n+\frac{1}{2}}) + B^{n+\frac{1}{2}}(u^{n+\frac{1}{2}}) \Delta W^{n+1}, \tag{4.12}$$

where $u^{n+\frac{1}{2}} = (u^{n+1} + u^n)/2$, $F^{n+\frac{1}{2}}(u^{n+\frac{1}{2}}) = F(t_{n+\frac{1}{2}}, u^{n+\frac{1}{2}})$ and $B^{n+\frac{1}{2}}(u^{n+\frac{1}{2}}) = B(t_{n+\frac{1}{2}}, u^{n+\frac{1}{2}})$. By introducing two discrete operators

$$S_\tau = \left(Id - \frac{\tau}{2} M \right)^{-1} \left(Id + \frac{\tau}{2} M \right), \quad T_\tau = \left(Id - \frac{\tau}{2} M \right)^{-1},$$

the stochastic midpoint method (4.12) can be further rewritten as

$$u^{n+1} = S_\tau u^n + \tau T_\tau F^{n+\frac{1}{2}}(u^{n+\frac{1}{2}}) + T_\tau B^{n+\frac{1}{2}}(u^{n+\frac{1}{2}}) \Delta W^{n+1} \tag{4.13}$$

for $n = 0, 1, \ldots, N - 1$.

4.1.1.1 Analysis of the Stochastic Midpoint Method

This subsection presents the analysis of the stochastic midpoint method. Both the additive and multiplicative noise cases are studied. Below we give the well-posedness of the stochastic midpoint method (4.12), and analyze the regularity of the numerical solution.

(a) The additive noise case

We restrict ourselves to (4.1) in the additive noise case, and the corresponding stochastic midpoint method reads as

$$u^{n+1} = u^n + \tau M u^{n+\frac{1}{2}} + \tau F^{n+\frac{1}{2}}(u^{n+\frac{1}{2}}) + B^{n+\frac{1}{2}} \Delta W^{n+1}, \tag{4.14}$$

where $B^{n+\frac{1}{2}} := B(t_{n+\frac{1}{2}})$. It follows from (4.13) that

$$u^{n+1} = S_\tau u^n + \tau T_\tau F^{n+\frac{1}{2}}(u^{n+\frac{1}{2}}) + T_\tau B^{n+\frac{1}{2}} \Delta W^{n+1}. \tag{4.15}$$

Proposition 4.2 *Let F satisfy Assumption 2.2, $B(t) \in HS(U_0, \mathbb{H})$ for $t \in [0, T]$, and $u_0 \in L^{2p}(\Omega, \mathbb{H})$ for $p \geq 1$. For sufficiently small $\tau > 0$, there exists a unique $\mathbb{H}$-valued $\{\mathscr{F}_{t_n}\}_{0 \leq n \leq N}$-adapted solution $\{u^n; \ n = 0, 1, \ldots, N\}$ of the stochastic midpoint method (4.14). Moreover, there exists a positive constant $C = C(p, T)$ such that*

$$\max_{1 \leq n \leq N} \|u^n\|_{L^{2p}(\Omega, \mathbb{H})} \leq C\left(1 + \|u_0\|_{L^{2p}(\Omega, \mathbb{H})}\right). \tag{4.16}$$

Proof

Step 1: *Existence and uniqueness.* Fix a set $\Omega' \subset \Omega$ with $\mathbb{P}(\Omega') = 1$ such that $W(t, \omega) \in U = L^2(D)$ for all $t \in [0, T]$ and $\omega \in \Omega'$. Below let us assume that $\omega \in \Omega'$. Combining with (4.16), the existence of $\{u^n; \ n = 0, 1, \ldots, N\}$ follows from the standard Galerkin method and the Brouwer fixed point theorem.

In fact, define a mapping

$$\Lambda : \mathbb{H} \times U \to \mathscr{T}(\mathbb{H}), \quad (u^n, \Delta W^{n+1}) \mapsto \Lambda(u^n, \Delta W^{n+1}),$$

where $\mathscr{T}(\mathbb{H})$ is the set of all subsets of $\mathbb{H}$, and $\Lambda(u^n, \Delta W^{n+1})$ is the set of solutions u^{n+1} of (4.14). The inequality (4.16) implies that Λ is bounded and its graph is closed by the closed graph theorem. Hence, there exists a Borel measurable mapping $\lambda_n : \mathbb{H} \times U \to \mathbb{H}$ such that $\lambda_n(s_1, s_2) \in \Lambda(s_1, s_2)$ for all $(s_1, s_2) \in \mathbb{H} \times U$. Therefore, the $\mathscr{F}_{t_{n+1}}$-measurability of u^{n+1} follows from the Doob–Dynkin lemma (see e.g., [146, Proposition 3]).

To show the uniqueness of the numerical solution, we assume that there are two different solutions u^{n+1} and v^{n+1} satisfying (4.15) with $u^n = v^n$. Then it follows that

$$u^{n+1} - v^{n+1} = \tau T_\tau \left(F^{n+\frac{1}{2}}(u^{n+\frac{1}{2}}) - F^{n+\frac{1}{2}}(v^{n+\frac{1}{2}}) \right),$$

which combining the Lipschitz continuity of F and Lemma C.3 (i) leads to

$$\|u^{n+1} - v^{n+1}\|_{\mathbb{H}} \leq C\tau \|u^{n+1} - v^{n+1}\|_{\mathbb{H}}$$

with C independent of τ. Consequently, for sufficiently small τ, the uniqueness of the solution for (4.14) is obtained.

Step 2:   *Proof of* (4.16). We only present the proof for the case $p = 1$. The proof for $p > 1$ follows the same procedure. From (4.15), we utilize Lemma C.3 (i)–(ii) to derive that

$$\mathbb{E}\left[\|u^{n+1}\|_{\mathbb{H}}^2\right] = \mathbb{E}\left[\|S_\tau u^n\|_{\mathbb{H}}^2\right]$$
$$+ 2\mathbb{E}\left[\langle S_\tau u^n, \ \tau T_\tau F^{n+\frac{1}{2}}(u^{n+\frac{1}{2}}) + T_\tau B^{n+\frac{1}{2}} \Delta W^{n+1}\rangle_{\mathbb{H}}\right]$$
$$+ \mathbb{E}\left[\|\tau T_\tau F^{n+\frac{1}{2}}(u^{n+\frac{1}{2}}) + T_\tau B^{n+\frac{1}{2}} \Delta W^{n+1}\|_{\mathbb{H}}^2\right]$$
$$= \mathbb{E}\left[\|S_\tau u^n\|_{\mathbb{H}}^2\right] + 2\mathbb{E}\left[\langle S_\tau u^n, \ \tau T_\tau F^{n+\frac{1}{2}}(u^{n+\frac{1}{2}})\rangle_{\mathbb{H}}\right]$$
$$+ \mathbb{E}\left[\|\tau T_\tau F^{n+\frac{1}{2}}(u^{n+\frac{1}{2}}) + T_\tau B^{n+\frac{1}{2}} \Delta W^{n+1}\|_{\mathbb{H}}^2\right]$$
$$\leq C\tau \mathbb{E}\left[\|u^{n+1}\|_{\mathbb{H}}^2\right] + (1 + C\tau)\mathbb{E}\left[\|u^n\|_{\mathbb{H}}^2\right] + C\tau$$

due to the independence between u^n and ΔW^{n+1}, the linear growth of F, and the assumption $B(t) \in HS(U_0, \mathbb{H})$. Therefore, for sufficiently small τ, by the Grönwall inequality, one arrives at the assertion (4.16) when $p = 1$. $\square$

Similarly, one can obtain the following $\mathscr{D}(M^k)$-regularity ($k \in \mathbb{N}_+$) of the numerical solution $\{u^n; n = 0, 1, \ldots, N\}$ for the stochastic midpoint method (4.14).

Proposition 4.3 *Let F satisfy Assumption 2.4, $B(t) \in HS(U_0, \mathscr{D}(M^k))$ for any $t \in [0, T]$, and $u_0 \in L^{2p}(\Omega, \mathscr{D}(M^k))$ for $p \geq 1$. Then there exists a positive constant $C = C(p, T, F, B)$ such that*

(i) $\displaystyle\max_{1 \leq n \leq N} \|u^n\|_{L^{2p}(\Omega, \mathscr{D}(M^k))} \leq C\Big(1 + \|u_0\|_{L^{2p}(\Omega, \mathscr{D}(M^k))}\Big);$

(ii) $\displaystyle\max_{0 \leq n \leq N-1} \mathbb{E}\Big[\|u^{n+1} - u^n\|_{\mathscr{D}(M^{k-1})}^{2p}\Big] \leq C\tau^p;$

(iii) $\displaystyle\max_{0 \leq n \leq N-1} \Big\|\mathbb{E}[u^{n+1} - u^n]\Big\|_{\mathscr{D}(M^{k-1})} \leq C\tau.$

(b) The multiplicative noise case

When (4.1) is driven by multiplicative noise, the analysis of the regularity of the numerical solution will be more complicated than that of the additive noise case. Below we focus on the analysis of the stochastic midpoint method for (4.1) driven by linear multiplicative noise, i.e.,

$$\begin{cases} \mathrm{d}u(t) = \Big[Mu(t) + F(t, u(t))\Big]\mathrm{d}t + \lambda \mathbb{J}^{-1}u(t) \circ \mathrm{d}W(t), & t \in (0, T], \\ u(0) = u_0 \end{cases}$$

(4.17)

with λ being a nonzero constant. Then, the stochastic midpoint method (4.12) becomes

$$u^{n+1} = u^n + \tau M u^{n+\frac{1}{2}} + \tau F^{n+\frac{1}{2}}(u^{n+\frac{1}{2}}) + \lambda \mathbb{J}^{-1} u^{n+\frac{1}{2}} \Delta W^{n+1}.$$

(4.18)

In this case, the well-posedness of u^n in the $\mathbb{H}$-norm is stated in the following proposition.

Proposition 4.4 *Let F satisfy Assumption 2.2 and $u_0 \in \mathbb{H}$. For sufficiently small $\tau > 0$, there exists a unique $\mathbb{H}$-valued and $\{\mathscr{F}_{t_n}\}_{0 \leq n \leq N}$-adapted solution $\{u^n; n = 0, 1, \ldots, N\}$ of the stochastic midpoint method (4.18). Moreover,*

$$\max_{1 \leq n \leq N} \|u^n\|_{\mathbb{H}}^2 \leq C\big(1 + \|u_0\|_{\mathbb{H}}^2\big), \quad \mathbb{P}\text{-}a.s.,$$

(4.19)

where $C = C(T, \lambda)$ is a positive constant.

Proof The proof of the existence and uniqueness of the solution of (4.18) is similar to that of Proposition 4.2, and hence is omitted here.

Applying $\langle \cdot, u^n + u^{n+1}\rangle_{\mathbb{H}}$ to both sides of (4.18) leads to

$$\|u^{n+1}\|_{\mathbb{H}}^2 = \|u^n\|_{\mathbb{H}}^2 + \tau \langle F^{n+\frac{1}{2}}(u^{n+\frac{1}{2}}), u^n + u^{n+1}\rangle_{\mathbb{H}}$$

$$\leq (1 + C\tau)\|u^n\|_{\mathbb{H}}^2 + C\tau\|u^{n+1}\|_{\mathbb{H}}^2 + C\tau,$$

which implies that $\|u^n\|_{\mathbb{H}}^2 \leq C(1 + \|u_0\|_{\mathbb{H}}^2)$ for sufficiently small τ, based on the Grönwall inequality. $\qquad\qquad\qquad\qquad\qquad\qquad\qquad\qquad\qquad\qquad\qquad\qquad\square$

In particular, if F satisfies $\langle F(t, v), v \rangle_{\mathbb{H}} = 0$ for $t \in [0, T]$ and $v \in \mathbb{H}$, then we have

$$\|u^n\|_{\mathbb{H}}^2 = \|u_0\|_{\mathbb{H}}^2, \quad \mathbb{P}\text{-}a.s.$$

In addition, the $\mathscr{D}(M)$-regularity of the numerical solution of (4.18) can also be obtained, which is stated in the following proposition. Note that (4.18) is fully implicit, and then the increments $\Delta W^{n+1}, n \in \mathbb{N}$ of the Wiener process are generally substituted by some truncated random variables (see [137]). Let

$$\Delta \overline{W}^{n+1} := \sqrt{\tau} \sum_{i \in \mathbb{N}} \xi_i^{n+1} Q^{\frac{1}{2}} e_i, \qquad (4.20)$$

where

$$\xi_i^{n+1} = \begin{cases} A_\tau^b, & \vartheta_i^{n+1} > A_\tau^b, \\ \vartheta_i^{n+1}, & |\vartheta_i^{n+1}| \leq A_\tau^b, \\ -A_\tau^b, & \vartheta_i^{n+1} < -A_\tau^b \end{cases}$$

with $A_\tau^b = (2b|\ln \tau|)^{1/2}$ for some $b \geq 2$, and $\{\vartheta_i^{n+1}\}_{i \in \mathbb{N}}, n = 0, 1, \ldots, N-1$ being a family of independent standard normal random variables. One can check that

$$\mathbb{E}\big[(\xi_i^{n+1} - \vartheta_i^{n+1})^2\big] \leq \tau^b, \quad \mathbb{E}\big[(\xi_i^{n+1})^2 - (\vartheta_i^{n+1})^2\big] \leq (1 + 2A_\tau^b)\tau^b.$$

Therefore for any $Q^{\frac{1}{2}} \in HS(U, H^\gamma)$ with $\gamma \geq 0$, it holds that

$$\mathbb{E}\big[\|\Delta \overline{W}^{n+1} - \Delta W^{n+1}\|_{H^\gamma}^2\big] \leq C\tau^{b+1},$$

$$\mathbb{E}\big[\|(\Delta \overline{W}^{n+1})^2 - (\Delta W^{n+1})^2\|_{H^\gamma}^2\big] \leq C\tau^{b+1}. \qquad (4.21)$$

In the following proposition, we only give the proof of the case $F \equiv 0$. The proof is also suitable for the case that $F \not\equiv 0$ under certain conditions on F. For the sake of simplicity, we denote $\zeta^{n+1} := \Delta \overline{W}^{n+1}$.

Proposition 4.5 *Assume that $\varepsilon = \mu \equiv 1$, $\mathbb{E}\big[\|Mu_0\|_{\mathbb{H}}^2 + \|\nabla \cdot \mathbf{E}_0\|_U^2 + \|\nabla \cdot \mathbf{H}_0\|_U^2\big] < \infty$, $Q^{\frac{1}{2}} \in HS(U, H^{2+\gamma}(D))$ and $\sum_{i \in \mathbb{N}} \|Q^{\frac{1}{2}} e_i\|_{H^{1+\gamma}(D)} < \infty$ with $\gamma > 3/2$. Then for sufficiently small τ, there exists a positive constant $C = C(\lambda, T, u_0, Q)$ such* *that*

$$\sup_{1\leq n\leq N} \mathbb{E}\Big[\|Mu^n\|_{\mathbb{H}}^2 + \|\nabla \cdot \mathbf{E}^n\|_U^2 + \|\nabla \cdot \mathbf{H}^n\|_U^2\Big] \leq C. \tag{4.22}$$

Proof

Step 1. *Estimates of the divergence of the numerical solution.* From (4.18), we have

$$\mathbf{E}^{n+1} - \mathbf{E}^n = \tau \nabla \times \mathbf{H}^{n+\frac{1}{2}} - \lambda \mathbf{H}^{n+\frac{1}{2}}\zeta^{n+1}, \tag{4.23}$$

$$\mathbf{H}^{n+1} - \mathbf{H}^n = -\tau \nabla \times \mathbf{E}^{n+\frac{1}{2}} + \lambda \mathbf{E}^{n+\frac{1}{2}}\zeta^{n+1}. \tag{4.24}$$

Applying $\nabla\cdot$ to both sides of (4.23)–(4.24) and using the equality $\nabla \cdot (\nabla \times \boldsymbol{v}) \equiv 0$ for vector function $\boldsymbol{v}$ lead to

$$\nabla \cdot \mathbf{E}^{n+1} - \nabla \cdot \mathbf{E}^n = -\lambda \nabla \cdot (\mathbf{H}^{n+\frac{1}{2}}\zeta^{n+1}), \tag{4.25}$$

$$\nabla \cdot \mathbf{H}^{n+1} - \nabla \cdot \mathbf{H}^n = \lambda \nabla \cdot (\mathbf{E}^{n+\frac{1}{2}}\zeta^{n+1}). \tag{4.26}$$

Applying $\langle \cdot, \nabla \cdot \mathbf{E}^{n+\frac{1}{2}}\rangle_U$ and $\langle \cdot, \nabla \cdot \mathbf{H}^{n+\frac{1}{2}}\rangle_U$ to both sides of (4.25) and (4.26), respectively, and adding up the two derived equations, we obtain

$$\|\nabla \cdot \mathbf{E}^{n+1}\|_U^2 + \|\nabla \cdot \mathbf{H}^{n+1}\|_U^2 - \|\nabla \cdot \mathbf{E}^n\|_U^2 - \|\nabla \cdot \mathbf{H}^n\|_U^2$$

$$= -2\lambda\langle\nabla \cdot (\mathbf{H}^{n+\frac{1}{2}}\zeta^{n+1}), \nabla \cdot \mathbf{E}^{n+\frac{1}{2}}\rangle_U + 2\lambda\langle\nabla \cdot (\mathbf{E}^{n+\frac{1}{2}}\zeta^{n+1}), \nabla \cdot \mathbf{H}^{n+\frac{1}{2}}\rangle_U$$

$$= -2\lambda\Big\langle\zeta^{n+1}\nabla \cdot \mathbf{H}^{n+\frac{1}{2}} + \mathbf{H}^{n+\frac{1}{2}} \cdot \nabla\zeta^{n+1}, \nabla \cdot \mathbf{E}^{n+\frac{1}{2}}\Big\rangle_U$$

$$\quad + 2\lambda\Big\langle\zeta^{n+1}\nabla \cdot \mathbf{E}^{n+\frac{1}{2}} + \mathbf{E}^{n+\frac{1}{2}} \cdot \nabla\zeta^{n+1}, \nabla \cdot \mathbf{H}^{n+\frac{1}{2}}\Big\rangle_U$$

$$= -2\lambda\Big\langle\mathbf{H}^{n+\frac{1}{2}} \cdot \nabla\zeta^{n+1}, \nabla \cdot \mathbf{E}^{n+\frac{1}{2}}\Big\rangle_U + 2\lambda\Big\langle\mathbf{E}^{n+\frac{1}{2}} \cdot \nabla\zeta^{n+1}, \nabla \cdot \mathbf{H}^{n+\frac{1}{2}}\Big\rangle_U. \tag{4.27}$$

By the fact that ζ^{n+1} is independent of $\mathscr{F}_{t_n}$, one has

$$\mathbb{E}\Big[\|\nabla \cdot \mathbf{E}^{n+1}\|_U^2 + \|\nabla \cdot \mathbf{H}^{n+1}\|_U^2\Big] - \mathbb{E}\Big[\|\nabla \cdot \mathbf{E}^n\|_U^2 + \|\nabla \cdot \mathbf{H}^n\|_U^2\Big]$$

$$= -\lambda\mathbb{E}\Big[\Big\langle\mathbf{H}^{n+1} \cdot \nabla\zeta^{n+1}, \nabla \cdot \mathbf{E}^{n+\frac{1}{2}}\Big\rangle_U\Big] - \frac{\lambda}{2}\mathbb{E}\Big[\Big\langle\mathbf{H}^n \cdot \nabla\zeta^{n+1}, \nabla \cdot \mathbf{E}^{n+1}\Big\rangle_U\Big]$$

$$+ \lambda \mathbb{E}\Big[\Big\langle \mathbf{E}^{n+1} \cdot \nabla \zeta^{n+1}, \nabla \cdot \mathbf{H}^{n+\frac{1}{2}} \Big\rangle_U\Big] + \frac{\lambda}{2}\mathbb{E}\Big[\Big\langle \mathbf{E}^n \cdot \nabla \zeta^{n+1}, \nabla \cdot \mathbf{H}^{n+1} \Big\rangle_U\Big]$$

$$=: I_1^n + I_2^n + I_3^n + I_4^n.$$

For the term I_1^n,

$$I_1^n = -\frac{\lambda}{2}\mathbb{E}\Big[\Big\langle \mathbf{H}^{n+1} \cdot \nabla \zeta^{n+1}, \nabla \cdot \mathbf{E}^{n+1} - \nabla \cdot \mathbf{E}^n \Big\rangle_U\Big]$$

$$- \lambda \mathbb{E}\Big[\Big\langle (\mathbf{H}^{n+1} - \mathbf{H}^n) \cdot \nabla \zeta^{n+1}, \nabla \cdot \mathbf{E}^n \Big\rangle_U\Big]$$

$$=: I_{11}^n + I_{12}^n.$$

Plugging (4.25) into the term I_{11}^n and using the Sobolev embedding $H^\gamma(D) \hookrightarrow L^\infty(D)$ with $\gamma > 3/2$, we derive

$$I_{11}^n = -\frac{\lambda}{2}\mathbb{E}\Big[\Big\langle \mathbf{H}^{n+1} \cdot \nabla \zeta^{n+1}, -\lambda \nabla \cdot (\mathbf{H}^{n+\frac{1}{2}} \zeta^{n+1}) \Big\rangle_U\Big]$$

$$= \frac{\lambda^2}{4}\mathbb{E}\Big[\Big\langle \mathbf{H}^{n+1} \cdot (\nabla \zeta^{n+1}), \zeta^{n+1} \nabla \cdot \mathbf{H}^{n+1} \Big\rangle_U + \|\mathbf{H}^{n+1} \cdot \nabla \zeta^{n+1}\|_U^2\Big]$$

$$+ \frac{\lambda^2}{4}\mathbb{E}\Big[\Big\langle \mathbf{H}^{n+1} \cdot \nabla \zeta^{n+1}, \zeta^{n+1} \nabla \cdot \mathbf{H}^n + \mathbf{H}^n \cdot \nabla \zeta^{n+1} \Big\rangle_U\Big]$$

$$\leq C\,\mathbb{E}\Big[\|\zeta^{n+1}\|_{H^{1+\gamma}(D)}^2 \|\mathbf{H}^{n+1}\|_{L^2(D)^3}\big(\|\mathbf{H}^{n+1}\|_{L^2(D)^3} + \|\nabla \cdot \mathbf{H}^{n+1}\|_U\big)\Big]$$

$$+ C\,\mathbb{E}\Big[\|\zeta^{n+1}\|_{H^{1+\gamma}(D)}^2 \|\mathbf{H}^{n+1}\|_{L^2(D)^3}\big(\|\mathbf{H}^n\|_{L^2(D)^3} + \|\nabla \cdot \mathbf{H}^n\|_U\big)\Big].$$

This, combining the Young inequality and Proposition 4.4, yields

$$I_{11}^n \leq \frac{\tau}{10}\mathbb{E}\Big[\|\nabla \cdot \mathbf{H}^{n+1}\|_U^2\Big] + \frac{\tau}{15}\mathbb{E}\Big[\|\nabla \cdot \mathbf{H}^n\|_U^2\Big] + C\tau.$$

Similarly, we plug (4.24) into the term I_{12}^n and obtain

$$I_{12}^n \leq \frac{\tau}{5}\mathbb{E}\Big[\|\nabla \times \mathbf{E}^{n+1}\|_{L^2(D)^3}^2\Big] + \frac{\tau}{15}\mathbb{E}\Big[\|\nabla \cdot \mathbf{E}^n\|_U^2\Big] + C\tau.$$

Therefore,

$$I_1^n \leq \frac{\tau}{5}\mathbb{E}\Big[\|\nabla \times \mathbf{E}^{n+1}\|_{L^2(D)^3}^2\Big] + \frac{\tau}{10}\mathbb{E}\Big[\|\nabla \cdot \mathbf{H}^{n+1}\|_U^2\Big]$$

$$+ \frac{\tau}{15}\mathbb{E}\Big[\|\nabla \cdot \mathbf{E}^n\|_U^2 + \|\nabla \cdot \mathbf{H}^n\|_U^2\Big] + C\tau.$$

For the term I_2^n, similar to the estimate of I_{11}^n, one has

$$I_2^n = -\frac{\lambda}{2}\mathbb{E}\Big[\big\langle \mathbf{H}^n \cdot \nabla \zeta^{n+1}, \nabla \cdot \mathbf{E}^{n+1} - \nabla \cdot \mathbf{E}^n \big\rangle_U\Big]$$

$$= \frac{\lambda^2}{4}\mathbb{E}\Big[\big\langle \mathbf{H}^n \cdot \nabla \zeta^{n+1}, \zeta^{n+1}\nabla \cdot \mathbf{H}^{n+1} + \mathbf{H}^{n+1} \cdot \nabla \zeta^{n+1} \big\rangle_U\Big]$$

$$+ \frac{\lambda^2}{4}\mathbb{E}\Big[\big\langle \mathbf{H}^n \cdot \nabla \zeta^{n+1}, \zeta^n\nabla \cdot \mathbf{H}^n + \mathbf{H}^n \cdot \nabla \zeta^n \big\rangle_U\Big]$$

$$\leq \frac{\tau}{10}\mathbb{E}\Big[\|\nabla \cdot \mathbf{H}^{n+1}\|_U^2\Big] + \frac{\tau}{15}\mathbb{E}\Big[\|\nabla \cdot \mathbf{H}^n\|_U^2\Big] + C\tau.$$

The estimates of I_3^n and I_4^n are similar to those of I_1^n and I_2^n, respectively, which satisfy

$$I_3^n \leq \frac{\tau}{5}\mathbb{E}\Big[\|\nabla \times \mathbf{H}^{n+1}\|_{L^2(D)^3}^2\Big] + \frac{\tau}{10}\mathbb{E}\Big[\|\nabla \cdot \mathbf{E}^{n+1}\|_U^2\Big]$$

$$+ \frac{\tau}{5}\mathbb{E}\Big[\|\nabla \cdot \mathbf{H}^n\|_U^2 + \|\nabla \cdot \mathbf{E}^n\|_U^2\Big] + C\tau,$$

$$I_4^n \leq \frac{\tau}{10}\mathbb{E}\Big[\|\nabla \cdot \mathbf{E}^{n+1}\|_U^2\Big] + \frac{\tau}{15}\mathbb{E}\Big[\|\nabla \cdot \mathbf{E}^n\|_U^2\Big] + C\tau.$$

Combining the estimates of I_1^n, I_2^n, I_3^n, and I_4^n together, we derive

$$(1 - \frac{\tau}{5})\mathbb{E}\Big[\|\nabla \cdot \mathbf{E}^{n+1}\|_U^2 + \|\nabla \cdot \mathbf{H}^{n+1}\|_U^2\Big]$$

$$\leq (1 + \frac{\tau}{5})\mathbb{E}\Big[\|\nabla \cdot \mathbf{E}^n\|_U^2 + \|\nabla \cdot \mathbf{H}^n\|_U^2\Big] + \frac{\tau}{5}\mathbb{E}\big[\|Mu^{n+1}\|_{\mathbb{H}}^2\big] + C\tau. \qquad (4.28)$$

Step 2. Estimates of the curl of the numerical solution. Applying $\langle \cdot, M(u^{n+1} - u^n)\rangle_{\mathbb{H}}$ to both sides of (4.18), we obtain

$$\|Mu^{n+1}\|_{\mathbb{H}}^2 = \|Mu^n\|_{\mathbb{H}}^2 - \frac{2\lambda}{\tau}\big\langle \mathbb{J}^{-1}u^{n+\frac{1}{2}}\zeta^{n+1}, M(u^{n+1} - u^n)\big\rangle_{\mathbb{H}}. \qquad (4.29)$$

Using the skew-adjointness of the Maxwell operator M and substituting (4.18) into the right-hand side of (4.29) yield

$$\mathbb{E}\Big[\|Mu^{n+1}\|_{\mathbb{H}}^2\Big] = \mathbb{E}\Big[\|Mu^n\|_{\mathbb{H}}^2\Big] + 2\lambda\mathbb{E}\Big[\big\langle M(\mathbb{J}^{-1}u^{n+\frac{1}{2}}\zeta^{n+1}), Mu^{n+\frac{1}{2}}\big\rangle_{\mathbb{H}}\Big]. \qquad (4.30)$$

Denote

$$R_n^\tau := \begin{bmatrix} (\nabla \zeta^{n+1}) \times & 0 \\ 0 & (\nabla \zeta^{n+1}) \times \end{bmatrix}.$$

Then

$$2\lambda \mathbb{E}\Big[\Big\langle M(\mathbb{J}^{-1} u^{n+\frac{1}{2}} \zeta^{n+1}), M u^{n+\frac{1}{2}} \Big\rangle_{\mathbb{H}}\Big]$$

$$= 2\lambda \mathbb{E}\Big[\Big\langle \zeta^{n+1} \mathbb{J}^{-1} M u^{n+\frac{1}{2}}, M u^{n+\frac{1}{2}} \Big\rangle_{\mathbb{H}}\Big] + 2\lambda \mathbb{E}\Big[\Big\langle R_n^\tau u^{n+\frac{1}{2}}, M u^{n+\frac{1}{2}} \Big\rangle_{\mathbb{H}}\Big]$$

$$= \frac{\lambda}{2} \mathbb{E}\Big[\Big\langle R_n^\tau (u^{n+1} - u^n), M(u^{n+1} + u^n) \Big\rangle_{\mathbb{H}}\Big] + \lambda \mathbb{E}\Big[\Big\langle R_n^\tau u^n, M(u^{n+1} - u^n) \Big\rangle_{\mathbb{H}}\Big]$$

$$=: II_1^n + II_2^n.$$

We substitute (4.18) into the term II_1^n and obtain

$$II_1^n = \frac{\lambda \tau}{4} \mathbb{E}\Big[\Big\langle R_n^\tau M(u^{n+1} + u^n), M(u^{n+1} + u^n) \Big\rangle_{\mathbb{H}}\Big]$$

$$+ \frac{\lambda^2}{4} \mathbb{E}\Big[\Big\langle R_n^\tau \mathbb{J}^{-1}(u^{n+1} + u^n) \zeta^{n+1}, M(u^{n+1} + u^n) \Big\rangle_{\mathbb{H}}\Big]$$

$$=: II_{11}^n + II_{12}^n.$$

It follows from the Sobolev embedding $H^\gamma(D) \hookrightarrow L^\infty(D)$ with $\gamma > 3/2$ and the Young inequality that

$$II_{11}^n = \frac{\lambda \tau}{4} \mathbb{E}\Big[\Big\langle R_n^\tau M u^{n+1}, M u^{n+1} \Big\rangle_{\mathbb{H}}\Big] + \frac{\lambda \tau}{4} \mathbb{E}\Big[\Big\langle R_n^\tau M u^{n+1}, M u^n \Big\rangle_{\mathbb{H}}\Big]$$

$$+ \frac{\lambda \tau}{4} \mathbb{E}\Big[\Big\langle R_n^\tau M u^n, M u^{n+1} \Big\rangle_{\mathbb{H}}\Big]$$

$$\leq C\tau \, \mathbb{E}\Big[\|\zeta^{n+1}\|_{H^{1+\gamma}(D)} \|M u^{n+1}\|_{\mathbb{H}}^2\Big]$$

$$+ C\tau \mathbb{E}\Big[\|\zeta^{n+1}\|_{H^{1+\gamma}(D)} \|M u^{n+1}\|_{\mathbb{H}} \|M u^n\|_{\mathbb{H}}\Big]$$

$$\leq C\tau \, \mathbb{E}\Big[\|\zeta^{n+1}\|_{H^{1+\gamma}(D)} \|M u^{n+1}\|_{\mathbb{H}}^2\Big] + \frac{\tau}{5} \mathbb{E}\big[\|M u^{n+1}\|_{\mathbb{H}}^2\big] + C\tau^2 \mathbb{E}\big[\|M u^n\|_{\mathbb{H}}^2\big].$$

The assumption $\sum_{i \in \mathbb{N}} \|Q^{\frac{1}{2}} e_i\|_{H^{1+\gamma}(D)} < \infty$ implies $\|\zeta^{n+1}\|_{H^{1+\gamma}(D)} \leq C\tau^{\frac{1}{2}} A_\tau^b$, which yields

$$II_{11}^n \leq C\tau^{\frac{3}{2}} A_\tau \mathbb{E}\big[\|M u^{n+1}\|_{\mathbb{H}}^2\big] + \frac{\tau}{5} \mathbb{E}\big[\|M u^{n+1}\|_{\mathbb{H}}^2\big] + C\tau^2 \mathbb{E}\big[\|M u^n\|_{\mathbb{H}}^2\big].$$

For the term II^n_{12}, we utilize Proposition 4.4 to obtain

$$II^n_{12} \leq \frac{\tau}{5}\mathbb{E}\Big[\|Mu^{n+1}\|^2_{\mathbb{H}} + \|Mu^n\|^2_{\mathbb{H}}\Big] + C\tau.$$

Combining the estimates of II^n_{11} and II^n_{12} gives

$$II^n_1 \leq \Big(C\sqrt{\tau}A^b_\tau + \frac{2}{5}\Big)\tau\mathbb{E}[\|Mu^{n+1}\|^2_{\mathbb{H}}] + \Big(C\tau + \frac{1}{5}\Big)\tau\mathbb{E}[\|Mu^n\|^2_{\mathbb{H}}] + C\tau.$$

For the term II^n_2, the skew-adjointness of the Maxwell operator M leads to

$$II^n_2 = -\lambda\mathbb{E}\Big[\big\langle M(R^\tau_n u^n), u^{n+1} - u^n\big\rangle_{\mathbb{H}}\Big]$$

$$= -\frac{\lambda\tau}{2}\mathbb{E}\Big[\big\langle M(R^\tau_n u^n), Mu^{n+1}\big\rangle_{\mathbb{H}}\Big] - \lambda^2\mathbb{E}\Big[\big\langle M(R^\tau_n u^n), \mathbb{J}^{-1}u^{n+\frac{1}{2}}\zeta^{n+1}\big\rangle_{\mathbb{H}}\Big]$$

$$=: II^n_{21} + II^n_{22},$$

where we substituted the expression of $u^{n+1} - u^n$ by (4.18) in the second step. Note that

$$M(R^\tau_n u^n) = \begin{bmatrix} (\nabla \cdot \mathbf{H}^n)\nabla\zeta^{n+1} \\ -(\nabla \cdot \mathbf{E}^n)\nabla\zeta^{n+1} \end{bmatrix} + \begin{bmatrix} (\mathbf{H}^n \cdot \nabla)\nabla\zeta^{n+1} \\ -(\mathbf{E}^n \cdot \nabla)\nabla\zeta^{n+1} \end{bmatrix}$$

$$- (\nabla \cdot (\nabla\zeta^{n+1}))\mathbb{J}u^n + \begin{bmatrix} -(\nabla\zeta^{n+1} \cdot \nabla)\mathbf{H}^n \\ (\nabla\zeta^{n+1} \cdot \nabla)\mathbf{E}^n \end{bmatrix},$$

which implies that

$$\|M(R^\tau_n u^n)\|_{\mathbb{H}} \leq C\|\zeta^{n+1}\|_{H^{2+\gamma}(D)}\Big(\|\nabla \cdot \mathbf{H}^n\|_U + \|\nabla \cdot \mathbf{E}^n\|_U + \|u^n\|_{\mathbb{H}} + \|Mu^n\|_{\mathbb{H}}\Big)$$

due to the Sobolev embedding $H^\gamma(D) \hookrightarrow L^\infty(D)$ with $\gamma > 3/2$ and the Young inequality. Thus, for the term II^n_{21}, it follows from the independence of ζ^{n+1} and $\mathscr{F}_{t_n}$ that

$$II^n_{21} \leq \frac{\tau}{5}\mathbb{E}[\|Mu^{n+1}\|_{\mathbb{H}}] + C\tau\mathbb{E}[\|M(R^\tau_n u^n)\|^2_{\mathbb{H}}]$$

$$\leq \frac{\tau}{5}\mathbb{E}[\|Mu^{n+1}\|_{\mathbb{H}}] + C\tau^2\mathbb{E}[\|Mu^n\|_{\mathbb{H}}]$$

$$+ C\tau^2\mathbb{E}[\|\nabla \cdot \mathbf{H}^n\|^2_U + \|\nabla \cdot \mathbf{E}^n\|^2_U] + C\tau^2.$$

For the term II_{22}^n, by the Sobolev embedding $H^\gamma(D) \hookrightarrow L^\infty(D)$ with $\gamma > 3/2$, it holds

$$II_{22}^n \leq C\mathbb{E}\left[\|M(R_n^\tau u^n)\|_{\mathbb{H}}\|\zeta^{n+1}\|_{H^\gamma(D)}\Big(\|u^{n+1}\|_{\mathbb{H}} + \|u^n\|_{\mathbb{H}}\Big)\right]$$

$$\leq \frac{\tau}{5}\mathbb{E}\left[\|\nabla \cdot \mathbf{H}^n\|_U^2 + \|\nabla \cdot \mathbf{E}^n\|_U^2 + \|u^n\|_{\mathbb{H}}^2 + \|Mu^n\|_{\mathbb{H}}^2\right]$$

$$+ C\tau\mathbb{E}\left[\|u^{n+1}\|_{\mathbb{H}}^2 + \|u^n\|_{\mathbb{H}}^2\right] + C\frac{1}{\tau^3}\mathbb{E}\left[\|\zeta^{n+1}\|_{H^{2+\gamma}(D)}^4\|\zeta^{n+1}\|_{H^\gamma(D)}^4\right]$$

$$\leq \frac{\tau}{5}\mathbb{E}\left[\|\nabla \cdot \mathbf{E}^n\|_U^2 + \|\nabla \cdot \mathbf{H}^n\|_U^2\right] + \frac{\tau}{5}\mathbb{E}\left[\|Mu^n\|_{\mathbb{H}}^2\right] + C\tau.$$

Hence,

$$II_2^n \leq \left(C\tau + \frac{1}{5}\right)\tau\mathbb{E}\left[\|\nabla \cdot \mathbf{E}^n\|_U^2 + \|\nabla \cdot \mathbf{H}^n\|_U^2\right] + \frac{\tau}{5}\mathbb{E}\left[\|Mu^{n+1}\|_{\mathbb{H}}^2\right]$$

$$+ \left(C\tau + \frac{1}{5}\right)\tau\mathbb{E}\left[\|Mu^n\|_{\mathbb{H}}^2\right] + C\tau.$$

Combining the estimates of II_1^n and II_2^n, we have

$$2\lambda\mathbb{E}\left[\Big\langle M(\mathbb{J}^{-1}u^{n+\frac{1}{2}}\zeta^{n+1}), Mu^{n+\frac{1}{2}}\Big\rangle_{\mathbb{H}}\right]$$

$$\leq \left(C\sqrt{\tau}A_\tau^b + \frac{3}{5}\right)\tau\mathbb{E}[\|Mu^{n+1}\|_{\mathbb{H}}^2] + \left(C\tau + \frac{2}{5}\right)\tau\mathbb{E}[\|Mu^n\|_{\mathbb{H}}^2] \qquad (4.31)$$

$$+ \left(C\tau + \frac{1}{5}\right)\tau\mathbb{E}\left[\|\nabla \cdot \mathbf{E}^n\|_U^2 + \|\nabla \cdot \mathbf{H}^n\|_U^2\right] + C\tau.$$

Plugging (4.31) into (4.30) yields

$$\mathbb{E}[\|Mu^{n+1}\|_{\mathbb{H}}^2] \leq \mathbb{E}[\|Mu^n\|_{\mathbb{H}}^2] + \left(C\sqrt{\tau}A_\tau^b + \frac{3}{5}\right)\tau\mathbb{E}[\|Mu^{n+1}\|_{\mathbb{H}}^2]$$

$$+ \left(C\tau + \frac{2}{5}\right)\tau\mathbb{E}[\|Mu^n\|_{\mathbb{H}}^2]$$

$$+ \left(C\tau + \frac{1}{5}\right)\tau\mathbb{E}\left[\|\nabla \cdot \mathbf{E}^n\|_U^2 + \|\nabla \cdot \mathbf{H}^n\|_U^2\right] + C\tau.$$

Combining (4.28), we conclude that

$$\mathbb{E}\left[\|Mu^{n+1}\|_{\mathbb{H}}^2 + \|\nabla \cdot \mathbf{E}^{n+1}\|_U^2 + \|\nabla \cdot \mathbf{H}^{n+1}\|_U^2\right]$$

$$\leq \mathbb{E}\left[\|Mu^n\|_{\mathbb{H}}^2 + \|\nabla \cdot \mathbf{E}^n\|_U^2 + \|\nabla \cdot \mathbf{H}^n\|_U^2\right]$$

$$+ \left(C\tau + \frac{2}{5} \right) \tau \, \mathbb{E} \left[\| Mu^n \|_{\mathbb{H}}^2 + \| \nabla \cdot \mathbf{E}^n \|_U^2 + \| \nabla \cdot \mathbf{H}^n \|_U^2 \right]$$

$$+ \left(C\sqrt{\tau} A_\tau + \frac{4}{5} \right) \tau \, \mathbb{E} \left[\| Mu^{n+1} \|_{\mathbb{H}}^2 + \| \nabla \cdot \mathbf{E}^{n+1} \|_U^2 + \| \nabla \cdot \mathbf{H}^{n+1} \|_U^2 \right] + C\tau.$$

For sufficiently small τ, the Grönwall inequality gives the desired result. Thus we finish the proof of Proposition 4.5. □

Remark 4.2 By multiplying $\| \nabla \cdot \mathbf{E}^{n+1} \|_U^2 + \| \nabla \cdot \mathbf{H}^{n+1} \|_U^2$ and $\| Mu^{n+1} \|_{\mathbb{H}}^2$ on both sides of (4.27) and (4.29), respectively, one can further derive that

$$\sup_{1 \le n \le N} \mathbb{E} \left[\| Mu^n \|_{\mathbb{H}}^4 + \| \nabla \cdot \mathbf{E}^n \|_U^4 + \| \nabla \cdot \mathbf{H}^n \|_U^4 \right] \le C. \qquad (4.32)$$

If we impose the following PEC boundary conditions

$$\mathbf{n} \times \mathbf{E} = 0, \quad \mathbf{n} \cdot \mathbf{H} = 0, \quad \text{on} \quad [0, T] \times \partial D, \qquad (4.33)$$

the H^1-regularity of the solution of (4.18) is implied by Lemma B.2 and Proposition 4.5, that is, there exists a positive constant C such that

$$\sup_{1 \le n \le N} \mathbb{E} \left[\| u^n \|_{H^1(D)^6}^2 \right] \le C.$$

Moreover, if we turn our attention to the damped stochastic Maxwell equations (3.34), then we consider the following modified stochastic midpoint method

$$u^{n+1} = e^{-\sigma\tau} u^n + \frac{\tau}{2} M \left(u^{n+1} + e^{-\sigma\tau} u^n \right) + \theta \Delta W_2^{n+1}$$

$$+ \frac{\lambda}{2} \mathbb{J}^{-1} \left(u^{n+1} + e^{-\sigma\tau} u^n \right) \Delta \overline{W}_1^{n+1}, \quad n \in \mathbb{N}. \qquad (4.34)$$

Using similar arguments as in the proof of Proposition 4.5, we can obtain that the solution of (4.34) is bounded uniformly in the $L^2(\Omega, H^1(D)^6)$-norm, i.e.,

$$\sup_{n \in \mathbb{N}} \mathbb{E} \left[\| u^n \|_{H^1(D)^6}^2 \right] \le C.$$

We refer to [45] for details. It can be shown that this result ensures the existence of the invariant measure for (4.34) while the uniqueness of the invariant measure is obtained by using the general Harris Theorem (see [91, Theorem 4.8]).

Theorem 4.1 *Let conditions in Proposition 4.5 and the PEC boundary conditions (4.33) hold. Suppose that $u_0 \in L^2(\Omega, H^1(D)^6)$ and $F_{Q_1} \in W^{1,\infty}(D)$. In addition, let $Q_i^{\frac{1}{2}} \in HS(U, H^{\gamma_i}(D))$, $i = 1, 2$, $\sum_{i \in \mathbb{N}} \| Q_1^{\frac{1}{2}} e_i \|_{H^{\gamma_1 - 1}(D)} < \infty$, for any $\gamma_1 > 7/2$, $\gamma_2 \ge 2$, and let $\sigma \in W^{1,\infty}(D)$, $\sigma \ge \sigma_0 > 0$ with a constant σ_0.*

(i) Let $\{u^n; n \in \mathbb{N}\}$ and $\{\widetilde{u}^n; n \in \mathbb{N}\}$ be solutions of (4.34) with initial data u_0 and $\widetilde{u}_0$, respectively. Then

$$\mathbb{E}\big[\|u^n - \widetilde{u}^n\|_{\mathbb{H}}^2\big] \leq e^{-2\sigma_0 t_n}\mathbb{E}\big[\|u_0 - \widetilde{u}_0\|_{\mathbb{H}}^2\big].$$

(ii) For sufficiently small τ, the numerical solution $\{u^n; n \in \mathbb{N}\}$ of (4.34) has a unique invariant measure $\pi^\tau \in \mathscr{P}_2(\mathbb{H})$. Thus, $\{u^n; n \in \mathbb{N}\}$ is ergodic. Moreover, $\{u^n; n \in \mathbb{N}\}$ is exponentially mixing.

(iii) For arbitrary two distributions $\pi_1, \pi_2 \in \mathscr{P}_2(\mathbb{H})$ and $n \in \mathbb{N}$,

$$\mathscr{W}_2((P_n^\tau)^*\pi_1, (P_n^\tau)^*\pi_2) \leq e^{-\sigma_0 t_n}\mathscr{W}_2(\pi_1, \pi_2),$$

where $(P_n^\tau)^*\pi, n \in \mathbb{N}$ denotes the probability distribution of u^n with initial probability distribution π. Moreover, for any distribution $\pi \in \mathscr{P}_2(\mathbb{H})$, $n \in \mathbb{N}$, we have

$$\mathscr{W}_2((P_n^\tau)^*\pi, \pi^\tau) \leq e^{-\sigma_0 t_n}\mathscr{W}_2(\pi, \pi^\tau).$$

Proof It follows from (4.34) that

$$(u^n - \widetilde{u}^n) - e^{-\sigma\tau}(u^{n-1} - \widetilde{u}^{n-1}) = \tau M \frac{(u^n - \widetilde{u}^n) + e^{-\sigma\tau}(u^{n-1} - \widetilde{u}^{n-1})}{2}$$
$$+ \lambda \mathbb{J}^{-1}\frac{(u^n - \widetilde{u}^n) + e^{-\sigma\tau}(u^{n-1} - \widetilde{u}^{n-1})}{2}\Delta\overline{W}_1^n.$$

Applying $\langle \cdot, (u^n - \widetilde{u}^n) + e^{-\sigma\tau}(u^{n-1} - \widetilde{u}^{n-1})\rangle_{\mathbb{H}}$ to both sides of the above equation and taking the expectation, we obtain

$$\mathbb{E}\big[\|u^n - \widetilde{u}^n\|_{\mathbb{H}}^2\big] \leq e^{-2\sigma_0\tau}\mathbb{E}\big[\|u^{n-1} - \widetilde{u}^{n-1}\|_{\mathbb{H}}^2\big] \leq \cdots \leq e^{-2\sigma_0 t_n}\mathbb{E}\big[\|u_0 - \widetilde{u}_0\|_{\mathbb{H}}^2\big]$$

for all $n \in \mathbb{N}$. Thus we have proved assertion (i). The proofs of assertions (ii) and (iii) of this theorem are analogous to those of Theorem 3.6. □

4.1.1.2 Analysis of Stochastic Symplectic Runge–Kutta Methods

This part is devoted to studying the well-posedness and regularity of the numerical solution for the stochastic symplectic Runge–Kutta method (4.3)–(4.4) to the stochastic Maxwell equations driven by additive noise. More precisely, consider the following temporal semi-discretization

$$U_i^n = u^n + \tau \sum_{j=1}^{s} a_{ij} \left(M U_j^n + F^{nj}(U_j^n) \right) + \sum_{j=1}^{s} \tilde{a}_{ij} B^{nj} \Delta W^{n+1}, \qquad (4.35)$$

$$u^{n+1} = u^n + \tau \sum_{i=1}^{s} b_i \left(M U_i^n + F^{ni}(U_i^n) \right) + \sum_{i=1}^{s} \tilde{b}_i B^{ni} \Delta W^{n+1}, \qquad (4.36)$$

where $B^{ni} = B(t_n + \tilde{c}_i \tau)$ and coefficients a_{ij}, b_i satisfy the symplectic condition (4.11).

To prove the existence of a numerical solution given by the stochastic symplectic Runge–Kutta method (4.35)–(4.36), we need the following coercivity condition: the matrix $A = (a_{ij})_{i,j=1}^{s}$ is invertible, and there exists a diagonal positive definite matrix $\mathcal{K} = \mathrm{diag}(k_1, k_2, \ldots, k_s)$ and a constant $\alpha > 0$ such that

$$u^\top \mathcal{K} A^{-1} u \geq \alpha u^\top \mathcal{K} u \quad \forall u \in \mathbb{R}^s. \qquad (4.37)$$

It is obvious that the stochastic midpoint method (4.14) satisfies the coercivity condition. For (4.35)–(4.36), we have the following well-posedness result.

Theorem 4.2 *Assume that the coefficients of the stochastic Runge–Kutta method (4.35)–(4.36) satisfy the symplectic condition (4.11) and the coercivity condition (4.37). If in addition F satisfies Assumption 2.2, and $B(t) \in HS(U_0, \mathscr{D}(M))$ for any $t \in [0, T]$, then for sufficiently small τ, there exists a unique $\mathbb{H}$-valued $\{\mathscr{F}_{t_n}\}_{0 \leq n \leq N}$-adapted numerical solution $\{u^n; \ n = 0, 1, \ldots, N\}$ of the stochastic symplectic Runge–Kutta method. Moreover, for any $p \geq 1$, there is a positive constant $C = C(s, p, T)$ such that*

$$\max_{1 \leq i \leq s} \mathbb{E}\left[\|U_i^n\|_{\mathbb{H}}^{2p} \right] \leq C \left(\mathbb{E}\left[\|u^n\|_{\mathbb{H}}^{2p} \right] + \tau \right), \quad n = 0, 1, \ldots, N, \qquad (4.38)$$

$$\max_{1 \leq n \leq N} \|u^n\|_{L^{2p}(\Omega, \mathbb{H})} \leq C \left(1 + \|u_0\|_{L^{2p}(\Omega, \mathbb{H})} \right). \qquad (4.39)$$

Proof

Step 1. Existence and uniqueness. The existence of the numerical solution is similar to that in *Step 1* of Proposition 4.2, which is omitted here.

The uniqueness of the numerical solution follows from the uniqueness of U_i^n, $i = 1, 2, \ldots, s$. Assume that (4.35) has two different numerical solutions $U_n = \left((U_1^n)^\top, \ldots, (U_s^n)^\top \right)^\top$ and $V_n = \left((V_1^n)^\top, \ldots, (V_s^n)^\top \right)^\top$ with $U_{n-1} = V_{n-1}$, then it follows that the compact form $U_n - V_n$ is given by

$$U_n - V_n = \tau \left[Id - \tau (A \otimes M) \right]^{-1} (A \otimes Id) \left(F^n(U_n) - F^n(V_n) \right).$$

Lemma C.2 (i) and the linear growth of F imply

$$\|U_n - V_n\|_{\mathbb{H}^{\otimes s}} \leq C\tau \|U_n - V_n\|_{\mathbb{H}^{\otimes s}},$$

where $\mathbb{H}^{\otimes s} = \mathbb{H} \times \mathbb{H} \times \cdots \times \mathbb{H}$. Consequently, if τ is sufficiently small, the uniqueness of U_n holds.

Step 2.   *Proof of* (4.38). We only present the proof for $p = 1$. The proof for $p > 1$ follows a similar procedure and is omitted. From the compact formula of (4.35), we have

$$U_n = \left[Id - \tau (A \otimes M) \right]^{-1} \left[\mathbf{1}_s \otimes u^n + \tau (A \otimes Id) F^n (U_n) \right. $$
$$\left. + (\widetilde{A} \otimes Id) B^n \Delta W^{n+1} \right],$$

which, along with Lemma C.2 (i) and the linear growth of F, implies

$$\|U_n\|_{\mathbb{H}^{\otimes s}}^2 \leq C \|\mathbf{1}_s \otimes u^n + \tau (A \otimes Id) F^n (U_n) + (\widetilde{A} \otimes Id) B^n \Delta W^{n+1}\|_{\mathbb{H}^{\otimes s}}^2$$

$$\leq C \|u^n\|_{\mathbb{H}}^2 + C\tau^2 \sum_{i=1}^{s} \|F^{ni}(U_i^n)\|_{\mathbb{H}}^2 + C \sum_{i=1}^{s} \|B^{ni} \Delta W^{n+1}\|_{\mathbb{H}}^2$$

$$\leq C \|u^n\|_{\mathbb{H}}^2 + C\tau^2 \sum_{i=1}^{s} \left(1 + \|U_i^n\|_{\mathbb{H}}^2 \right) + C \sum_{i=1}^{s} \|B^{ni} \Delta W^{n+1}\|_{\mathbb{H}}^2$$

$$\leq C \|u^n\|_{\mathbb{H}}^2 + C\tau^2 + C\tau^2 \|U_n\|_{\mathbb{H}^{\otimes s}}^2 + C \sum_{i=1}^{s} \|B^{ni} \Delta W^{n+1}\|_{\mathbb{H}}^2.$$

By the assumption $B(t) \in HS(U_0, \mathscr{D}(M))$, it holds that

$$\mathbb{E}\left[\|U_n\|_{\mathbb{H}^{\otimes s}}^2 \right] \leq C \mathbb{E}\left[\|u^n\|_{\mathbb{H}}^2 \right] + C\tau \qquad (4.40)$$

for sufficiently small τ. The proof of (4.38) is completed by $\|U_n\|_{\mathbb{H}^{\otimes s}}^2 = \sum_{i=1}^{s} \|U_i^n\|_{\mathbb{H}}^2$.

Step 3. *Proof of* (4.39). We start from (4.36) to obtain

$$\|u^{n+1}\|_{\mathbb{H}}^2 = \|u^n\|_{\mathbb{H}}^2 + \tau^2 \Big\| \sum_{i=1}^{s} b_i \big(M U_i^n + F^{ni}(U_i^n)\big) \Big\|_{\mathbb{H}}^2 + \Big\| \sum_{i=1}^{s} \widetilde{b}_i B^{ni} \Delta W^{n+1} \Big\|_{\mathbb{H}}^2$$

$$+ 2\tau \sum_{i=1}^{s} b_i \langle u^n, M U_i^n + F^{ni}(U_i^n) \rangle_{\mathbb{H}} + 2 \sum_{i=1}^{s} \widetilde{b}_i \langle u^n, B^{ni} \Delta W^{n+1} \rangle_{\mathbb{H}}$$

$$+ 2\tau \sum_{i,j=1}^{s} b_i \widetilde{b}_j \langle M U_i^n + F^{ni}(U_i^n), B^{nj} \Delta W^{n+1} \rangle_{\mathbb{H}}.$$

For the fourth term of the right-hand side of the above equality, we note that by (4.35) and the skew-adjointness of M,

$$2\tau \sum_{i=1}^{s} b_i \langle u^n, M U_i^n + F^{ni}(U_i^n) \rangle_{\mathbb{H}}$$

$$= 2\tau \sum_{i=1}^{s} b_i \langle U_i^n, F^{ni}(U_i^n) \rangle_{\mathbb{H}} - 2\tau \sum_{i,j=1}^{s} b_i \widetilde{a}_{ij} \langle B^{nj} \Delta W^{n+1}, M U_i^n + F^{ni}(U_i^n) \rangle_{\mathbb{H}}$$

$$- \tau^2 \sum_{i,j=1}^{s} \big(b_i a_{ij} + b_j a_{ji}\big) \langle M U_j^n + F^{nj}(U_j^n), M U_i^n + F^{ni}(U_i^n) \rangle_{\mathbb{H}}.$$

It is now clear that

$$\|u^{n+1}\|_{\mathbb{H}}^2 = \|u^n\|_{\mathbb{H}}^2 + \Big\| \sum_{i=1}^{s} \widetilde{b}_i B^{ni} \Delta W^{n+1} \Big\|_{\mathbb{H}}^2 + 2\tau \sum_{i=1}^{s} b_i \langle U_i^n, F^{ni}(U_i^n) \rangle_{\mathbb{H}}$$

$$- \tau^2 \sum_{i,j=1}^{s} m_{ij} \langle M U_j^n + F^{nj}(U_j^n), M U_i^n + F^{ni}(U_i^n) \rangle_{\mathbb{H}} \qquad (4.41)$$

$$+ 2\tau \sum_{i,j=1}^{s} \big(b_i \widetilde{b}_j - b_i \widetilde{a}_{ij}\big) \langle B^{nj} \Delta W^{n+1}, M U_i^n + F^{ni}(U_i^n) \rangle_{\mathbb{H}}$$

$$+ 2 \sum_{i=1}^{s} \widetilde{b}_i \langle u^n, B^{ni} \Delta W^{n+1} \rangle_{\mathbb{H}}.$$

Due to the symplectic condition (4.11), we arrive at

$$\|u^{n+1}\|_{\mathbb{H}}^2 \le \|u^n\|_{\mathbb{H}}^2 + C(1+\tau)\sum_{i=1}^s \|B^{ni}\,\Delta W^{n+1}\|_{\mathbb{H}}^2 + C\tau\sum_{i=1}^s \|M(B^{ni}\,\Delta W^{n+1})\|_{\mathbb{H}}^2$$

$$+ C\tau\sum_{i=1}^s \|U_i^n\|_{\mathbb{H}}^2 + C\tau\sum_{i=1}^s \|F^{ni}(U_i^n)\|_{\mathbb{H}}^2 + 2\sum_{i=1}^s \widetilde{b}_i\langle u^n, B^{ni}\,\Delta W^{n+1}\rangle_{\mathbb{H}}.$$

Taking the expectation, together with the linear growth of F and the assumption $B(t) \in HS(U_0, \mathscr{D}(M))$, we have

$$\mathbb{E}\big[\|u^{n+1}\|_{\mathbb{H}}^2\big] \le \mathbb{E}\big[\|u^n\|_{\mathbb{H}}^2\big] + C\tau + C\tau\mathbb{E}\big[\|U_n\|_{\mathbb{H}^{\otimes s}}^2\big].$$

Plugging (4.40) into the above inequality yields

$$\mathbb{E}\big[\|u^{n+1}\|_{\mathbb{H}}^2\big] \le (1 + C\tau)\mathbb{E}\big[\|u^n\|_{\mathbb{H}}^2\big] + C\tau,$$

which implies the assertion (4.39) by the Grönwall inequality. The proof of Theorem 4.2 is thus finished.   □

Remark 4.3

(i) Theorem 4.2 still holds if we replace the symplectic condition (4.11) by the one that

$$\mathscr{M} = \big(m_{ij}\big)_{i,j=1}^s \quad \text{with} \quad m_{ij} := b_i a_{ij} + b_j a_{ji} - b_i b_j \tag{4.42}$$

is semi-positive definite, which can ensure that the fourth term of the right-hand side of (4.41) is non-positive. It is obvious that the stochastic symplectic Runge–Kutta method satisfies (4.42) naturally.

(ii) Note that for the well-posedness of the stochastic symplectic Runge–Kutta method, a technical assumption $\sup_{t\in[0,T]}\|B(t)\|_{HS(U_0,\mathscr{D}(M))} < \infty$ is required to bound the term $\|M(B^{ni}\,\Delta W^{n+1})\|_{\mathbb{H}}$. One may eliminate this assumption for certain concrete numerical methods, for instance, see Proposition 4.2 for the stochastic midpoint method.

Remark 4.4

(i) If we plug (4.35) again into the sixth term on the right-hand side of (4.41), then we have

$$\|u^{n+1}\|_{\mathbb{H}}^2 = \|u^n\|_{\mathbb{H}}^2 + 2\tau\sum_{i=1}^s \langle U_i^n, F^{ni}(U_i^n)\rangle_{\mathbb{H}} + 2\sum_{i=1}^s \widetilde{b}_i\langle U_i^n, B^{ni}\,\Delta W^{n+1}\rangle_{\mathbb{H}}$$

$$- \tau^2 \sum_{i,j=1}^s m_{ij}\langle MU_j^n + F^{nj}(U_j^n), MU_i^n + F^{ni}(U_i^n)\rangle_{\mathbb{H}}$$

$$- 2\tau \sum_{i,j=1}^{s} \widetilde{m}_{ij} \langle B^{ni} \Delta W^{n+1}, M U_j^n + F^{nj}(U_j^n) \rangle_{\mathbb{H}}$$

$$- \sum_{i,j=1}^{s} \widetilde{\widetilde{m}}_{ij} \langle B^{nj} \Delta W^{n+1}, B^{ni} \Delta W^{n+1} \rangle_{\mathbb{H}}.$$

Under the condition (4.7), we can obtain the following discrete energy evolution law

$$\|u^{n+1}\|_{\mathbb{H}}^2 = \|u^n\|_{\mathbb{H}}^2 + 2\tau \sum_{i=1}^{s} b_i \langle U_i^n, F^{ni}(U_i^n) \rangle_{\mathbb{H}}$$

$$+ 2 \sum_{i=1}^{s} \widetilde{b}_i \langle U_i^n, B^{ni} \Delta W^{n+1} \rangle_{\mathbb{H}}, \quad \mathbb{P}\text{-}a.s.$$

for all $n = 0, 1, \ldots, N - 1$, which can be considered as the discrete version of (3.23).

(ii) Denote $F^{ni}(U_i^n) =: (F_e^{ni}(U_i^n)^\top, F_m^{ni}(U_i^n)^\top)^\top$. From (4.36), after straightforward calculations we end up with the discrete averaged divergence evolution laws

$$\mathbb{E}[\nabla \cdot \mathbf{E}^{n+1}] = \mathbb{E}[\nabla \cdot \mathbf{E}^n] + \tau \mathbb{E}\Big[\sum_{i=1}^{s} b_i \nabla \cdot F_e^{ni}(U_i^n) \Big],$$

$$\mathbb{E}[\nabla \cdot \mathbf{H}^{n+1}] = \mathbb{E}[\nabla \cdot \mathbf{H}^n] + \tau \mathbb{E}\Big[\sum_{i=1}^{s} b_i \nabla \cdot F_m^{ni}(U_i^n) \Big],$$

which can be considered as the discrete version of (3.26).

Similar to Theorem 4.2, we can obtain the $\mathscr{D}(M^k)$-regularity ($k \in \mathbb{N}_+$) for the numerical solution $\{u^n; n = 0, 1, \ldots, N\}$ of the stochastic symplectic Runge–Kutta method (4.35)–(4.36). As a consequence, the discrete versions of the Hölder continuity in $\mathscr{D}(M^{k-1})$ of the numerical solution are also obtained.

Proposition 4.6 *Suppose that conditions in Theorem 4.2 hold, $u_0 \in L^{2p}(\Omega, \mathscr{D}(M^k))$ for $p \geq 1$, $B(t) \in HS(U_0, \mathscr{D}(M^{k+1}))$ for any $t \in [0, T]$, and that F satisfies Assumption 2.4. Then there exists a constant $C = C(s, p, T, F, B) > 0$ such that*

(i) $\displaystyle \max_{1 \leq i \leq s} \mathbb{E}[\|U_i^n\|_{\mathscr{D}(M^k)}^{2p}] \leq C\Big(\tau + \mathbb{E}[\|u^n\|_{\mathscr{D}(M^k)}^{2p}]\Big)$, *for $n = 0, 1, \ldots, N$;*

(ii) $\displaystyle \max_{1 \leq n \leq N} \|u^n\|_{L^{2p}(\Omega, \mathscr{D}(M^k))} \leq C\Big(1 + \|u_0\|_{L^{2p}(\Omega, \mathscr{D}(M^k))}\Big)$;

(iii) $\displaystyle\max_{0\leq n\leq N-1} \mathbb{E}\left[\|u^{n+1} - u^n\|^{2p}_{\mathscr{D}(M^{k-1})}\right] \leq C\tau^p;$

(iv) $\displaystyle\max_{0\leq n\leq N-1} \|\mathbb{E}[u^{n+1} - u^n]\|_{\mathscr{D}(M^{k-1})} \leq C\tau.$

4.1.2 Exponential-Type Methods

As we know, the Itô form of (4.1) reads as follows

$$
\begin{cases}
\mathrm{d}u(t) = \left[Mu(t) + \widetilde{F}(t, u(t))\right]\mathrm{d}t + B(t, u(t))\mathrm{d}W(t), & t \in (0, T], \\
u(0) = u_0,
\end{cases}
$$

$$(4.43)$$

where

$$
\widetilde{F}(t, u(t)) := F(t, u(t)) + \frac{1}{2}B_u(t, u(t))B(t, u(t))F_Q, \quad F_Q(\mathbf{x}) = \sum_{k\in\mathbb{N}}\left(Q^{\frac{1}{2}}e_j(\mathbf{x})\right)^2.
$$

Under conditions in Remark 2.4 of Sect. 2.2, the mild solution of (4.43) exists globally and satisfies

$$
u(t_n) = S(\tau)u(t_{n-1}) + \int_{t_{n-1}}^{t_n} S(t_n - s)\widetilde{F}(s, u(s))\mathrm{d}s
$$

$$
+ \int_{t_{n-1}}^{t_n} S(t - s)B(s, u(s))\mathrm{d}W(s), \quad \mathbb{P}\text{-}a.s.
$$

for $0 \leq t_{n-1} < t_n \leq T$.

We now turn our attention to the exponential-type methods. One is the exponential Euler method for (4.43), which is defined recurrently by

$$
u^n = S(\tau)u^{n-1} + \tau S(\tau)\widetilde{F}(t_{n-1}, u^{n-1})
$$

$$
+ S(\tau)B(t_{n-1}, u^{n-1})\Delta W^n, \quad n = 1, 2, \ldots, N \tag{4.44}
$$

with $u^0 = u_0$. We refer to [53] for more details on the analysis of the exponential Euler method for the stochastic Maxwell equations. Another one is the accelerated exponential Euler method of (4.43),

$$
u^n = S(\tau)u^{n-1} + \int_{t_{n-1}}^{t_n} S(t_n - s)\widetilde{F}(t_{n-1}, u^{n-1})\mathrm{d}s
$$

$$
+ \int_{t_{n-1}}^{t_n} S(t_n - s)B(t_{n-1}, u^{n-1})\mathrm{d}W(s) \tag{4.45}
$$

for $n = 1, 2, \ldots, N$, where $u^0 = u_0$. For the above two exponential-type methods, we have the following *a priori* estimates of the numerical solutions.

Proposition 4.7 *Let $u_0 \in L^{2p}(\Omega, \mathscr{D}(M))$, $p \geq 1$. Assume that the drift term $\widetilde{F}(t, \cdot)$ satisfies Assumption 2.4 with $k = 1$ and the diffusion term $B(t, \cdot)$ satisfies Assumption 2.5 with $k = 1$ for any $t \in [0, T]$. Then solutions of (4.44) and (4.45) satisfy*

$$\max_{0 \leq n \leq N} \mathbb{E}\left[\|u^n\|_{\mathscr{D}(M)}^{2p}\right] \leq C,$$

where $C = C(u_0, T, p)$ is a positive constant.

Proof We only present the proof of the exponential Euler method, and the assertion for the accelerated exponential Euler method follows a similar procedure.

It follows from (4.44) that

$$u^n = S(t_n)u_0 + \tau \sum_{k=0}^{n-1} S(t_n - t_k)\widetilde{F}(t_k, u^k) + \sum_{k=0}^{n-1} S(t_n - t_k)B(t_k, u^k)\Delta W^{k+1}$$

for all $n = 1, 2, \ldots, N$. Taking the $\mathscr{D}(M)$-norm yields that for $p \geq 1$,

$$\max_{1 \leq n \leq N} \mathbb{E}\left[\|u^n\|_{\mathscr{D}(M)}^{2p}\right] \leq C \max_{1 \leq n \leq N} \mathbb{E}\left[\|S(t_n)u_0\|_{\mathscr{D}(M)}^{2p}\right]$$

$$+ C \max_{1 \leq n \leq N} \mathbb{E}\left[\left\|\tau \sum_{k=0}^{n-1} S(t_n - t_k)\widetilde{F}(t_k, u^k)\right\|_{\mathscr{D}(M)}^{2p}\right]$$

$$+ C \max_{1 \leq n \leq N} \mathbb{E}\left[\left\|\sum_{k=0}^{n-1} S(t_n - t_k)B(t_k, u^k)\Delta W^{k+1}\right\|_{\mathscr{D}(M)}^{2p}\right]$$

$$=: I_1 + I_2 + I_3.$$

For the first term I_1, using the unitarity of the operator S, we have

$$I_1 = C\mathbb{E}\left[\|u_0\|_{\mathscr{D}(M)}^{2p}\right].$$

For the second term I_2, the linear growth of $\widetilde{F}$ and the Hölder inequality lead to

$$I_2 \leq C + C\tau \sum_{k=0}^{N-1} \mathbb{E}\left[\|u^k\|_{\mathscr{D}(M)}^{2p}\right].$$

For the third term I_3, Proposition D.4 (ii) and the linear growth of B yield

$$I_3 = C \max_{0 \le t_n \le T} \mathbb{E}\Big[\Big\| \int_0^{t_n} \Big(S\Big(t_n - \lfloor \tfrac{s}{\tau}\rfloor\tau\Big) B(t_{\lfloor \frac{s}{\tau}\rfloor\tau}, u^{\lfloor \frac{s}{\tau}\rfloor\tau})\Big)\mathrm{d}W(s)\Big\|^{2p}_{\mathscr{D}(M)}\Big]$$

$$\le C\, \mathbb{E}\Big[\Big(\int_0^T \|B(t_{\lfloor \frac{s}{\tau}\rfloor\tau}, u^{\lfloor \frac{s}{\tau}\rfloor\tau})\|^2_{HS(U_0,\mathscr{D}(M))}\mathrm{d}s\Big)^p\Big]$$

$$\le C + C\, \mathbb{E}\Big[\Big(\int_0^T \|u^{\lfloor \frac{s}{\tau}\rfloor\tau}\|^2_{\mathscr{D}(M)}\mathrm{d}s\Big)^p\Big]$$

$$= C + C\, \mathbb{E}\Big[\Big(\tau \sum_{k=0}^{N-1} \|u^k\|^2_{\mathscr{D}(M)}\Big)^p\Big]$$

$$\le C + C\tau \sum_{k=0}^{N-1} \mathbb{E}\Big[\|u^k\|^{2p}_{\mathscr{D}(M)}\Big],$$

where $\lfloor \tfrac{s}{\tau}\rfloor$ is the integer part of $\tfrac{s}{\tau}$. Combining the above estimates, we have

$$\max_{1 \le n \le N} \mathbb{E}\Big[\|u^n\|^{2p}_{\mathscr{D}(M)}\Big] \le C + C\, \mathbb{E}\Big[\|u_0\|^{2p}_{\mathscr{D}(M)}\Big] + C\tau \sum_{k=0}^{N-1} \mathbb{E}\Big[\|u^k\|^{2p}_{\mathscr{D}(M)}\Big].$$

Then the Grönwall inequality finishes the proof of Proposition 4.7.   □

Particularly, if $\varepsilon = \mu \equiv 1$, $F \equiv 0$ and $B = (\lambda_1^\top, \lambda_2^\top)^\top$, the exponential Euler method (4.44) and the accelerated exponential Euler method (4.45) become

$$u^n = S(\tau)u^{n-1} + S(\tau)B\Delta W^n, \quad n = 1, 2, \dots, N \tag{4.46}$$

and

$$u^n = S(\tau)u^{n-1} + \int_{t_{n-1}}^{t_n} S(t_n - s)B\mathrm{d}W(s), \quad n = 1, 2, \dots, N, \tag{4.47}$$

respectively.

The following propositions state that these two methods preserve the stochastic symplectic structure, and fulfill both the averaged energy evolution law and the averaged divergence conservation laws. In the sequel, we sketch only the proof for (4.46), and the proof for (4.47) is similar.

Proposition 4.8 *The methods* (4.46) *and* (4.47) *preserve the discrete stochastic symplectic structure*

$$\varpi_{n+1} := \int_D d\mathbf{E}^{n+1}(\mathbf{x}) \wedge d\mathbf{H}^{n+1}(\mathbf{x})d\mathbf{x} = \int_D d\mathbf{E}^n(\mathbf{x}) \wedge d\mathbf{H}^n(\mathbf{x})d\mathbf{x} = \varpi_n, \quad \mathbb{P}\text{-}a.s.$$

for all $n = 0, 1, \ldots, N - 1$.

Proof Taking the exterior derivative of (4.46) leads to

$$du^{n+1} = d\big(S(\tau)u^n\big).$$

Therefore the stochastic symplecticity follows from properties of the deterministic linear Maxwell equations (see Theorem 1.1 in Sect. 1.1.2). □

Proposition 4.9 *The methods* (4.46) *and* (4.47) *fulfill the averaged energy evolution law*

$$\mathbb{E}\big[\|u^n\|_{\mathbb{H}}^2\big] = \mathbb{E}\big[\|u^{n-1}\|_{\mathbb{H}}^2\big] + \varkappa\tau, \quad n = 1, 2, \ldots, N$$

with $\varkappa = (\lambda_1^2 + \lambda_2^2)\text{Tr}(Q)$.

Proof Applying $\mathbb{E}\big[\|\cdot\|_{\mathbb{H}}^2\big]$ to both sides of (4.46) leads to

$$\mathbb{E}\big[\|u^n\|_{\mathbb{H}}^2\big] = \mathbb{E}\Big[\|S(\tau)u^{n-1}\|_{\mathbb{H}}^2\Big] + \mathbb{E}\Big[\|S(\tau)B\Delta W^n\|_{\mathbb{H}}^2\Big]$$

$$+ 2\mathbb{E}\Big[\big\langle S(\tau)u^{n-1}, S(\tau)B\Delta W^n\big\rangle_{\mathbb{H}}\Big]$$

$$= \mathbb{E}\big[\|u^{n-1}\|_{\mathbb{H}}^2\big] + \mathbb{E}\big[\|B\Delta W^n\|_{\mathbb{H}}^2\big].$$

We note that

$$\mathbb{E}\big[\|B\Delta W^n\|_{\mathbb{H}}^2\big] = (\lambda_1^2 + \lambda_2^2)\int_D \mathbb{E}\Big[\big\|\int_{t_{n-1}}^{t_n} dW(s)\big\|_{\mathbb{H}}^2\Big]d\mathbf{x}$$

$$= (\lambda_1^2 + \lambda_2^2)\tau \int_D \Big[\sum_{i\in\mathbb{N}}\big(Q^{\frac{1}{2}}e_i(\mathbf{x})\big)^2\Big]d\mathbf{x}$$

$$= (\lambda_1^2 + \lambda_2^2)\text{Tr}(Q)\tau$$

due to the Itô isometry. □

Proposition 4.10 *The methods* (4.46) *and* (4.47) *fulfill the discrete averaged divergence conservation laws*

$$\mathbb{E}\big[\nabla \cdot \mathbf{E}^n\big] = \mathbb{E}\big[\nabla \cdot \mathbf{E}^{n-1}\big], \quad \mathbb{E}\big[\nabla \cdot \mathbf{H}^n\big] = \mathbb{E}\big[\nabla \cdot \mathbf{H}^{n-1}\big], \quad n = 1, 2, \ldots, N.$$

Proof Denote $(\mathrm{div}, \mathrm{div})(\mathbf{E}^\top, \mathbf{H}^\top)^\top := (\nabla \cdot \mathbf{E}^\top, \nabla \cdot \mathbf{H}^\top)^\top$. Performing the operator $(\mathrm{div}, \mathrm{div})$ and taking the expectation on both sides of (4.46) yield

$$\mathbb{E}\big[(\mathrm{div}, \mathrm{div})u^n\big] = \mathbb{E}\big[(\mathrm{div}, \mathrm{div})S(\tau)u^{n-1}\big].$$

Notice that $S(\tau)u^{n-1}$ is the exact solution of the following deterministic linear Maxwell equations

$$\begin{cases} \frac{\mathrm{d}}{\mathrm{d}t}u(t) = Mu(t), & t \in (t_{n-1}, t_n], \\ u(t_{n-1}) = u^{n-1}. \end{cases}$$

These equations possess the divergence conservation laws, that is,

$$(\mathrm{div}, \mathrm{div})S(\tau)u^{n-1} = (\mathrm{div}, \mathrm{div})u^{n-1}$$

for all $n = 1, 2, \ldots, N+1$. $\qquad\square$

4.2 Fully Discrete Algorithms

Section 4.1 proposes and analyzes the temporal semi-discretizations for the stochastic Maxwell equations. This section discretizes these temporal semi-discretizations further in the spatial direction to obtain several fully discrete algorithms, including the stochastic multi-symplectic algorithms, the stochastic multi-symplectic wavelet algorithm, the stochastic symplectic and multi-symplectic dG algorithms. The geometric structures and physical properties of these algorithms are also analyzed.

For the considered spatial domain $D := [a_1^-, a_1^+] \times [a_2^-, a_2^+] \times [a_3^-, a_3^+]$, we introduce a uniform partition with $\Delta x = (a_1^+ - a_1^-)/I$, $\Delta y = (a_2^+ - a_2^-)/J$, and $\Delta z = (a_3^+ - a_3^-)/K$ being the mesh sizes along x, y, and z-directions, respectively. Here I, J, and K are positive integers. Recall that $\tau = T/N$ is the uniform time step size of $[0, T]$ with N being a positive integer. For $n = 1, \ldots, N$, $i = 1, \ldots, I$, $j = 1, \ldots, J$, and $k = 1, \ldots, K$, denote grid points by $t_n = n\tau$, $x_i = a_1^- + i\Delta x$, $y_j = a_2^- + j\Delta y$, $z_k = a_3^- + k\Delta z$. Let $u_{i,j,k}^n$ be an approximation of $u(t_n, x_i, y_j, z_k)$. The difference operators in t-direction and x-direction are defined by

$$\delta_t u_{i,j,k}^n = \frac{u_{i,j,k}^{n+1} - u_{i,j,k}^n}{\tau}, \qquad \bar{\delta}_t u_{i,j,k}^n = \frac{u_{i,j,k}^{n+1} - u_{i,j,k}^{n-1}}{2\tau},$$

$$\delta_x u_{i,j,k}^n = \frac{u_{i+1,j,k}^n - u_{i,j,k}^n}{\Delta x}, \qquad \bar{\delta}_x u_{i,j,k}^n = \frac{u_{i+1,j,k}^n - u_{i-1,j,k}^n}{2\Delta x}.$$

And the average operators in t-direction and x-direction are defined by

$$A_t u_{i,j,k}^n = \frac{u_{i,j,k}^{n+1} + u_{i,j,k}^n}{2}, \qquad A_x u_{i,j,k}^n = \frac{u_{i+1,j,k}^n + u_{i,j,k}^n}{2}.$$

For the simplicity of notations, we may also use $u_{i,j,k}^{n+\frac{1}{2}}$ and $u_{i+\frac{1}{2},j,k}^n$ to replace $A_t u_{i,j,k}^n$ and $A_x u_{i,j,k}^n$, respectively. Similar to δ_x, $\bar{\delta}_x$, and A_x, we can define operators δ_y, $\bar{\delta}_y$, and A_y in y-direction and operators δ_z, $\bar{\delta}_z$, and A_z in z-direction.

4.2.1 Stochastic Multi-Symplectic Algorithms

As shown in Sect. 3.2, the stochastic Maxwell equations, including the multiplicative noise case (3.10) and the additive noise case (3.21), can be rewritten in the formulation of the stochastic Hamiltonian partial differential equation, i.e.,

$$\mathbb{F}du + \mathbb{K}_1\partial_x u \, dt + \mathbb{K}_2\partial_y u \, dt + \mathbb{K}_3\partial_z u \, dt = \nabla_u S(u) \circ dW(t), \qquad (4.48)$$

where $S(u) = \frac{\lambda_2}{2}|\mathbf{E}|^2 + \frac{\lambda_1}{2}|\mathbf{H}|^2$ for (3.10) and $S(u) = \lambda_2 \cdot \mathbf{E} - \lambda_1 \cdot \mathbf{H}$ for (3.21). Matrices $\mathbb{F}$ and $\mathbb{K}_p$ ($p = 1, 2, 3$) are defined in (3.12). And in the previous chapter, we have shown that (4.48) possesses the stochastic multi-symplectic conservation law

$$d_t \varpi + \partial_x \kappa_1 dt + \partial_y \kappa_2 dt + \partial_z \kappa_3 dt = 0, \qquad \mathbb{P}\text{-}a.s.$$

with ϖ and κ_p ($p = 1, 2, 3$) being the differential 2-forms associated with the skew-symmetric matrices $\mathbb{F}$ and $\mathbb{K}_p$ ($p = 1, 2, 3$), respectively.

We first give the definition of the stochastic multi-symplectic algorithm.

Definition 4.2 A stochastic algorithm

$$\mathbb{F}\partial_t^{n,i,j,k} u_{i,j,k}^n + \mathbb{K}_1\partial_x^{n,i,j,k} u_{i,j,k}^n$$
$$+ \mathbb{K}_2\partial_y^{n,i,j,k} u_{i,j,k}^n + \mathbb{K}_3\partial_z^{n,i,j,k} u_{i,j,k}^n = (\nabla_u S(u))_{i,j,k}^n \gamma_{i,j,k}^n$$

of (4.48) is said to be a stochastic multi-symplectic algorithm if the discrete stochastic multi-symplectic conservation law

$$\partial_t^{n,i,j,k} \varpi_{i,j,k}^n + \partial_x^{n,i,j,k} (\kappa_1)_{i,j,k}^n + \partial_y^{n,i,j,k} (\kappa_2)_{i,j,k}^n + \partial_z^{n,i,j,k} (\kappa_3)_{i,j,k}^n = 0, \qquad \mathbb{P}\text{-}a.s.$$

is fulfilled. Here, $\partial_t^{n,i,j,k}$, $\partial_x^{n,i,j,k}$, $\partial_y^{n,i,j,k}$, and $\partial_z^{n,i,j,k}$ are certain discretizations of the derivatives ∂_t, ∂_x, ∂_y, and ∂_z, respectively. And $\gamma_{i,j,k}^n$ is an approximation of the noise.

We are going to propose several structure-preserving algorithms which are stochastic multi-symplectic.

MS Method-I *The first algorithm is constructed by applying the midpoint method in both spatial and temporal directions of* (4.48). *Namely, for* $n = 0, 1, \ldots, N, i = 0, 1, \ldots, I, j = 0, 1, \ldots, J$, *and* $k = 0, 1, \ldots, K$,

$$
\mathbb{F}\delta_t u_{i+\frac{1}{2},j+\frac{1}{2},k+\frac{1}{2}}^n + \mathbb{K}_1 \delta_x A_t u_{i,j+\frac{1}{2},k+\frac{1}{2}}^n + \mathbb{K}_2 \delta_y A_t u_{i+\frac{1}{2},j,k+\frac{1}{2}}^n
$$
$$
+ \mathbb{K}_3 \delta_z A_t u_{i+\frac{1}{2},j+\frac{1}{2},k}^n = \nabla_u S(A_t u_{i+\frac{1}{2},j+\frac{1}{2},k+\frac{1}{2}}^n)\gamma_{i,j,k}^n, \quad \mathbb{P}\text{-}a.s.,
$$
(4.49)

where

$$
\gamma_{i,j,k}^n := \frac{1}{\tau}(\Delta W)_{i,j,k}^{n+1} \quad and \quad (\Delta W)_{i,j,k}^{n+1} := W(t_{n+1}, x_i, y_j, z_k) - W(t_n, x_i, y_j, z_k).
$$

MS Method-II *The second algorithm is constructed based on the central finite difference method in both spatial and temporal directions of* (4.48). *Namely, for* $n = 1, 2, \ldots, N, i = 0, 1, \ldots, I, j = 0, 1, \ldots, J$, *and* $k = 0, 1, \ldots, K$,

$$
\mathbb{F}\bar{\delta}_t u_{i,j,k}^n + \left(\mathbb{K}_1 \bar{\delta}_x + \mathbb{K}_2 \bar{\delta}_y + \mathbb{K}_3 \bar{\delta}_z\right) u_{i,j,k}^n = \nabla_u S(u_{i,j,k}^n)\widetilde{\gamma}_{i,j,k}^{n+1}, \quad \mathbb{P}\text{-}a.s.,
$$
(4.50)

where

$$
\widetilde{\gamma}_{i,j,k}^{n+1} := \frac{1}{2\tau}\left(W(t_{n+1}, x_i, y_j, z_k) - W(t_{n-1}, x_i, y_j, z_k)\right).
$$

MS Method-III *The third algorithm is constructed by applying the central finite difference method in spatial direction and the midpoint method in temporal direction of* (4.48). *Namely, for* $n = 0, 1, \ldots, N, i = 0, 1, \ldots, I, j = 0, 1, \ldots, J$, *and* $k = 0, 1, \ldots, K$,

$$
\mathbb{F}\delta_t u_{i,j,k}^n + \left(\mathbb{K}_1 \bar{\delta}_x A_t + \mathbb{K}_2 \bar{\delta}_y + \mathbb{K}_3 \bar{\delta}_z A_t\right) u_{i,j,k}^n = \nabla_u S(A_t u_{i,j,k}^n)\gamma_{i,j,k}^n, \quad \mathbb{P}\text{-}a.s.
$$
(4.51)

4.2.1.1   Stochastic Multi-Symplectic Conservation Law

This part is devoted to studying the stochastic multi-symplectic conservation laws for the above three structure-preserving algorithms.

Proposition 4.11 *The MS Method-I (4.49) possesses the discrete stochastic multi-symplectic conservation law*

$$\delta_t A_x A_y A_z \varpi^n_{i,j,k} + \delta_x A_t A_y A_z (\kappa_1)^n_{i,j,k}$$

$$+ \delta_y A_t A_z A_x (\kappa_2)^n_{i,j,k} + \delta_z A_t A_x A_y (\kappa_3)^n_{i,j,k} = 0, \quad \mathbb{P}\text{-}a.s.,$$

where $\varpi^n_{i,j,k} = \frac{1}{2} du^n_{i,j,k} \wedge \mathbb{F} du^n_{i,j,k}$ *and* $(\kappa_p)^n_{i,j,k} = \frac{1}{2} du^n_{i,j,k} \wedge \mathbb{K}_p du^n_{i,j,k}$, $p = 1, 2, 3$.

Proof Taking the exterior derivative on both sides of (4.49) yields

$$\mathbb{F} \delta_t du^n_{i+\frac{1}{2},j+\frac{1}{2},k+\frac{1}{2}} + \mathbb{K}_1 \delta_x A_t du^n_{i,j+\frac{1}{2},k+\frac{1}{2}} + \mathbb{K}_2 \delta_y A_t du^n_{i+\frac{1}{2},j,k+\frac{1}{2}}$$

$$+ \mathbb{K}_3 \delta_z A_t du^n_{i+\frac{1}{2},j+\frac{1}{2},k} = \nabla^2 S(A_t u^n_{i+\frac{1}{2},j+\frac{1}{2},k+\frac{1}{2}}) A_t du^n_{i+\frac{1}{2},j+\frac{1}{2},k+\frac{1}{2}} \gamma^n_{i,j,k}.$$

$$(4.52)$$

Next, we take the wedge product between $A_t du^n_{i+\frac{1}{2},j+\frac{1}{2},k+\frac{1}{2}}$ and (4.52). The term $A_t du^n_{i+\frac{1}{2},j+\frac{1}{2},k+\frac{1}{2}} \wedge \nabla^2 S(A_t u^n_{i+\frac{1}{2},j+\frac{1}{2},k+\frac{1}{2}}) A_t du^n_{i+\frac{1}{2},j+\frac{1}{2},k+\frac{1}{2}} \gamma^n_{i,j,k}$ vanishes due to the symmetry of $\nabla^2 S(\cdot)$. For the first term on the left-hand side of (4.52), we have

$$A_t du^n_{i+\frac{1}{2},j+\frac{1}{2},k+\frac{1}{2}} \wedge \mathbb{F} \delta_t du^n_{i+\frac{1}{2},j+\frac{1}{2},k+\frac{1}{2}}$$

$$= \frac{1}{2\tau} du^{n+1}_{i+\frac{1}{2},j+\frac{1}{2},k+\frac{1}{2}} \wedge \mathbb{F} du^{n+1}_{i+\frac{1}{2},j+\frac{1}{2},k+\frac{1}{2}} - \frac{1}{2\tau} du^n_{i+\frac{1}{2},j+\frac{1}{2},k+\frac{1}{2}} \wedge \mathbb{F} du^n_{i+\frac{1}{2},j+\frac{1}{2},k+\frac{1}{2}}$$

$$= \delta_t \varpi^n_{i+\frac{1}{2},j+\frac{1}{2},k+\frac{1}{2}}.$$

Similarly, for the rest terms on the left-hand side of (4.52), we have

$$A_t du^n_{i+\frac{1}{2},j+\frac{1}{2},k+\frac{1}{2}} \wedge \mathbb{K}_1 \delta_x A_t du^n_{i,j+\frac{1}{2},k+\frac{1}{2}} = \delta_x A_t (\kappa_1)^n_{i,j+\frac{1}{2},k+\frac{1}{2}},$$

$$A_t du^n_{i+\frac{1}{2},j+\frac{1}{2},k+\frac{1}{2}} \wedge \mathbb{K}_2 \delta_y A_t du^n_{i+\frac{1}{2},j,k+\frac{1}{2}} = \delta_y A_t (\kappa_2)^n_{i+\frac{1}{2},j,k+\frac{1}{2}},$$

$$A_t du^n_{i+\frac{1}{2},j+\frac{1}{2},k+\frac{1}{2}} \wedge \mathbb{K}_3 \delta_z A_t du^n_{i+\frac{1}{2},j+\frac{1}{2},k} = \delta_z A_t (\kappa_3)^n_{i+\frac{1}{2},j+\frac{1}{2},k}.$$

Combining the above results completes the proof. □

Proposition 4.12 *The MS Method-II* (4.50) *possesses the discrete stochastic multi-symplectic conservation law*

$$
\frac{\widehat{\varpi}_{i,j,k}^{n+\frac{1}{2}} - \widehat{\varpi}_{i,j,k}^{n-\frac{1}{2}}}{\tau} + \frac{(\widehat{\kappa}_1)_{i+\frac{1}{2},j,k}^{n} - (\widehat{\kappa}_1)_{i-\frac{1}{2},j,k}^{n}}{\Delta x} + \frac{(\widehat{\kappa}_2)_{i,j+\frac{1}{2},k}^{n} - (\widehat{\kappa}_2)_{i,j-\frac{1}{2},k}^{n}}{\Delta y}
$$

$$
+ \frac{(\widehat{\kappa}_3)_{i,j,k+\frac{1}{2}}^{n} - (\widehat{\kappa}_3)_{i,j,k-\frac{1}{2}}^{n}}{\Delta z} = 0, \quad \mathbb{P}\text{-}a.s.,
$$

where

$$
\widehat{\varpi}_{i,j,k}^{n+\frac{1}{2}} = \frac{1}{2} du_{i,j,k}^{n} \wedge \mathbb{F} du_{i,j,k}^{n+1}, \qquad (\widehat{\kappa}_1)_{i+\frac{1}{2},j,k}^{n} = \frac{1}{2} du_{i,j,k}^{n} \wedge \mathbb{K}_1 du_{i+1,j,k}^{n},
$$

$$
(\widehat{\kappa}_2)_{i,j+\frac{1}{2},k}^{n} = \frac{1}{2} du_{i,j,k}^{n} \wedge \mathbb{K}_2 du_{i,j+1,k}^{n}, \qquad (\widehat{\kappa}_3)_{i,j,k+\frac{1}{2}}^{n} = \frac{1}{2} du_{i,j,k}^{n} \wedge \mathbb{K}_3 du_{i,j,k+1}^{n}.
$$

Proof Similar to the proof of Proposition 4.11, we take the exterior derivative on both sides of (4.50) to obtain

$$
\Delta x \Delta y \Delta z \big(\mathbb{F} du_{i,j,k}^{n+1} - \mathbb{F} du_{i,j,k}^{n-1} \big) + \tau \Delta y \Delta z \big(\mathbb{K}_1 du_{i+1,j,k}^{n} - \mathbb{K}_1 du_{i-1,j,k}^{n} \big)
$$

$$
+ \tau \Delta x \Delta z \big(\mathbb{K}_2 du_{i,j+1,k}^{n} - \mathbb{K}_2 du_{i,j-1,k}^{n} \big) + \tau \Delta x \Delta y \big(\mathbb{K}_3 du_{i,j,k+1}^{n+1} - \mathbb{K}_3 du_{i,j,k-1}^{n+1} \big)
$$

$$
= \Delta x \Delta y \Delta z \nabla^2 S(u_{i,j,k}^{n}) du_{i,j,k}^{n} (W_{i,j,k}^{n+1} - W_{i,j,k}^{n-1}).
$$

Then taking the wedge product between $du_{i,j,k}^{n}$ and the above equation, one obtains the conclusion via the symmetry of $\nabla^2 S(\cdot)$. □

Proposition 4.13 *The MS Method-III* (4.51) *possesses the discrete stochastic multi-symplectic conservation law*

$$
\frac{\widetilde{\varpi}_{i,j,k}^{n+1} - \widetilde{\varpi}_{i,j,k}^{n}}{\tau} + \frac{(\widetilde{\kappa}_1)_{i+\frac{1}{2},j,k}^{n+\frac{1}{2}} - (\widetilde{\kappa}_1)_{i-\frac{1}{2},j,k}^{n+\frac{1}{2}}}{\Delta x} + \frac{(\widetilde{\kappa}_2)_{i,j+\frac{1}{2},k}^{n+\frac{1}{2}} - (\widetilde{\kappa}_2)_{i,j-\frac{1}{2},k}^{n+\frac{1}{2}}}{\Delta y}
$$

$$
+ \frac{(\widetilde{\kappa}_3)_{i,j,k+\frac{1}{2}}^{n+\frac{1}{2}} - (\widetilde{\kappa}_3)_{i,j,k-\frac{1}{2}}^{n+\frac{1}{2}}}{\Delta z} = 0, \quad \mathbb{P}\text{-}a.s.,
$$

where

$$
\widetilde{\varpi}_{i,j,k}^{n+1} = \frac{1}{2} du_{i,j,k}^{n+1} \wedge \mathbb{F} du_{i,j,k}^{n+1}, \qquad (\widetilde{\kappa}_1)_{i+\frac{1}{2},j,k}^{n+\frac{1}{2}} = \frac{1}{2} du_{i,j,k}^{n+\frac{1}{2}} \wedge \mathbb{K}_1 du_{i+1,j,k}^{n+\frac{1}{2}},
$$

$$
(\widetilde{\kappa}_2)_{i,j+\frac{1}{2},k}^{n+\frac{1}{2}} = \frac{1}{2} du_{i,j,k}^{n+\frac{1}{2}} \wedge \mathbb{K}_2 du_{i,j+1,k}^{n+\frac{1}{2}}, \qquad (\widetilde{\kappa}_3)_{i,j,k+\frac{1}{2}}^{n+\frac{1}{2}} = \frac{1}{2} du_{i,j,k}^{n+\frac{1}{2}} \wedge \mathbb{K}_3 du_{i,j,k+1}^{n+\frac{1}{2}}.
$$

Proof The proof is similar to that of Proposition 4.12 and thus is omitted here. □

4.2.1.2 Energy Evolution Law

In Sect. 3.3, we have shown that for the stochastic Maxwell equations, the energy is conserved for the multiplicative noise case (i.e., $S(u) = \frac{\lambda_2}{2}|\mathbf{E}|^2 + \frac{\lambda_1}{2}|\mathbf{H}|^2$ in (4.48) with $\lambda_1 = \lambda_2$), whereas the averaged energy evolutes linearly with the rate $\tilde{\varkappa} = (|\lambda_1|^2 + |\lambda_2|^2)\mathrm{Tr}(Q)$ for the additive noise case (i.e., $S(u) = \lambda_2 \cdot \mathbf{E} - \lambda_1 \cdot \mathbf{H}$ in (4.48); see Remark 3.2). In this part, we investigate the evolution of the discrete energies for the three stochastic multi-symplectic algorithms (4.49)–(4.51).

Proposition 4.14 *Denote by* $u_{i,j,k}^n = ((\mathbf{E}_{i,j,k}^n)^\top, (\mathbf{H}_{i,j,k}^n)^\top)^\top$ *the solution of the MS Method-I* (4.49) *with* $S(u) = \frac{\lambda_2}{2}|\mathbf{E}|^2 + \frac{\lambda_1}{2}|\mathbf{H}|^2$. *Assume that* $\lambda_1 = \lambda_2 = \lambda$. *Then under the periodic boundary condition, the discrete energy satisfies*

$$\Phi^{[\mathrm{I}]}(t_{n+1}) = \Phi^{[\mathrm{I}]}(t_n), \quad \mathbb{P}\text{-}a.s.$$

for all $n = 0, 1, \ldots, N - 1$. *Here, the discrete energy* $\Phi^{[\mathrm{I}]}$ *is defined as*

$$\Phi^{[\mathrm{I}]}(t_n) := \Delta x \Delta y \Delta z \sum_{i=0}^{I-1} \sum_{j=0}^{J-1} \sum_{k=0}^{K-1} \left| u_{i+\frac{1}{2}, j+\frac{1}{2}, k+\frac{1}{2}}^n \right|^2.$$

Proof Recall $S(u) = \frac{\lambda}{2}|\mathbf{E}|^2 + \frac{\lambda}{2}|\mathbf{H}|^2$ in (4.49), and the MS Method-I reads as

$$\mathbb{F}\delta_t u_{i+\frac{1}{2}, j+\frac{1}{2}, k+\frac{1}{2}}^n + \mathbb{K}_1 \delta_x A_t u_{i, j+\frac{1}{2}, k+\frac{1}{2}}^n + \mathbb{K}_2 \delta_y A_t u_{i+\frac{1}{2}, j, k+\frac{1}{2}}^n$$
$$+ \mathbb{K}_3 \delta_z A_t u_{i+\frac{1}{2}, j+\frac{1}{2}, k}^n = \lambda A_t u_{i+\frac{1}{2}, j+\frac{1}{2}, k+\frac{1}{2}}^n \gamma_{i,j,k}^n, \quad \mathbb{P}\text{-}a.s. \tag{4.53}$$

Next, we multiply (4.53) by $2\tau \Delta x \Delta y \Delta z \mathbb{F} A_t u_{i+\frac{1}{2}, j+\frac{1}{2}, k+\frac{1}{2}}^n$ and then sum up over all spatial indices i, j, k. For the first term on the left-hand side of (4.53), we have

$$2\tau \Delta x \Delta y \Delta z \sum_{i=0}^{I-1} \sum_{j=0}^{J-1} \sum_{k=0}^{K-1} \left(\mathbb{F}\delta_t u_{i+\frac{1}{2}, j+\frac{1}{2}, k+\frac{1}{2}}^n \right) \cdot \left(\mathbb{F} A_t u_{i+\frac{1}{2}, j+\frac{1}{2}, k+\frac{1}{2}}^n \right)$$
$$= \Phi^{[\mathrm{I}]}(t_{n+1}) - \Phi^{[\mathrm{I}]}(t_n).$$

Using the periodic boundary condition, the other corresponding terms on the left-hand side of (4.53) vanish. For the term on the right-hand side of (4.53), the skew-symmetry of $\mathbb{F}$ implies

$$\lambda \left(A_t u_{i+\frac{1}{2}, j+\frac{1}{2}, k+\frac{1}{2}}^n \gamma_{i,j,k}^n \right) \cdot \left(\mathbb{F} A_t u_{i+\frac{1}{2}, j+\frac{1}{2}, k+\frac{1}{2}}^n \right) = 0.$$

Hence, combining the above results gives the conclusion. $\qquad\square$

Similarly, we can prove the following result for the additive noise case.

Proposition 4.15 *Denote by $u_{i,j,k}^n = ((E_{i,j,k}^n)^\top, (H_{i,j,k}^n)^\top)^\top$ the solution of the MS Method-I (4.49) with $S(u) = \lambda_2 \cdot E - \lambda_1 \cdot H$. Then under the periodic boundary condition, the discrete energy satisfies*

$$\Phi^{[I]}(t_{n+1}) = \Phi^{[I]}(t_n)$$

$$+ 2\Delta x \Delta y \Delta z \sum_{i=0}^{I-1} \sum_{j=0}^{J-1} \sum_{k=0}^{K-1} \left(\Upsilon_{i+\frac{1}{2},j+\frac{1}{2},k+\frac{1}{2}}^{n+\frac{1}{2}} (\Delta W)_{i,j,k}^{n+1} \right), \quad \mathbb{P}\text{-}a.s.$$

for $n = 0, 1, \ldots, N - 1$, where

$$\Upsilon_{i+\frac{1}{2},j+\frac{1}{2},k+\frac{1}{2}}^{n+\frac{1}{2}} := \lambda_1 \cdot E_{i+\frac{1}{2},j+\frac{1}{2},k+\frac{1}{2}}^{n+\frac{1}{2}} + \lambda_2 \cdot H_{i+\frac{1}{2},j+\frac{1}{2},k+\frac{1}{2}}^{n+\frac{1}{2}}.$$

Proof When $S(u) = \lambda_2 \cdot E - \lambda_1 \cdot H$, (4.49) reads as

$$\mathbb{F}\delta_t u_{i+\frac{1}{2},j+\frac{1}{2},k+\frac{1}{2}}^n + \mathbb{K}_1 \delta_x A_t u_{i,j+\frac{1}{2},k+\frac{1}{2}}^n + \mathbb{K}_2 \delta_y A_t u_{i+\frac{1}{2},j,k+\frac{1}{2}}^n$$

$$+ \mathbb{K}_3 \delta_z A_t u_{i+\frac{1}{2},j+\frac{1}{2},k}^n = \lambda \gamma_{i,j,k}^n, \quad \mathbb{P}\text{-}a.s.,$$

(4.54)

where $\lambda := (\lambda_2^\top, -\lambda_1^\top)^\top \in \mathbb{R}^6$. The rest of the proof is similar to that of Proposition 4.14, and the only difference lies in the treatment of the term on the right-hand side of (4.54). Note that

$$2\tau \Delta x \Delta y \Delta z \sum_{i=0}^{I-1} \sum_{j=0}^{J-1} \sum_{k=0}^{K-1} \left(\lambda \cdot \mathbb{F} A_t u_{i+\frac{1}{2},j+\frac{1}{2},k+\frac{1}{2}}^n \right) \gamma_{i,j,k}^n$$

$$= 2\Delta x \Delta y \Delta z \sum_{i=0}^{I-1} \sum_{j=0}^{J-1} \sum_{k=0}^{K-1}$$

$$\times \left[\lambda_1 \cdot A_t E_{i+\frac{1}{2},j+\frac{1}{2},k+\frac{1}{2}}^n + \lambda_2 \cdot A_t H_{i+\frac{1}{2},j+\frac{1}{2},k+\frac{1}{2}}^n \right] (\Delta W)_{i,j,k}^{n+1}.$$

Thus, the proof is finished.   □

Specially, we can obtain the estimate of the discrete averaged energy in the case that W depends only on the time variable.

Corollary 4.1 *Let conditions in Proposition 4.15 hold. If* $W : [0, T] \times \Omega \to \mathbb{R}$ *is a Brownian motion, then there exists a constant* $\widetilde{\varkappa} = (|\lambda_1|^2 + |\lambda_2|^2)|D|$ *such that*

$$\mathbb{E}\big[\Phi^{[I]}(t_{n+1})\big] = \mathbb{E}\big[\Phi^{[I]}(t_n)\big] + \widetilde{\varkappa}\tau, \quad n = 0, 1, \dots, N - 1.$$

Here and hereafter $|D|$ *denotes the volume of the spatial domain D.*

Proof The proof is similar to that of Proposition 4.15. The main difference concerns the treatment of the random term. We observe that

$$\Upsilon^{n+\frac{1}{2}}_{i+\frac{1}{2},j+\frac{1}{2},k+\frac{1}{2}}\Delta W^{n+1} \tag{4.55}$$

$$= \left[\Big(\lambda_1 \cdot \mathbf{E}^n_{i+\frac{1}{2},j+\frac{1}{2},k+\frac{1}{2}} + \frac{1}{2}\lambda_1 \cdot \big(\mathbf{E}^{n+1}_{i+\frac{1}{2},j+\frac{1}{2},k+\frac{1}{2}} - \mathbf{E}^n_{i+\frac{1}{2},j+\frac{1}{2},k+\frac{1}{2}}\big)\Big)\Delta W^{n+1}\right]$$

$$+ \left[\Big(\lambda_2 \cdot \mathbf{H}^n_{i+\frac{1}{2},j+\frac{1}{2},k+\frac{1}{2}} + \frac{1}{2}\lambda_2 \cdot \big(\mathbf{H}^{n+1}_{i+\frac{1}{2},j+\frac{1}{2},k+\frac{1}{2}} - \mathbf{H}^n_{i+\frac{1}{2},j+\frac{1}{2},k+\frac{1}{2}}\big)\Big)\Delta W^{n+1}\right].$$

Since the increments of the Wiener process W are independent, we have

$$\mathbb{E}\Big[\lambda_1 \cdot \mathbf{E}^n_{i+\frac{1}{2},j+\frac{1}{2},k+\frac{1}{2}}\Delta W^{n+1}\Big] = \mathbb{E}\Big[\lambda_2 \cdot \mathbf{H}^n_{i+\frac{1}{2},j+\frac{1}{2},k+\frac{1}{2}}\Delta W^{n+1}\Big] = 0.$$

Thus, taking the expectation on both sides of (4.55), we arrive at

$$\mathbb{E}\Big[\Upsilon^{n+\frac{1}{2}}_{i+\frac{1}{2},j+\frac{1}{2},k+\frac{1}{2}}\Delta W^{n+1}\Big] = \frac{1}{2}\mathbb{E}\Big[\lambda_1 \cdot \big(\mathbf{E}^{n+1}_{i+\frac{1}{2},j+\frac{1}{2},k+\frac{1}{2}} - \mathbf{E}^n_{i+\frac{1}{2},j+\frac{1}{2},k+\frac{1}{2}}\big)\Delta W^{n+1}\Big]$$

$$+ \frac{1}{2}\mathbb{E}\Big[\lambda_2 \cdot \big(\mathbf{H}^{n+1}_{i+\frac{1}{2},j+\frac{1}{2},k+\frac{1}{2}} - \mathbf{H}^n_{i+\frac{1}{2},j+\frac{1}{2},k+\frac{1}{2}}\big)\Delta W^{n+1}\Big]$$

$$= \frac{\tau}{2}\mathbb{E}\Big[\Delta W^{n+1}\lambda_1 \cdot \delta_t \mathbf{E}^n_{i+\frac{1}{2},j+\frac{1}{2},k+\frac{1}{2}}\Big] + \frac{\tau}{2}\mathbb{E}\Big[\Delta W^{n+1}\lambda_2 \cdot \delta_t \mathbf{H}^n_{i+\frac{1}{2},j+\frac{1}{2},k+\frac{1}{2}}\Big]$$

$$= \frac{\tau}{2}\sum_{p=1}^{3}\mathbb{E}\Big[\Delta W^{n+1}\big(\lambda_2^{(p)}\delta_t (H_p)^n_{i+\frac{1}{2},j+\frac{1}{2},k+\frac{1}{2}} + \lambda_1^{(p)}\delta_t (E_p)^n_{i+\frac{1}{2},j+\frac{1}{2},k+\frac{1}{2}}\big)\Big],$$

$$\tag{4.56}$$

where $\lambda_1 = (\lambda_1^{(1)}, \lambda_1^{(2)}, \lambda_1^{(3)})^\top$ and $\lambda_2 = (\lambda_2^{(1)}, \lambda_2^{(2)}, \lambda_2^{(3)})^\top$. To estimate (4.56), let us consider the componentwise form of (4.54),

$$\delta_t(E_1)^n_{i+\frac{1}{2},j+\frac{1}{2},k+\frac{1}{2}} = \delta_y A_t(H_3)^n_{i+\frac{1}{2},j,k+\frac{1}{2}} - \delta_z A_t(H_2)^n_{i+\frac{1}{2},j+\frac{1}{2},k} + \lambda_1^{(1)} \gamma^n_{i,j,k},$$

$$\delta_t(E_2)^n_{i+\frac{1}{2},j+\frac{1}{2},k+\frac{1}{2}} = \delta_z A_t(H_1)^n_{i+\frac{1}{2},j+\frac{1}{2},k} - \delta_x A_t(H_3)^n_{i,j+\frac{1}{2},k+\frac{1}{2}} + \lambda_1^{(2)} \gamma^n_{i,j,k},$$

$$\delta_t(E_3)^n_{i+\frac{1}{2},j+\frac{1}{2},k+\frac{1}{2}} = \delta_x A_t(H_2)^n_{i,j+\frac{1}{2},k+\frac{1}{2}} - \delta_y A_t(H_1)^n_{i+\frac{1}{2},j,k+\frac{1}{2}} + \lambda_1^{(3)} \gamma^n_{i,j,k},$$

$$\delta_t(H_1)^n_{i+\frac{1}{2},j+\frac{1}{2},k+\frac{1}{2}} = \delta_z A_t(E_2)^n_{i+\frac{1}{2},j+\frac{1}{2},k} - \delta_y A_t(E_3)^n_{i+\frac{1}{2},j,k+\frac{1}{2}} + \lambda_2^{(1)} \gamma^n_{i,j,k},$$

$$\delta_t(H_2)^n_{i+\frac{1}{2},j+\frac{1}{2},k+\frac{1}{2}} = \delta_x A_t(E_3)^n_{i,j+\frac{1}{2},k+\frac{1}{2}} - \delta_z A_t(E_1)^n_{i+\frac{1}{2},j+\frac{1}{2},k} + \lambda_2^{(2)} \gamma^n_{i,j,k},$$

$$\delta_t(H_3)^n_{i+\frac{1}{2},j+\frac{1}{2},k+\frac{1}{2}} = \delta_y A_t(E_1)^n_{i+\frac{1}{2},j,k+\frac{1}{2}} - \delta_x A_t(E_2)^n_{i,j+\frac{1}{2},k+\frac{1}{2}} + \lambda_2^{(3)} \gamma^n_{i,j,k}.$$
$$(4.57)$$

When W is a Brownian motion, $\gamma^n_{i,j,k}$ in (4.57) is defined by $\gamma^n_{i,j,k} = W(t_{n+1}) - W(t_n)$. Plugging (4.57) into the term on the right-hand side of (4.56) and using the periodic boundary condition, we obtain

$$2\Delta x \Delta y \Delta z \sum_{i=0}^{I-1} \sum_{j=0}^{J-1} \sum_{k=0}^{K-1} \mathbb{E}\left[\gamma^{n+\frac{1}{2}}_{i+\frac{1}{2},j+\frac{1}{2},k+\frac{1}{2}} \Delta W^{n+1} \right] = \tilde{\varkappa}\tau.$$

Therefore,

$$\mathbb{E}\left[\Phi^{[I]}(t_{n+1})\right] = \mathbb{E}\left[\Phi^{[I]}(t_n)\right] + \tilde{\varkappa}\tau, \quad n = 0, 1, \ldots, N-1,$$

which finishes the proof. $\qquad\square$

For the MS Method-II, we have the following two discrete energy evolution properties. Proofs are similar to those of Propositions 4.14 and 4.15.

Proposition 4.16 *Denote by* $u^n_{i,j,k} = ((E^n_{i,j,k})^\top, (H^n_{i,j,k})^\top)^\top$ *the solution of the MS Method-II* (4.50) *with* $S(u) = \frac{\lambda_2}{2}|E|^2 + \frac{\lambda_1}{2}|H|^2$. *Assume that* $\lambda_1 = \lambda_2 = \lambda$. *Then under the periodic boundary condition, the discrete energy satisfies*

$$\Phi^{[II]}(t_{n+1}) = \Phi^{[II]}(t_n), \quad \mathbb{P}\text{-}a.s.$$

for all $n = 1, 2, \ldots, N-1$. *Here, the discrete energy* $\Phi^{[II]}$ *is defined as*

$$\Phi^{[II]}(t_{n+1}) := \Delta x \Delta y \Delta z \sum_{i=1}^{I} \sum_{j=1}^{J} \sum_{k=1}^{K} u^{n+1}_{i,j,k} \cdot u^n_{i,j,k}.$$

Proposition 4.17 *Denote by* $u_{i,j,k}^n = ((\mathbf{E}_{i,j,k}^n)^\top, (\mathbf{H}_{i,j,k}^n)^\top)^\top$ *the solution of the MS Method-II (4.50) with* $S(u) = \lambda_2 \cdot \mathbf{E} - \lambda_1 \cdot \mathbf{H}$. *Then under the periodic boundary condition, the discrete energy satisfies*

$$\Phi^{[\mathrm{II}]}(t_{n+1}) = \Phi^{[\mathrm{II}]}(t_n) + \Delta x \Delta y \Delta z \sum_{i=1}^I \sum_{j=1}^J \sum_{k=1}^K \Upsilon_{i,j,k}^n (W_{i,j,k}^{n+1} - W_{i,j,k}^{n-1}), \quad \mathbb{P}\text{-a.s.}$$

for $n = 1, 2, \ldots, N - 1$, *where*

$$\Upsilon_{i,j,k}^n := \lambda_1 \cdot \mathbf{E}_{i,j,k}^n + \lambda_2 \cdot \mathbf{H}_{i,j,k}^n.$$

Moreover, we have the following discrete averaged energy linear growth of the MS Method-II.

Corollary 4.2 *Under conditions in Proposition 4.17, there exists a constant* $\widehat{\varkappa} = (|\lambda_1|^2 + |\lambda_2|^2) \bar{V}^Q(D)$ *such that*

$$\mathbb{E}\Big[\Phi^{[\mathrm{II}]}(t_{n+1})\Big] = \mathbb{E}\Big[\Phi^{[\mathrm{II}]}(t_n)\Big] + \widehat{\varkappa}\tau, \quad n = 0, 1, \ldots, N - 1,$$

where $\bar{V}^Q(D) := \Delta x \Delta y \Delta z \sum_{i,j,k} \sum_{m \in \mathbb{N}} \left(Q^{\frac{1}{2}} e_m(x_i, y_j, z_k)\right)^2$.

Proof Similar to Corollary 4.1, using the independence of the increments of Wiener process, we have the following result

$$\mathbb{E}\Big[\Upsilon_{i,j,k}^n (W_{i,j,k}^{n+1} - W_{i,j,k}^{n-1})\Big]$$

$$= \mathbb{E}\Big[\lambda_1 \cdot \Big(\mathbf{E}_{i,j,k}^n - \mathbf{E}_{i,j,k}^{n-2}\Big)(\Delta W)_{i,j,k}^n + \lambda_2 \cdot \Big(\mathbf{H}_{i,j,k}^n - \mathbf{H}_{i,j,k}^{n-2}\Big)(\Delta W)_{i,j,k}^n\Big]$$

$$= \tau \mathbb{E}\Big[(\Delta W)_{i,j,k}^n \lambda_1 \cdot \bar{\delta}_t \mathbf{E}_{i,j,k}^{n-1}\Big] + \tau \mathbb{E}\Big[(\Delta W)_{i,j,k}^n \lambda_2 \cdot \bar{\delta}_t \mathbf{H}_{i,j,k}^{n-1}\Big]$$

$$= \tau \sum_{p=1}^3 \mathbb{E}\Big[(\Delta W)_{i,j,k}^n \Big(\lambda_1^{(p)} \bar{\delta}_t (E_p)_{i,j,k}^{n-1} + \lambda_2^{(p)} \bar{\delta}_t (H_p)_{i,j,k}^{n-1}\Big)\Big].$$

$$(4.58)$$

In the same way, plugging the following componentwise form of (4.50)

$$\bar{\delta}_t(E_1)_{i,j,k}^n = \bar{\delta}_y(H_3)_{i,j,k}^n - \bar{\delta}_z(H_2)_{i,j,k}^n + \lambda_1^{(1)} \widetilde{\gamma}_{i,j,k}^{n+1},$$

$$\bar{\delta}_t(E_2)_{i,j,k}^n = \bar{\delta}_z(H_1)_{i,j,k}^n - \bar{\delta}_x(H_3)_{i,j,k}^n + \lambda_1^{(2)} \widetilde{\gamma}_{i,j,k}^{n+1},$$

$$\bar{\delta}_t(E_3)_{i,j,k}^n = \bar{\delta}_x(H_2)_{i,j,k}^n - \bar{\delta}_y(H_1)_{i,j,k}^n + \lambda_1^{(3)} \widetilde{\gamma}_{i,j,k}^{n+1},$$

$$\bar{\delta}_t(H_1)^n_{i,j,k} = \bar{\delta}_z(E_2)^n_{i,j,k} - \bar{\delta}_y(E_3)^n_{i,j,k} + \lambda^{(1)}_2 \tilde{\gamma}^{n+1}_{i,j,k},$$

$$\bar{\delta}_t(H_2)^n_{i,j,k} = \bar{\delta}_x(E_3)^n_{i,j,k} - \bar{\delta}_z(E_1)^n_{i,j,k} + \lambda^{(2)}_2 \tilde{\gamma}^{n+1}_{i,j,k},$$

$$\bar{\delta}_t(H_3)^n_{i,j,k} = \bar{\delta}_y(E_1)^n_{i,j,k} - \bar{\delta}_x(E_2)^n_{i,j,k} + \lambda^{(3)}_2 \tilde{\gamma}^{n+1}_{i,j,k}$$

into (4.58) leads to

$$\Delta x \Delta y \Delta z \sum_{i=1}^{I} \sum_{j=1}^{J} \sum_{k=1}^{K} \mathbb{E}\Big[\Upsilon^n_{i,j,k}(W^{n+1}_{i,j,k} - W^{n-1}_{i,j,k})\Big]$$

$$= \Delta x \Delta y \Delta z \sum_{i=1}^{I} \sum_{j=1}^{J} \sum_{k=1}^{K} \mathbb{E}\Big[(|\lambda_1|^2 + |\lambda_2|^2)(W^n_{i,j,k} - W^{n-2}_{i,j,k})(\Delta W)^n_{i,j,k}\Big]$$

$$= (|\lambda_1|^2 + |\lambda_2|^2)\Delta x \Delta y \Delta z \sum_{i=1}^{I} \sum_{j=1}^{J} \sum_{k=1}^{K} \mathbb{E}\Big[((\Delta W)^n_{i,j,k})^2\Big]$$

$$= \widehat{\varkappa}\tau,$$

which completes the proof.   □

Remark 4.5 Notice that

$$\bar{V}^Q(D) \approx \sum_{m\in\mathbb{N}} \eta_m \int_D e^2_m(\mathbf{x})\mathrm{d}\mathbf{x} = \sum_{m\in\mathbb{N}} \eta_m = \mathrm{Tr}(Q).$$

Thus $\bar{V}^Q(D)$ can be regarded as an approximation of $\mathrm{Tr}(Q)$.

Proposition 4.18 Denote by $u^n_{i,j,k} = ((\mathbf{E}^n_{i,j,k})^\top, (\mathbf{H}^n_{i,j,k})^\top)^\top$ the solution of the MS Method-III (4.51) with $S(u) = \frac{\lambda_2}{2}|\mathbf{E}|^2 + \frac{\lambda_1}{2}|\mathbf{H}|^2$. Assume that $\lambda_1 = \lambda_2 = \lambda$. Then under the periodic boundary condition, the discrete energy satisfies

$$\Phi^{[\mathrm{III}]}(t_{n+1}) = \Phi^{[\mathrm{III}]}(t_n), \quad \mathbb{P}\text{-a.s.}$$

for all $n = 0, 1, \ldots, N - 1$. Here, the discrete energy $\Phi^{[\mathrm{III}]}$ is defined as

$$\Phi^{[\mathrm{III}]}(t_n) := \Delta x \Delta y \Delta z \sum_{i=1}^{I} \sum_{j=1}^{J} \sum_{k=1}^{K} |u^n_{i,j,k}|^2.$$

Proof The proof is similar to that of Proposition 4.14.   □

Proposition 4.19 *Denote by* $u_{i,j,k}^n = ((\mathbf{E}_{i,j,k}^n)^\top, (\mathbf{H}_{i,j,k}^n)^\top)^\top$ *the solution of the MS method-III (4.51) with* $S(u) = \lambda_2 \cdot \mathbf{E} - \lambda_1 \cdot \mathbf{H}$. *Then under the periodic boundary condition, the discrete energy satisfies*

$$\Phi^{[\text{III}]}(t_{n+1}) = \Phi^{[\text{III}]}(t_n) + 2\Delta x \Delta y \Delta z \sum_{i=1}^{I}\sum_{j=1}^{J}\sum_{k=1}^{K} \Upsilon_{i,j,k}^{n+\frac{1}{2}} (\Delta W)_{i,j,k}^{n+1}, \quad \mathbb{P}\text{-}a.s.$$

$$(4.59)$$

for all $n = 0, 1, \ldots, N-1$, *where*

$$\Upsilon_{i,j,k}^{n+\frac{1}{2}} := \lambda_1 \cdot \mathbf{E}_{i,j,k}^{n+\frac{1}{2}} + \lambda_2 \cdot \mathbf{H}_{i,j,k}^{n+\frac{1}{2}}.$$

Proof The proof is similar to that of Proposition 4.15.   □

Corollary 4.3 *Let conditions in Proposition 4.18 hold. If* $W : [0, T] \times \Omega \to \mathbb{R}$ *is a Brownian motion, then there exists a constant* $\tilde{\varkappa} = (|\lambda_1|^2 + |\lambda_2|^2)|D|$ *such that*

$$\mathbb{E}\Big[\Phi^{[\text{III}]}(t_{n+1})\Big] = \mathbb{E}\Big[\Phi^{[\text{III}]}(t_n)\Big] + \tilde{\varkappa}\tau, \quad n = 0, 1, \ldots, N-1.$$

Proof The proof is similar to that of Corollary 4.1 and thus is omitted here.   □

4.2.1.3 Averaged Divergence Conservation Laws

It follows from Proposition 3.2 that, for $S(u) = \lambda_2 \cdot \mathbf{E} - \lambda_1 \cdot \mathbf{H}$, (4.48) possesses the averaged divergence conservation laws. In order to show that the divergence conservation laws also hold in the discrete sense, we first define two discrete divergence operators $\bar{\nabla}_{i,j,k}^{[\text{I}]}$ and $\bar{\nabla}_{i,j,k}^{[\text{II}]}$ as

$$\bar{\nabla}_{i,j,k}^{[\text{I}]} \cdot V := \delta_x\big((V_1)_{i-\frac{1}{2},j^*,k^*}\big) + \delta_y\big((V_2)_{i^*,j-\frac{1}{2},k^*}\big) + \delta_z\big((V_3)_{i^*,j^*,k-\frac{1}{2}}\big),$$

$$\bar{\nabla}_{i,j,k}^{[\text{II}]} \cdot V := \bar{\delta}_x(V_1)_{i,j,k} + \bar{\delta}_y(V_2)_{i,j,k} + \bar{\delta}_z(V_3)_{i,j,k}$$

$$(4.60)$$

with

$$(V_1)_{i-\frac{1}{2},j^*,k^*} := (V_1)_{i-\frac{1}{2},j+\frac{1}{2},k+\frac{1}{2}} + (V_1)_{i-\frac{1}{2},j+\frac{1}{2},k-\frac{1}{2}}$$
$$+ (V_1)_{i-\frac{1}{2},j-\frac{1}{2},k-\frac{1}{2}} + (V_1)_{i-\frac{1}{2},j-\frac{1}{2},k+\frac{1}{2}},$$

$$(V_2)_{i^*,j-\frac{1}{2},k^*} := (V_2)_{i+\frac{1}{2},j-\frac{1}{2},k+\frac{1}{2}} + (V_2)_{i+\frac{1}{2},j-\frac{1}{2},k-\frac{1}{2}}$$
$$+ (V_2)_{i-\frac{1}{2},j-\frac{1}{2},k+\frac{1}{2}} + (V_2)_{i-\frac{1}{2},j-\frac{1}{2},k-\frac{1}{2}},$$

$$(V_3)_{i*,j*,k-\frac{1}{2}} := (V_3)_{i+\frac{1}{2},j+\frac{1}{2},k-\frac{1}{2}} + (V_3)_{i+\frac{1}{2},j-\frac{1}{2},k-\frac{1}{2}}$$

$$+ (V_3)_{i-\frac{1}{2},j+\frac{1}{2},k-\frac{1}{2}} + (V_3)_{i-\frac{1}{2},j-\frac{1}{2},k-\frac{1}{2}},$$

for $V = (V_1, V_2, V_3)^\top \in \mathbb{R}^3$, $i = 1, 2, \ldots, I$, $j = 1, 2, \ldots, J$, and $k = 1, 2, \ldots, K$.

Proposition 4.20 *The MS Method-I (4.49) with $S(u) = \lambda_2 \cdot \mathbf{E} - \lambda_1 \cdot \mathbf{H}$ preserves the following discrete averaged divergence conservation laws*

$$\mathbb{E}\left[\bar{\nabla}_{i,j,k}^{[I]} \cdot \mathbf{E}^{n+1}\right] = \mathbb{E}\left[\bar{\nabla}_{i,j,k}^{[I]} \cdot \mathbf{E}^n\right], \quad \mathbb{E}\left[\bar{\nabla}_{i,j,k}^{[I]} \cdot \mathbf{H}^{n+1}\right] = \mathbb{E}\left[\bar{\nabla}_{i,j,k}^{[I]} \cdot \mathbf{H}^n\right]$$

for all $n = 0, 1, \ldots, N - 1$, $i = 1, 2, \ldots, I$, $j = 1, 2, \ldots, J$, and $k = 1, 2, \ldots, K$.

Proof We sketch only the proof for the discrete divergence of $\mathbf{E}$. The proof for $\mathbf{H}$ is similar and is omitted. It follows from the definition (4.60) that

$$\bar{\nabla}_{i,j,k}^{[I]} \cdot \mathbf{E}^{n+1} - \bar{\nabla}_{i,j,k}^{[I]} \cdot \mathbf{E}^n$$

$$= \tau\left[\delta_x\delta_t(E_1)_{i-\frac{1}{2},j*,k*}^n + \delta_y\delta_t(E_2)_{i*,j-\frac{1}{2},k*}^n + \delta_z\delta_t(E_3)_{i*,j*,k-\frac{1}{2}}^n\right].$$

Plugging (4.57) into the above equation leads to

$$\bar{\nabla}_{i,j,k}^{[I]} \cdot \mathbf{E}^{n+1} - \bar{\nabla}_{i,j,k}^{[I]} \cdot \mathbf{E}^n$$

$$= \tau\delta_x\left[\delta_y A_t(H_3)_{i-\frac{1}{2},(j-\frac{1}{2})*,k*}^n - \delta_z A_t(H_2)_{i-\frac{1}{2},j*,(k-\frac{1}{2})*}^n\right]$$

$$+ \tau\delta_y\left[\delta_z A_t(H_1)_{i*,j-\frac{1}{2},(k-\frac{1}{2})*}^n - \delta_x A_t(H_3)_{(i-\frac{1}{2})*,j-\frac{1}{2},k*}^n\right]$$

$$+ \tau\delta_z\left[\delta_x A_t(H_2)_{(i-\frac{1}{2})*,j*,k-\frac{1}{2}}^n - \delta_y A_t(H_1)_{i*,(j-\frac{1}{2})*,k-\frac{1}{2}}^n\right]$$

$$+ \lambda_1^{(1)}\delta_x(\Delta W)_{i-1,(j-\frac{1}{2})*,(k-\frac{1}{2})*}^{n+1} + \lambda_1^{(2)}\delta_y(\Delta W)_{(i-\frac{1}{2})*,j-1,(k-\frac{1}{2})*}^{n+1}$$

$$+ \lambda_1^{(3)}\delta_z(\Delta W)_{(i-\frac{1}{2})*,(j-\frac{1}{2})*,k-1}^{n+1}$$

$$= \lambda_1^{(1)}\delta_x(\Delta W)_{i-1,(j-\frac{1}{2})*,(k-\frac{1}{2})*}^{n+1} + \lambda_1^{(2)}\delta_y(\Delta W)_{(i-\frac{1}{2})*,j-1,(k-\frac{1}{2})*}^{n+1}$$

$$+ \lambda_1^{(3)}\delta_z(\Delta W)_{(i-\frac{1}{2})*,(j-\frac{1}{2})*,k-1}^{n+1},$$

which implies that

$$\mathbb{E}\left[\bar{\nabla}^{[\mathrm{I}]}_{i,j,k} \cdot \mathbf{E}^{n+1} - \bar{\nabla}^{[\mathrm{I}]}_{i,j,k} \cdot \mathbf{E}^{n}\right] = 0.$$

Thus the proof is finished. □

In a similar way, we can obtain the discrete averaged divergence conservation laws for the MS Method-II (4.50) and the MS Method-III (4.51), which are stated in the following two propositions.

Proposition 4.21 *The MS method-II (4.50) with* $S(u) = \lambda_2 \cdot \mathbf{E} - \lambda_1 \cdot \mathbf{H}$ *preserves the following discrete averaged divergence conservation laws*

$$\mathbb{E}\left[\bar{\nabla}^{[\mathrm{II}]} \cdot \mathbf{E}^{n+\frac{1}{2}}_{i,j,k}\right] = \mathbb{E}\left[\bar{\nabla}^{[\mathrm{II}]} \cdot \mathbf{E}^{n-\frac{1}{2}}_{i,j,k}\right], \quad \mathbb{E}\left[\bar{\nabla}^{[\mathrm{II}]} \cdot \mathbf{H}^{n+\frac{1}{2}}_{i,j,k}\right] = \mathbb{E}\left[\bar{\nabla}^{[\mathrm{II}]} \cdot \mathbf{H}^{n-\frac{1}{2}}_{i,j,k}\right]$$

for all $n = 1, 2, \ldots, N - 1$, $i = 1, 2, \ldots, I$, $j = 1, 2, \ldots, J$, *and* $k = 1, 2, \ldots, K$.

Proposition 4.22 *The MS method-III (4.51) with* $S(u) = \lambda_2 \cdot \mathbf{E} - \lambda_1 \cdot \mathbf{H}$ *preserves the following discrete averaged divergence conservation laws*

$$\mathbb{E}\left[\bar{\nabla}^{[\mathrm{II}]} \cdot \mathbf{E}^{n+1}_{i,j,k}\right] = \mathbb{E}\left[\bar{\nabla}^{[\mathrm{II}]} \cdot \mathbf{E}^{n}_{i,j,k}\right], \quad \mathbb{E}\left[\bar{\nabla}^{[\mathrm{II}]} \cdot \mathbf{H}^{n+1}_{i,j,k}\right] = \mathbb{E}\left[\bar{\nabla}^{[\mathrm{II}]} \cdot \mathbf{H}^{n}_{i,j,k}\right]$$

for all $n = 0, 1, \ldots, N - 1$, $i = 1, 2, \ldots, I$, $j = 1, 2, \ldots, J$, *and* $k = 1, 2, \ldots, K$.

4.2.2 Stochastic Multi-Symplectic Wavelet Algorithm

In this section, we study the stochastic multi-symplectic wavelet algorithm for

$$\begin{cases} d\mathbf{E}(t) = \nabla \times \mathbf{H}(t)dt - \lambda \mathbf{H} \circ dW(t), & t \in (0, T], \\ d\mathbf{H}(t) = -\nabla \times \mathbf{E}(t)dt + \lambda \mathbf{E} \circ dW(t), & t \in (0, T], \\ \mathbf{E}(0) = \mathbf{E}_0, \quad \mathbf{H}(0) = \mathbf{H}_0 \end{cases} \tag{4.61}$$

with the nonzero constant $\lambda \in \mathbb{R}$, aiming to inherit the stochastic multi-symplectic conservation law and the energy conservation law of the original equations simultaneously. The algorithm combines the midpoint method in the temporal direction and the wavelet interpolation method in the spatial direction. Moreover, the system of algebraic equations obtained under discretization has some nice features, such as skew-symmetry and sparsity. These features can lead to a large reduction in the computational cost.

We first give a brief introduction to the autocorrelation function and the interpolation operator. See [15, 16, 170] for more details. For the simplicity of notations, we restrict ourselves to the one-dimensional case. The approach can also be generalized to the multi-dimensional case. Define the Daubechies scaling function $\phi : \mathbb{R} \to \mathbb{R}$ of order r as

$$\phi(x) := \sum_{k=0}^{r-1} h_k \phi(2x - k), \quad x \in \mathbb{R}, \tag{4.62}$$

where r is a positive even integer and $\{h_k\}_{k=0}^{r-1}$ is a sequence of non-vanishing "filter coefficients". The autocorrelation function $\theta(x)$ of $\phi(x)$ is given by

$$\theta(x) = \int_{-\infty}^{\infty} \phi(r)\phi(r - x)\mathrm{d}r, \quad x \in \mathbb{R},$$

which has the following nice properties.

(i) Compact support:

$$\mathrm{supp}(\theta(x)) = [-r + 1, r - 1].$$

(ii) Orthonormal property:

$$\theta(0) = 1, \quad \theta(l) = 0 \quad \forall l \in \mathbb{Z}.$$

(iii) Derivative property:

$$\theta^{(2k)}(-x) = \theta^{(2k)}(x), \quad \theta^{(2k+1)}(-x) = -\theta^{(2k+1)}(x), \quad k \in \mathbb{N},$$

where $\theta^{(p)}(x)$ denotes the p-th derivative of the function θ.

Let

$$\mathscr{V}_j := \mathrm{span}\left\{\theta_{j,k}(x) = 2^{j/2}\theta(2^j x - k), \ k \in \mathbb{Z}\right\}, \quad j \in \mathbb{Z}.$$

Then $\{\mathscr{V}_j\}_{j\in\mathbb{Z}}$ forms a multiresolution analysis, which, roughly speaking, describes the increment in the information needed from a coarser approximation to a higher resolution approximation. For $j \in \mathbb{Z}$, define the interpolation operator $\mathscr{I} : H^1(\mathbb{R}) \to \mathscr{V}_j$, with the mesh size $\Delta x = 2^{-j}$ as

$$\mathscr{I}f(x) := \sum_{l\in\mathbb{Z}} f(x_l)\theta(2^j x - l),$$

where the collocation points $x_l = 2^{-j}l, l \in \mathbb{Z}$.

For the general d-dimensional case, we can define the interpolation operator $\mathscr{I}$ on $V_{j_1} \otimes \cdots \otimes V_{j_d}$ similarly as

$$\mathscr{I} f(\mathbf{x}) := \sum_{l_1,\dots,l_d \in \mathbb{Z}} f((x_1)_{l_1}, \cdots, (x_d)_{l_d}) \theta(2^{j_1} x_1 - l_1) \dots \theta(2^{j_d} x_d - l_d),$$

$$(4.63)$$

where $j_p \in \mathbb{Z}$ and the collocation points $(x_p)_{l_p} = 2^{-j_p} l_p$ for $l_p \in \mathbb{Z}$ with the mesh sizes $\Delta x_p = 2^{-j_p}$, $p = 1, 2, \dots, d$.

We apply the stochastic midpoint method to (4.61) in the temporal direction and obtain the following temporal semi-discretization:

$$E_1^{n+1} = E_1^n - \tau \left(\partial_z H_2^{n+\frac{1}{2}} - \partial_y H_3^{n+\frac{1}{2}} \right) - \lambda H_1^{n+\frac{1}{2}} \Delta W^{n+1},$$

$$E_2^{n+1} = E_2^n - \tau \left(\partial_x H_3^{n+\frac{1}{2}} - \partial_z H_1^{n+\frac{1}{2}} \right) - \lambda H_2^{n+\frac{1}{2}} \Delta W^{n+1},$$

$$E_3^{n+1} = E_3^n - \tau \left(\partial_y H_1^{n+\frac{1}{2}} - \partial_x H_2^{n+\frac{1}{2}} \right) - \lambda H_3^{n+\frac{1}{2}} \Delta W^{n+1},$$

$$H_1^{n+1} = H_1^n + \tau \left(\partial_z E_2^{n+\frac{1}{2}} - \partial_y E_3^{n+\frac{1}{2}} \right) + \lambda E_1^{n+\frac{1}{2}} \Delta W^{n+1},$$

$$H_2^{n+1} = H_2^n + \tau \left(\partial_x E_3^{n+\frac{1}{2}} - \partial_z E_1^{n+\frac{1}{2}} \right) + \lambda E_2^{n+\frac{1}{2}} \Delta W^{n+1},$$

$$H_3^{n+1} = H_3^n + \tau \left(\partial_y E_1^{n+\frac{1}{2}} - \partial_x E_2^{n+\frac{1}{2}} \right) + \lambda E_3^{n+\frac{1}{2}} \Delta W^{n+1},$$

$$(4.64)$$

where $\Delta W^{n+1} = W(t_{n+1}) - W(t_n)$, $n = 0, 1, \dots, N - 1$.

Further, the wavelet interpolation technique is applied to (4.64) in the spatial direction to obtain the stochastic multi-symplectic wavelet algorithm. We choose the autocorrelation function θ as the test function. By properties of θ, we know that the obtained first-order differentiation matrix is skew-symmetric and sparse.

For the numerical implementation, we consider (4.61) in $D = [a_1^-, a_1^+] \times [a_2^-, a_2^+] \times [a_3^-, a_3^+]$ with $L_p := a_p^+ - a_p^-$ ($p = 1, 2, 3$) being integers. Assume that the periodic boundary condition holds. Fix integers J_1, J_2, J_3, and set $N_1 = L_1 \cdot 2^{J_1}$, $N_2 = L_2 \cdot 2^{J_2}$, $N_3 = L_3 \cdot 2^{J_3}$. The collocation points are $(x_i, y_j, z_k) = (i/2^{J_1}, j/2^{J_2}, k/2^{J_3})$, $i = 1, 2, \dots, N_1$, $j = 1, 2, \dots, N_2$, $k = 1, 2, \dots, N_3$. For $s = 1, 2, \dots, 6$, it follows from (4.63) that the interpolation of the s-th component u_s of $u = (\mathbf{E}^\top, \mathbf{H}^\top)^\top$ at these collocation points is

$$\mathscr{I} u_s(t, x, y, z) = \sum_{i=1}^{N_1} \sum_{j=1}^{N_2} \sum_{k=1}^{N_3} (u_s)_{i,j,k} \theta(2^{J_1} x - i) \theta(2^{J_2} y - j) \theta(2^{J_3} z - k)$$

with the mesh sizes $\Delta x = 2^{-J_1}$, $\Delta y = 2^{-J_2}$, and $\Delta z = 2^{-J_3}$. Taking the partial derivative with respect to the variable x and evaluating the resulting expression at collocation points (x_i, y_j, z_k), we obtain

$$\partial_x[\mathscr{I}u_s(t, x_i, y_j, z_k)]$$

$$= \sum_{i'=1}^{N_1}\sum_{j'=1}^{N_2}\sum_{k'=1}^{N_3}(u_s)_{i',j',k'}\theta(2^{J_2}y_j - j')\theta(2^{J_3}z_k - k')\frac{d\theta(2^{J_1}x - i')}{dx}\Big|_{x=x_i}$$

$$= 2^{J_1}\sum_{i'=1}^{N_1}(u_s)_{i',j,k}\theta'(i - i') = \sum_{i'=i-(r-1)}^{i+(r-1)}(u_s)_{i',j,k}(B^x)_{i,i'}$$

$$= ((B^x \otimes I_{N_2} \otimes I_{N_3})\mathbf{u_s})_{i,j,k},$$

where $\mathbf{u_s} = \big((u_s)_{1,1,1}, \ldots, (u_s)_{N_1,1,1}, (u_s)_{1,2,1}, \ldots, (u_s)_{N_1,2,1}, \ldots, (u_s)_{N_1,N_2,N_3}\big)^{\top}$ and B^x is an $N_1 \times N_1$ sparse skew-symmetric circulant matrix with entries

$$(B^x)_{m,M} = \begin{cases} 2^{J_1}\theta'(m - M), & m - (r - 1) \leq M \leq m + (r - 1), \\[2mm] 2^{J_1}\theta'(-l), & m - M = N_1 - l, \quad 1 \leq l \leq r - 1, \\[2mm] 2^{J_1}\theta'(l), & M - m = N_1 - l, \quad 1 \leq l \leq r - 1, \\[2mm] 0, & \text{otherwise.} \end{cases}$$

Obviously, B^x has $(2r - 1)$ nonzero elements in each row, and it can be rewritten as

$$B^x = 2^{J_1} \cdot \text{Circ}\Big(\theta'(0), \theta'(-1), \ldots, \theta'(-(r - 1)), 0, \ldots, 0, \theta'(r - 1), \ldots, \theta'(1)\Big).$$

Similarly, for $s = 1, 2, \ldots, 6$, we have

$$\partial_y[\mathscr{I}u_s(t, x_i, y_j, z_k)] = ((I_{N_1} \otimes B^y \otimes I_{N_3})\mathbf{u_s})_{i,j,k},$$

$$\partial_z[\mathscr{I}u_s(t, x_i, y_j, z_k)] = ((I_{N_1} \otimes I_{N_2} \otimes B^z)\mathbf{u_s})_{i,j,k}$$

with B^y and B^z being $N_2 \times N_2$ and $N_3 \times N_3$ circulant matrices, respectively.

Applying the wavelet interpolation technique to (4.64), we derive the fully discrete algorithm for (4.61), which is called the stochastic multi-symplectic wavelet algorithm:

$$\mathbf{E}_1^{n+1} - \mathbf{E}_1^n = \tau\left(A_2\mathbf{H}_3^{n+\frac{1}{2}} - A_3\mathbf{H}_2^{n+\frac{1}{2}}\right) - \lambda\mathbf{H}_1^{n+\frac{1}{2}}\Delta W^{n+1},$$

$$\mathbf{E}_2^{n+1} - \mathbf{E}_2^n = \tau\left(A_3\mathbf{H}_1^{n+\frac{1}{2}} - A_1\mathbf{H}_3^{n+\frac{1}{2}}\right) - \lambda\mathbf{H}_2^{n+\frac{1}{2}}\Delta W^{n+1},$$

$$\mathbf{E}_3^{n+1} - \mathbf{E}_3^n = \tau\left(A_1\mathbf{H}_2^{n+\frac{1}{2}} - A_2\mathbf{H}_1^{n+\frac{1}{2}}\right) - \lambda\mathbf{H}_3^{n+\frac{1}{2}}\Delta\mathbf{W}^{n+1},$$

$$\mathbf{H}_1^{n+1} - \mathbf{H}_1^n = \tau\left(A_3\mathbf{E}_2^{n+\frac{1}{2}} - A_2\mathbf{E}_3^{n+\frac{1}{2}}\right) + \lambda\mathbf{E}_1^{n+\frac{1}{2}}\Delta\mathbf{W}^{n+1}, \qquad (4.65)$$

$$\mathbf{H}_2^{n+1} - \mathbf{H}_2^n = \tau\left(A_1\mathbf{E}_3^{n+\frac{1}{2}} - A_3\mathbf{E}_1^{n+\frac{1}{2}}\right) + \lambda\mathbf{E}_2^{n+\frac{1}{2}}\Delta\mathbf{W}^{n+1},$$

$$\mathbf{H}_3^{n+1} - \mathbf{H}_3^n = \tau\left(A_2\mathbf{E}_1^{n+\frac{1}{2}} - A_1\mathbf{E}_2^{n+\frac{1}{2}}\right) + \lambda\mathbf{E}_3^{n+\frac{1}{2}}\Delta\mathbf{W}^{n+1},$$

where $A_1 = B^x \otimes I_{N_2} \otimes I_{N_3}$, $A_2 = I_{N_1} \otimes B^y \otimes I_{N_3}$, and $A_3 = I_{N_1} \otimes I_{N_2} \otimes B^z$ are skew-symmetric matrices corresponding to differential matrices B^x, B^y, and B^z, respectively. Here,

$$\Delta\mathbf{W}^{n+1} = \left(\Delta W_{1,1,1}^{n+1}, \ldots, \Delta W_{N_1,1,1}^{n+1}, \Delta W_{1,2,1}^{n+1}, \ldots, \Delta W_{N_1,2,1}^{n+1}, \ldots, \Delta W_{N_1,N_2,N_3}^{n+1}\right)^\top \tag{4.66}$$

with $\Delta W_{i,j,k}^{n+1} = W(t_{n+1}, x_i, y_j, z_k) - W(t_n, x_i, y_j, z_k)$, and $\mathbf{u}_s^{n+\frac{1}{2}}\Delta\mathbf{W}^{n+1}$ denotes the components multiplication between $\mathbf{u}_s^{n+\frac{1}{2}}$ and $\Delta\mathbf{W}^{n+1}$ for $s = 1, 2, \ldots, 6$.

4.2.2.1 Stochastic Multi-Symplectic Conservation Law

The following proposition presents the stochastic multi-symplectic conservation law for the full discretization (4.65).

Proposition 4.23 *The fully discrete algorithm* (4.65) *preserves the following discrete stochastic multi-symplectic conservation law*

$$\frac{\varpi_{i,j,k}^{n+1} - \varpi_{i,j,k}^n}{\tau} + \sum_{i'=i-(r-1)}^{i+(r-1)} (B^x)_{i,i'}(\kappa_1)_{i',j,k}^{n+\frac{1}{2}}$$

$$+ \sum_{j'=j-(r-1)}^{j+(r-1)} (B^y)_{j,j'}(\kappa_2)_{i,j',k}^{n+\frac{1}{2}} + \sum_{k'=k-(r-1)}^{k+(r-1)} (B^z)_{k,k'}(\kappa_3)_{i,j,k'}^{n+\frac{1}{2}} = 0, \quad \mathbb{P}\text{-a.s.} \tag{4.67}$$

for all $n = 0, 1, \ldots, N-1$, $i = 1, 2, \ldots, N_1$, $j = 1, 2, \ldots, N_2$, *and* $k = 1, 2, \ldots, N_3$. *Here,* r *is the order of the Daubechies scaling function* ϕ, *and*

$$\varpi_{i,j,k}^n = \frac{1}{2}du_{i,j,k}^n \wedge \mathbb{F}du_{i,j,k}^n, \qquad (\kappa_1)_{i',j,k}^{n+\frac{1}{2}} = \frac{1}{2}du_{i,j,k}^{n+\frac{1}{2}} \wedge \mathbb{K}_1 du_{i',j,k}^{n+\frac{1}{2}},$$

$$(\kappa_2)_{i,j',k}^{n+\frac{1}{2}} = \frac{1}{2}du_{i,j,k}^{n+\frac{1}{2}} \wedge \mathbb{K}_2 du_{i,j',k}^{n+\frac{1}{2}}, \qquad (\kappa_3)_{i,j,k'}^{n+\frac{1}{2}} = \frac{1}{2}du_{i,j,k}^{n+\frac{1}{2}} \wedge \mathbb{K}_3 du_{i,j,k'}^{n+\frac{1}{2}}.$$

Proof Note that (4.65) can be rewritten as

$$\mathbb{F}\frac{u_{i,j,k}^{n+1} - u_{i,j,k}^{n}}{\tau} + \sum_{i'=i-(r-1)}^{i+(r-1)} (B^x)_{i,i'}\left(\mathbb{K}_1 u_{i',j,k}^{n+\frac{1}{2}}\right) + \sum_{j'=j-(r-1)}^{j+(r-1)} (B^y)_{j,j'}\left(\mathbb{K}_2 u_{i,j',k}^{n+\frac{1}{2}}\right)$$

$$+ \sum_{k'=k-(r-1)}^{k+(r-1)} (B^z)_{k,k'}\left(\mathbb{K}_3 u_{i,j,k'}^{n+\frac{1}{2}}\right) = \nabla_u S(u_{i,j,k}^{n+\frac{1}{2}})\frac{\Delta W_{i,j,k}^{n+1}}{\tau},$$

where $S(u) = \frac{\lambda}{2}|u|^2$. Taking the exterior derivative on both sides of the above equation, we have

$$\mathbb{F}\frac{du_{i,j,k}^{n+1} - du_{i,j,k}^{n}}{\tau} + \sum_{i'=i-(r-1)}^{i+(r-1)} (B^x)_{i,i'}\left(\mathbb{K}_1 du_{i',j,k}^{n+\frac{1}{2}}\right)$$

$$+ \sum_{j'=j-(r-1)}^{j+(r-1)} (B^y)_{j,j'}\left(\mathbb{K}_2 du_{i,j',k}^{n+\frac{1}{2}}\right)$$

$$+ \sum_{k'=k-(r-1)}^{k+(r-1)} (B^z)_{k,k'}\left(\mathbb{K}_3 du_{i,j,k'}^{n+\frac{1}{2}}\right)$$

$$= \nabla^2 S(u_{i,j,k}^{n+\frac{1}{2}}) du_{i,j,k}^{n+\frac{1}{2}}\frac{\Delta W_{i,j,k}^{n+1}}{\tau}.$$

Then, performing the wedge product against $du_{i,j,k}^{n+\frac{1}{2}}$ and utilizing the symmetry of $\nabla^2 S(\cdot)$, we finish the proof.   □

4.2.2.2 Energy Conservation Law

The following proposition presents the energy conservation law for the full discretization (4.65).

Proposition 4.24 *The stochastic multi-symplectic wavelet algorithm* (4.65) *possesses the following discrete energy conservation law*

$$\|\mathbf{E}^n\|^2 + \|\mathbf{H}^n\|^2 = \|\mathbf{E}_0\|^2 + \|\mathbf{H}_0\|^2, \quad \mathbb{P}\text{-}a.s. \tag{4.68}$$

for all $n = 0, 1, \ldots, N$. Here,

$$\|\mathbf{E}^n\|^2 = \Delta x \, \Delta y \, \Delta z \sum_{i=1}^{N_1} \sum_{j=1}^{N_2} \sum_{k=1}^{N_3} \left((E_{1_{i,j,k}}^n)^2 + (E_{2_{i,j,k}}^n)^2 + (E_{3_{i,j,k}}^n)^2 \right),$$

$$\|\mathbf{H}^n\|^2 = \Delta x \, \Delta y \, \Delta z \sum_{i=1}^{N_1} \sum_{j=1}^{N_2} \sum_{k=1}^{N_3} \left((H_{1_{i,j,k}}^n)^2 + (H_{2_{i,j,k}}^n)^2 + (H_{3_{i,j,k}}^n)^2 \right).$$

Proof We multiply each equation in (4.65) by $\mathbf{E}_1^{n+\frac{1}{2}}, \mathbf{E}_2^{n+\frac{1}{2}}, \mathbf{E}_3^{n+\frac{1}{2}}, \mathbf{H}_1^{n+\frac{1}{2}}, \mathbf{H}_2^{n+\frac{1}{2}},$ and $\mathbf{H}_3^{n+\frac{1}{2}}$, respectively, to obtain

$$\frac{\|\mathbf{E}_1^{n+1}\|^2 - \|\mathbf{E}_1^n\|^2}{2} = \tau \left(A_2 \mathbf{H}_3^{n+\frac{1}{2}} - A_3 \mathbf{H}_2^{n+\frac{1}{2}} \right) \cdot \mathbf{E}_1^{n+\frac{1}{2}} - \lambda \left(\mathbf{H}_1^{n+\frac{1}{2}} \Delta W^{n+1} \right) \cdot \mathbf{E}_1^{n+\frac{1}{2}},$$

$$\frac{\|\mathbf{E}_2^{n+1}\|^2 - \|\mathbf{E}_2^n\|^2}{2} = \tau \left(A_3 \mathbf{H}_1^{n+\frac{1}{2}} - A_1 \mathbf{H}_3^{n+\frac{1}{2}} \right) \cdot \mathbf{E}_2^{n+\frac{1}{2}} - \lambda \left(\mathbf{H}_2^{n+\frac{1}{2}} \Delta W^{n+1} \right) \cdot \mathbf{E}_2^{n+\frac{1}{2}},$$

$$\frac{\|\mathbf{E}_3^{n+1}\|^2 - \|\mathbf{E}_3^n\|^2}{2} = \tau \left(A_1 \mathbf{H}_2^{n+\frac{1}{2}} - A_2 \mathbf{H}_1^{n+\frac{1}{2}} \right) \cdot \mathbf{E}_3^{n+\frac{1}{2}} - \lambda \left(\mathbf{H}_3^{n+\frac{1}{2}} \Delta W^{n+1} \right) \cdot \mathbf{E}_3^{n+\frac{1}{2}},$$

$$\frac{\|\mathbf{H}_1^{n+1}\|^2 - \|\mathbf{H}_1^n\|^2}{2} = \tau \left(A_3 \mathbf{E}_2^{n+\frac{1}{2}} - A_2 \mathbf{E}_3^{n+\frac{1}{2}} \right) \cdot \mathbf{H}_1^{n+\frac{1}{2}} + \lambda \left(\mathbf{E}_1^{n+\frac{1}{2}} \Delta W^{n+1} \right) \cdot \mathbf{H}_1^{n+\frac{1}{2}},$$

$$\frac{\|\mathbf{H}_2^{n+1}\|^2 - \|\mathbf{H}_2^n\|^2}{2} = \tau \left(A_1 \mathbf{E}_3^{n+\frac{1}{2}} - A_3 \mathbf{E}_1^{n+\frac{1}{2}} \right) \cdot \mathbf{H}_2^{n+\frac{1}{2}} + \lambda \left(\mathbf{E}_2^{n+\frac{1}{2}} \Delta W^{n+1} \right) \cdot \mathbf{H}_2^{n+\frac{1}{2}},$$

$$\frac{\|\mathbf{H}_3^{n+1}\|^2 - \|\mathbf{H}_3^n\|^2}{2} = \tau \left(A_2 \mathbf{E}_1^{n+\frac{1}{2}} - A_1 \mathbf{E}_2^{n+\frac{1}{2}} \right) \cdot \mathbf{H}_3^{n+\frac{1}{2}} + \lambda \left(\mathbf{E}_3^{n+\frac{1}{2}} \Delta W^{n+1} \right) \cdot \mathbf{H}_3^{n+\frac{1}{2}}.$$

By the fact that A_1, A_2, and A_3 are skew-symmetric matrices and adding up all terms in the above equations, we have

$$\frac{1}{2} \left[(\|\mathbf{E}^{n+1}\|^2 + \|\mathbf{H}^{n+1}\|^2) - (\|\mathbf{E}^n\|^2 + \|\mathbf{H}^n\|^2) \right] = 0,$$

which leads to the assertion. $\qquad\square$

4.2.3   *Stochastic Multi-Symplectic Discontinuous Galerkin Algorithms*

This section investigates the full discretization based on the dG method in space for the stochastic Maxwell equations with additive noise on $\mathbb{H}$:

$$\begin{cases} \mathrm{d}u(t) = Mu(t)\mathrm{d}t + \lambda \mathrm{d}W(t), & t \in (0, T], \\ u(0) = u_0, \end{cases} \tag{4.69}$$

where $\lambda = (\lambda_1^\top, \lambda_2^\top)^\top \in \mathbb{R}^6$ and

$$M = \begin{bmatrix} 0 & \varepsilon^{-1}\nabla \times \\ -\mu^{-1}\nabla \times & 0 \end{bmatrix} \quad \text{with} \quad \mathscr{D}(M) = H_0(\mathrm{curl}, D) \times H(\mathrm{curl}, D).$$

In this section, we suppose that the domain D is a polyhedron in $\mathbb{R}^3$ in order to cover the domain with a mesh consisting of polyhedral elements.

Let us first present a succinct introduction to the basic concepts of the dG method. For more details, we refer to [51, 143]. We consider a simplicial, shape- and contact-regular mesh $\mathscr{T}_h$ that partitions the domain D into disjoint polyhedral elements $\{K : K \in \mathscr{T}_h\}$, such that $\overline{D} = \bigcup_{K \in \mathscr{T}_h} K$. The index h refers to the maximum diameter of all elements of $\mathscr{T}_h$. We define $\mathbf{n}_K$ on ∂K as the unit outward normal to K. Denote by $P^r(K)$ the set of polynomials of total degree at most r defined on the element K for $r \geq 1$.

Furthermore, we denote the set of all interior faces by G_h^I, the set of all boundary faces by G_h^B, and the set of all faces by $G_h := G_h^I \cup G_h^B$. For an arbitrary interior face $G \in G_h^I$, we choose arbitrarily one of the unit outward normals of the two mesh elements composing the face G. We fix this face normal and denote it by $\mathbf{n}_G$. We use notations K and K_G for two neighboring elements $\partial K \cap \partial K_G = G \in G_h^I$, where the face normal $\mathbf{n}_G$ points from K to K_G. For a boundary face $G \in G_h^B$, $\mathbf{n}_G$ is always the unit outward normal. Fig. 4.1 illustrates some notations of the dG method for the reader's convenience.

Denote by $v_K := v_h|_K$ the restriction of the discrete function v_h on an element K. Furthermore, the jump and average of v_h on an interior face G are denoted by

$$[\![v_h]\!]_G := (v_{K_G})\big|_G - (v_K)\big|_G \quad \text{and} \quad \{\!\{v_h\}\!\}_G := \frac{1}{2}\big[(v_{K_G})\big|_G + (v_K)\big|_G\big],$$

respectively.

When $v_h : D \to \mathbb{R}^d$, the above jump and average operators act componentwise on v.

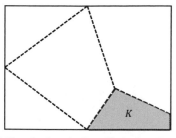
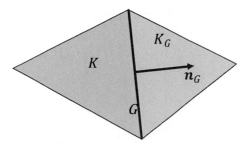

(a)Interior face (red) and boundary face (blue)

(b) Neighboring element and unit normal

Fig. 4.1 Some notations of the dG method. (**a**) Interior face (red) and boundary face (blue). (**b**) Neighboring element and unit normal

For $m \in \mathbb{N}$, we define the broken Sobolev space as

$$H^m(\mathscr{T}_h) := \left\{ v \in L^2(D) : \ v|_K \in H^m(K) \text{ for all } K \in \mathscr{T}_h \right\}^6,$$

which is a Hilbert space with the semi-norm and norm being

$$|v|^2_{H^m(\mathscr{T}_h)} := \sum_{K \in \mathscr{T}_h} |v|^2_{H^m(K)^6} \text{ and } \|v\|^2_{H^m(\mathscr{T}_h)} := \sum_{j=0}^{m} |v|^2_{H^j(\mathscr{T}_h)} \quad \forall \, v \in H^m(\mathscr{T}_h),$$

respectively. Clearly, the usual Sobolev spaces are subspaces of their broken versions, i.e., $H^m(D) \subset H^m(\mathscr{T}_h)$ for all $m \in \mathbb{N}$.

Define a finite element space consisting of piecewise polynomial

$$\mathbb{H}_{h,r} := \left\{ v_h \in L^2(D) : v_h|_K \in P^r(K) \text{ for all } K \in \mathscr{T}_h \right\}^6, \quad r \in \mathbb{N}. \tag{4.70}$$

Denote by Π_h the orthogonal projection on $\mathbb{H}_{h,r}$, which satisfies

$$\langle u - \Pi_h u, v_h \rangle_{\mathbb{H}} = 0 \quad \forall \, u \in \mathbb{H}, \ v_h \in \mathbb{H}_{h,r}.$$

We make the following assumption on coefficients ε and μ.

Assumption 4.1 *Suppose that the coefficients ε and μ are piecewise positive constants, i.e., $\varepsilon_K := \varepsilon|_K$ and $\mu_K := \mu|_K$ for each $K \in \mathscr{T}_h$.*

Under the above assumption, by the integration by parts formula, the dG method to discrete the spatial direction of (4.69) reads as follows: seek $u_h = (\mathbf{E}_h^\top, \mathbf{H}_h^\top)^\top \in$

$\mathbb{H}_{h,r}$ such that for every test function $v_h = (\psi_h^\top, \phi_h^\top)^\top \in \mathbb{H}_{h,r}$, it holds that

$$
\varepsilon_K \int_K d\mathbf{E}_h \cdot \psi_h d\mathbf{x} - \int_K (\nabla \times \mathbf{H}_h) \cdot \psi_h d\mathbf{x}dt
$$

$$
- \int_{\partial K} \left(\widehat{\mathbf{n}_K \times \mathbf{H}_h} - \mathbf{n}_K \times \mathbf{H}_K \right) \cdot \psi_K dSdt = \varepsilon_K \int_K \lambda_1 \cdot \psi_h d\mathbf{x}dW(t)
\tag{4.71}
$$

and

$$
\mu_K \int_K d\mathbf{H}_h \cdot \phi_h d\mathbf{x} + \int_K (\nabla \times \mathbf{E}_h) \cdot \phi_h d\mathbf{x}dt
$$

$$
+ \int_{\partial K} \left(\widehat{\mathbf{n}_K \times \mathbf{E}_h} - \mathbf{n}_K \times \mathbf{E}_K \right) \cdot \phi_K dSdt = \mu_k \int_K \lambda_2 \cdot \phi_h d\mathbf{x}dW(t).
\tag{4.72}
$$

Here, $\widehat{\mathbf{n}_K \times \mathbf{E}_h}$ and $\widehat{\mathbf{n}_K \times \mathbf{H}_h}$ are the so-called numerical fluxes. The numerical flux can be chosen according to the central, upwind, or hybrid principles.

(a) Stochastic symplecticity

To study the stochastic symplecticity of (4.71)–(4.72), we first derive the corresponding global formulation of the proposed dG algorithm. We take the central flux as an example. Let $G = \partial K \cap \partial K_G$. The central flux is defined by

$$
\widehat{\mathbf{n}_K \times \mathbf{E}_h}\Big|_G := \mathbf{n}_K \times \frac{\mathbf{E}_K + \mathbf{E}_{K_G}}{2}, \quad \widehat{\mathbf{n}_K \times \mathbf{H}_h}\Big|_G := \mathbf{n}_K \times \frac{\mathbf{H}_K + \mathbf{H}_{K_G}}{2}.
\tag{4.73}
$$

Note that for $u \in \mathscr{D}(M) \cap H^1(\mathscr{T}_h)$,

$$
\mathbf{n}_G \times [\![\mathbf{H}]\!]_G = \mathbf{n}_G \times [\![\mathbf{E}]\!]_G = 0 \quad \forall \, G \in G_h^I,
$$

$$
\mathbf{n}_G \times \mathbf{E} = 0 \quad \forall \, G \in G_h^B.
\tag{4.74}
$$

Therefore, plugging (4.73) into (4.71)–(4.72) and adding up over all elements $K \in \mathscr{T}_h$, we arrive at

$$
\int_D \varepsilon d\mathbf{E}_h \cdot \psi_h d\mathbf{x} - \int_D (\nabla \times \mathbf{H}_h) \cdot \psi_h d\mathbf{x}dt
$$

$$
- \sum_{G \in G_h^I} \int_G \mathbf{n}_G \times [\![\mathbf{H}_h]\!]_G \cdot \{\!\{\psi_h\}\!\}_G dSdt = \int_D \varepsilon \lambda_1 \cdot \psi_h d\mathbf{x}dW(t)
$$

and

$$\int_D \mu \mathrm{dH}_h \cdot \phi_h \mathrm{dx} + \int_D (\nabla \times \mathbf{E}_h) \cdot \phi_h \mathrm{dxd}t - \sum_{G \in G_h^B} \int_G \mathbf{n}_G \times \mathbf{E}_h \cdot \phi_h \mathrm{d}S \mathrm{d}t$$

$$+ \sum_{G \in G_h^I} \int_G \mathbf{n}_G \times [\![\mathbf{E}_h]\!]_G \cdot \{\!\{\phi_h\}\!\}_G \mathrm{d}S \mathrm{d}t = \int_D \mu \lambda_2 \cdot \phi_h \mathrm{dxd}W(t),$$

from which one can define the discrete version of the Maxwell operator.

Definition 4.3 For all $u_h = (\mathbf{E}_h^\top, \mathbf{H}_h^\top)^\top$, $v_h = (\psi_h^\top, \phi_h^\top)^\top \in \mathbb{H}_{h,r}$, the discrete Maxwell operator $M_h^{\mathrm{cf}} : \mathbb{H}_{h,r} \to \mathbb{H}_{h,r}$ with the central flux is defined as

$$\langle M_h^{\mathrm{cf}} u_h, v_h \rangle_{\mathbb{H}} := \sum_{K \in \mathscr{T}_h} \left[\left\langle \nabla \times \mathbf{H}_h, \psi_h \right\rangle_{L^2(K)^3} - \left\langle \nabla \times \mathbf{E}_h, \phi_h \right\rangle_{L^2(K)^3} \right]$$

$$+ \sum_{G \in G_h^I} \left[\left\langle \mathbf{n}_G \times [\![\mathbf{H}_h]\!]_G, \{\!\{\psi_h\}\!\}_G \right\rangle_{L^2(G)^3} - \left\langle \mathbf{n}_G \times [\![\mathbf{E}_h]\!]_G, \{\!\{\phi_h\}\!\}_G \right\rangle_{L^2(G)^3} \right]$$

$$+ \sum_{G \in G_h^B} \left\langle \mathbf{n}_G \times \mathbf{E}_h, \phi_h \right\rangle_{L^2(G)^3}.$$

Proceeding similarly, we can design the numerical flux by the upwind principle and obtain another discrete version of the Maxwell operator. See e.g. [97, Sect. 4] for more details.

Definition 4.4 For all $u_h = (\mathbf{E}_h^\top, \mathbf{H}_h^\top)^\top$, $v_h = (\psi_h^\top, \phi_h^\top)^\top \in \mathbb{H}_{h,r}$, the discrete Maxwell operator $M_h^{\mathrm{upw}} : \mathbb{H}_{h,r} \to \mathbb{H}_{h,r}$ with the upwind flux is defined as

$$\langle M_h^{\mathrm{upw}} u_h, v_h \rangle_{\mathbb{H}} := \sum_{K \in \mathscr{T}_h} \left[\left\langle \nabla \times \mathbf{H}_h, \psi_h \right\rangle_{L^2(K)^3} - \left\langle \nabla \times \mathbf{E}_h, \phi_h \right\rangle_{L^2(K)^3} \right]$$

$$+ \sum_{G \in G_h^I} \left[\left\langle \mathbf{n}_G \times [\![\mathbf{H}_h]\!]_G, \beta_K \psi_K + \beta_{K_G} \psi_{K_G} \right\rangle_{L^2(G)^3} \right.$$

$$- \left\langle \mathbf{n}_G \times [\![\mathbf{E}_h]\!]_G, \alpha_K \phi_K + \alpha_{K_G} \phi_{K_G} \right\rangle_{L^2(G)^3}$$

$$- \gamma_G \left\langle \mathbf{n}_G \times [\![\mathbf{E}_h]\!]_G, \mathbf{n}_G \times [\![\psi_h]\!]_G \right\rangle_{L^2(G)^3}$$

$$\left. - \delta_G \left\langle \mathbf{n}_G \times [\![\mathbf{H}_h]\!]_G, \mathbf{n}_G \times [\![\phi_h]\!]_G \right\rangle_{L^2(G)^3} \right]$$

$$+ \sum_{G \in G_h^B} \left[\left\langle \mathbf{n}_G \times \mathbf{E}_h, \phi_h \right\rangle_{L^2(G)^3} \right.$$

$$\left. - 2\gamma_G \left\langle \mathbf{n}_G \times \mathbf{E}_h, \mathbf{n}_G \times \psi_h \right\rangle_{L^2(G)^3} \right],$$

where

$$\alpha_K = \frac{C_{K_G} \varepsilon_{K_G}}{C_{K_G} \varepsilon_{K_G} + C_K \varepsilon_K}, \quad \beta_K = \frac{C_{K_G} \mu_{K_G}}{C_{K_G} \mu_{K_G} + C_K \mu_K},$$

$$\gamma_G = \frac{1}{C_{K_G} \mu_{K_G} + C_K \mu_K}, \quad \delta_G = \frac{1}{C_{K_G} \varepsilon_{K_G} + C_K \varepsilon_K}$$

with $C_K = (\varepsilon_K \mu_K)^{-1/2}$.

Applying a dG method with the central flux or upwind flux to (4.69) in the spatial direction, we have the following spatial semi-discretization

$$\mathrm{d}u_h(t) = M_h u_h(t)\mathrm{d}t + \Pi_h \lambda \mathrm{d}W(t), \quad t \in (0, T] \tag{4.75}$$

with $u_h(0) = \Pi_h u_0$ and $M_h \in \{M_h^{\mathrm{cf}}, M_h^{\mathrm{upw}}\}$. Furthermore, applying the stochastic midpoint method to (4.75) in the temporal direction leads to the following fully discrete algorithm of (4.69):

$$u_h^{n+1} = u_h^n + \frac{\tau}{2}\left(M_h u_h^n + M_h u_h^{n+1}\right) + \Pi_h \lambda \Delta W^{n+1}, \quad n = 0, 1, \ldots, N-1, \tag{4.76}$$

where $u_h^0 = \Pi_h u_0$ and $\Delta W^{n+1} = W(t_{n+1}) - W(t_n)$.

Denote

$$\mathbb{H}_{h,r} + \left(\mathscr{D}(M) \cap H^1(\mathscr{T}_h)\right) := \left\{v_h + u : v_h \in \mathbb{H}_{h,r}, \ u \in \mathscr{D}(M) \cap H^1(\mathscr{T}_h)\right\}.$$

Then the discrete Maxwell operators M_h^{cf} and M_h^{upw} are well-defined from $\mathbb{H}_{h,r} + \left(\mathscr{D}(M) \cap H^1(\mathscr{T}_h)\right)$ to $\mathbb{H}_{h,r}$. We have the following properties of the discrete Maxwell operators M_h^{cf} and M_h^{upw}.

Proposition 4.25 *For discrete Maxwell operators M_h^{cf} and M_h^{upw}, the following statements hold.*

(i) *For all $u \in \mathscr{D}(M) \cap H^1(\mathscr{T}_h)$,*

$$M_h^{\mathrm{cf}} u = \Pi_h M u \quad \text{and} \quad M_h^{\mathrm{upw}} u = \Pi_h M u.$$

(ii) *For all $u_h \in \mathbb{H}_{h,r}$,*

$$\langle M_h^{\mathrm{cf}} u_h, u_h \rangle_{\mathbb{H}} = 0 \quad and \quad \langle M_h^{\mathrm{upw}} u_h, u_h \rangle_{\mathbb{H}} \leq 0.$$

(iii) *For all $u \in \mathbb{H}_{h,r} + \big(\mathcal{D}(M) \cap H^1(\mathscr{T}_h)\big)$ and $v_h = (\psi_h^\top, \phi_h^\top)^\top \in \mathbb{H}_{h,r}$,*

$$\langle M_h^{\mathrm{cf}} u, v_h \rangle_{\mathbb{H}} = \sum_{K \in \mathscr{T}_h} \left[\big\langle \mathbf{H}, \nabla \times \psi_h \big\rangle_{L^2(K)^3} - \big\langle \mathbf{E}, \nabla \times \phi_h \big\rangle_{L^2(K)^3} \right]$$

$$+ \sum_{G \in G_h^I} \left[\big\langle \{\!\{\mathbf{H}\}\!\}_G, \mathbf{n}_G \times [\![\psi_h]\!]_G \big\rangle_{L^2(G)^3} - \big\langle \{\!\{\mathbf{E}\}\!\}_G, \mathbf{n}_G \times [\![\phi_h]\!]_G \big\rangle_{L^2(G)^3} \right]$$

$$- \sum_{G \in G_h^B} \big\langle \mathbf{H}, \mathbf{n}_G \times \psi_h \big\rangle_{L^2(G)^3}$$

and

$$\langle M_h^{\mathrm{upw}} u, v_h \rangle_{\mathbb{H}} = \sum_{K \in \mathscr{T}_h} \left[\big\langle \mathbf{H}, \nabla \times \psi_h \big\rangle_{L^2(K)^3} - \big\langle \mathbf{E}, \nabla \times \phi_h \big\rangle_{L^2(K)^3} \right]$$

$$+ \sum_{G \in G_h^I} \left[\big\langle \beta_K \mathbf{H}_{K_G} + \beta_{K_G} \mathbf{H}_K - \gamma_G \mathbf{n}_G \times [\![\mathbf{E}]\!]_G, \mathbf{n}_G \times [\![\psi_h]\!]_G \big\rangle_{L^2(G)^3} \right.$$

$$\left. - \big\langle \alpha_K \mathbf{E}_{K_G} + \alpha_{K_G} \mathbf{E}_K + \delta_G \mathbf{n}_G \times [\![\mathbf{H}]\!]_G, \mathbf{n}_G \times [\![\phi_h]\!]_G \big\rangle_{L^2(G)^3} \right]$$

$$- \sum_{G \in G_h^B} \left[\big\langle \mathbf{H}, \mathbf{n}_G \times \psi_h \big\rangle_{L^2(G)^3} + 2\gamma_G \big\langle \mathbf{n}_G \times \mathbf{E}, \mathbf{n}_G \times \psi_h \big\rangle_{L^2(G)^3} \right].$$

Proof We only give the proof for the operator M_h^{upw}. For the case of the operator M_h^{cf}, it can be proved similarly.

(i) By (4.74), the sum over the faces vanishes, and for any $v_h = (\psi_h^\top, \phi_h^\top)^\top \in \mathbb{H}_{h,r}$, we obtain

$$\langle M_h^{\mathrm{upw}} u, v_h \rangle_{\mathbb{H}} = \sum_{K \in \mathscr{T}_h} \left[\langle \nabla \times \mathbf{H}, \psi_h \rangle_{L^2(K)^3} - \langle \nabla \times \mathbf{E}, \phi_h \rangle_{L^2(K)^3} \right]$$

$$= \langle \nabla \times \mathbf{H}, \psi_h \rangle_{L^2(D)^3} - \langle \nabla \times \mathbf{E}, \phi_h \rangle_{L^2(D)^3}$$

$$= \langle M u, v_h \rangle_{\mathbb{H}},$$

which implies the assertion $M_h^{\mathrm{upw}} u = \Pi_h M u$.

(ii) The integration by parts formula yields

$$\sum_{K \in \mathscr{T}_h} \left[\left\langle \nabla \times \mathbf{H}_h, \mathbf{E}_h \right\rangle_{L^2(K)^3} - \left\langle \nabla \times \mathbf{E}_h, \mathbf{H}_h \right\rangle_{L^2(K)^3} \right]$$

$$= - \sum_{G \in G_h^I} \left[\left\langle \mathbf{n}_G \times \mathbf{E}_K, \mathbf{H}_K \right\rangle_{L^2(G)^3} + \left\langle \mathbf{n}_{KG} \times \mathbf{E}_{KG}, \mathbf{H}_{KG} \right\rangle_{L^2(G)^3} \right] \quad (4.77)$$

$$- \sum_{G \in G_h^B} \left\langle \mathbf{n}_G \times \mathbf{E}_h, \mathbf{H}_h \right\rangle_{L^2(G)^3}.$$

Plugging (4.77) into the definition of M_h^{upw} and using the fact $\alpha_K + \beta_K = 1$, we obtain

$$\langle M_h^{\mathrm{upw}} u_h, u_h \rangle_{\mathbb{H}} = \sum_{G \in G_h^I} \left[\alpha_{KG} \left\langle \mathbf{n}_G \times \mathbf{E}_K, \mathbf{H}_{KG} \right\rangle_{L^2(G)^3} \right.$$

$$- \alpha_K \left\langle \mathbf{n}_G \times \mathbf{E}_{KG}, \mathbf{H}_K \right\rangle_{L^2(G)^3}$$

$$+ \beta_K \left\langle \mathbf{n}_G \times \mathbf{H}_{KG}, \mathbf{E}_K \right\rangle_{L^2(G)^3}$$

$$- \beta_{KG} \left\langle \mathbf{n}_G \times \mathbf{H}_K, \mathbf{E}_{KG} \right\rangle_{L^2(G)^3}$$

$$\left. - \gamma_G \left\| \mathbf{n}_G \times [\![\mathbf{E}_h]\!]_G \right\|_{L^2(G)^3}^2 - \delta_G \left\| \mathbf{n}_G \times [\![\mathbf{H}_h]\!]_G \right\|_{L^2(G)^3}^2 \right]$$

$$- 2 \sum_{G \in G_h^B} \gamma_G \| \mathbf{n}_G \times \mathbf{E}_h \|_{L^2(G)^3}^2.$$

Due to the fact that $\alpha_K + \alpha_{KG} = 1$, $\beta_K + \beta_{KG} = 1$, and $\alpha_K = \beta_{KG}$, we have

$$\sum_{G \in G_h^I} \left[\alpha_{KG} \left\langle \mathbf{n}_G \times \mathbf{E}_K, \mathbf{H}_{KG} \right\rangle_{L^2(G)^3} + \beta_K \left\langle \mathbf{n}_G \times \mathbf{H}_{KG}, \mathbf{E}_K \right\rangle_{L^2(G)^3} \right] = 0,$$

$$\sum_{G \in G_h^I} \left[\alpha_K \left\langle \mathbf{n}_G \times \mathbf{E}_{KG}, \mathbf{H}_K \right\rangle_{L^2(G)^3} + \beta_{KG} \left\langle \mathbf{n}_G \times \mathbf{H}_K, \mathbf{E}_{KG} \right\rangle_{L^2(G)^3} \right] = 0.$$

Therefore,

$$\langle M_h^{\mathrm{upw}} u_h, u_h \rangle_{\mathbb{H}} = - \sum_{G \in G_h^I} \left[\gamma_G \left\| \mathbf{n}_G \times [\![\mathbf{E}_h]\!]_G \right\|^2_{L^2(G)^3} \right.$$

$$\left. + \delta_G \left\| \mathbf{n}_G \times [\![\mathbf{H}_h]\!]_G \right\|^2_{L^2(G)^3} \right]$$

$$- 2 \sum_{G \in G_h^B} \gamma_G \| \mathbf{n}_G \times \mathbf{E}_h \|^2_{L^2(G)^3} \leq 0.$$

(iii) The proof is analogous to the proof of (ii) and is omitted.

Combining (i)–(iii), we finish the proof.   □

Based on Proposition 4.25, it can be shown that the dG algorithm (4.76) with central flux preserves the stochastic symplectic structure numerically, which is stated as follows.

Proposition 4.26 *Suppose that Assumption 4.1 holds. Under the homogeneous boundary condition, the dG algorithm (4.76) with $M_h = M_h^{\mathrm{cf}}$ preserves the following discrete stochastic symplectic structure*

$$\int_D d\mathbf{E}_h^{n+1}(\mathbf{x}) \wedge d\mathbf{H}_h^{n+1}(\mathbf{x}) d\mathbf{x} = \int_D d\mathbf{E}_h^n(\mathbf{x}) \wedge d\mathbf{H}_h^n(\mathbf{x}) d\mathbf{x}, \quad \mathbb{P}\text{-}a.s.$$

for all $n = 0, 1, \ldots, N - 1$.

Proof By the property of M_h^{cf} in Proposition 4.25 (ii), and following a similar approach to the proof of Theorem 3.1, we can obtain the result.   □

Next we investigate the divergence conservation property of (4.76). Define the test space X_h as

$$X_h := \left\{ v \in C^0(\overline{D}) : v_h|_K \in P^{r+1}(K) \text{ for all } K \in \mathcal{T}_h \right\} \cap H_0^1(D).$$

By $\langle \cdot, \cdot \rangle_{-1}$ we denote the duality product between $H^{-1}(D)$ and $H_0^1(D)$, in which $\langle \nabla \cdot \mathbf{E}, \psi \rangle_{-1} = -\langle \mathbf{E}, \nabla \psi \rangle_{L^2(D)^3}$ for all $\psi \in H_0^1(D)$. Then (4.76) possesses the following discrete averaged divergence conservation laws.

Proposition 4.27 *Under Assumption 4.1, the dG algorithm (4.76) with $M_h \in \{M_h^{\mathrm{cf}}, M_h^{\mathrm{upw}}\}$ satisfies the following discrete averaged divergence conservation* *laws:*

$$\mathbb{E}\big[\langle \nabla \cdot (\varepsilon \mathbf{E}_h^{n+1}), \phi \rangle_{-1}\big] = \mathbb{E}\big[\langle \nabla \cdot (\varepsilon \mathbf{E}_h^{n}), \phi \rangle_{-1}\big] \qquad \forall \phi \in X_h,$$

$$\mathbb{E}\big[\langle \nabla \cdot (\mu \mathbf{H}_h^{n+1}), \phi \rangle_{-1}\big] = \mathbb{E}\big[\langle \nabla \cdot (\mu \mathbf{H}_h^{n}), \phi \rangle_{-1}\big] \qquad \forall \phi \in X_h$$

for all $n = 0, 1, \dots, N - 1$.

Proof For $\psi, \phi \in X_h$, using the definition of the duality product $\langle \cdot, \cdot \rangle_{-1}$, we obtain

$$\left\langle \begin{bmatrix} \nabla \cdot (\varepsilon \mathbf{E}_h^{n+1}) \\ \nabla \cdot (\mu \mathbf{H}_h^{n+1}) \end{bmatrix}, \begin{bmatrix} \psi \\ \phi \end{bmatrix} \right\rangle_{-1} = \langle \nabla \cdot (\varepsilon \mathbf{E}_h^{n+1}), \psi \rangle_{-1} + \langle \nabla \cdot (\varepsilon \mathbf{H}_h^{n+1}), \phi \rangle_{-1}$$

$$= -\left\langle \begin{bmatrix} \mathbf{E}_h^{n+1} \\ \mathbf{H}_h^{n+1} \end{bmatrix}, \begin{bmatrix} \nabla \psi \\ \nabla \phi \end{bmatrix} \right\rangle_{\mathbb{H}}.$$

It follows from (4.76) that

$$\left\langle \begin{bmatrix} \mathbf{E}_h^{n+1} \\ \mathbf{H}_h^{n+1} \end{bmatrix}, \begin{bmatrix} \nabla \psi \\ \nabla \phi \end{bmatrix} \right\rangle_{\mathbb{H}} = \left\langle \begin{bmatrix} \mathbf{E}_h^{n} \\ \mathbf{H}_h^{n} \end{bmatrix}, \begin{bmatrix} \nabla \psi \\ \nabla \phi \end{bmatrix} \right\rangle_{\mathbb{H}} + \frac{\tau}{2}\left\langle M_h \begin{bmatrix} \mathbf{E}_h^{n} + \mathbf{E}_h^{n+1} \\ \mathbf{H}_h^{n} + \mathbf{H}_h^{n+1} \end{bmatrix}, \begin{bmatrix} \nabla \psi \\ \nabla \phi \end{bmatrix} \right\rangle_{\mathbb{H}}$$

$$+ \left\langle \Pi_h \begin{bmatrix} \lambda_1 \Delta W^{n+1} \\ \lambda_2 \Delta W^{n+1} \end{bmatrix}, \begin{bmatrix} \nabla \psi \\ \nabla \phi \end{bmatrix} \right\rangle_{\mathbb{H}}. \tag{4.78}$$

From Proposition 4.25 (iii), the second term on the right-hand side of (4.78) vanishes, since for any function $\varphi \in X_h$, we have $\nabla \times (\nabla \varphi) = \mathbf{0}$, $\mathbf{n}_G \times [\![\nabla \varphi]\!]_G = \mathbf{0}$ for $G \in G_h^I$, and $\mathbf{n}_G \times (\nabla \varphi) = \mathbf{0}$ for $G \in G_h^B$. For the third term on the right-hand side of (4.78), it is easy to see that the expectation is zero. Thus,

$$\mathbb{E}\left[\left\langle \begin{bmatrix} \mathbf{E}_h^{n+1} \\ \mathbf{H}_h^{n+1} \end{bmatrix}, \begin{bmatrix} \nabla \psi \\ \nabla \phi \end{bmatrix} \right\rangle_{\mathbb{H}}\right] = \mathbb{E}\left[\left\langle \begin{bmatrix} \mathbf{E}_h^{n} \\ \mathbf{H}_h^{n} \end{bmatrix}, \begin{bmatrix} \nabla \psi \\ \nabla \phi \end{bmatrix} \right\rangle_{\mathbb{H}}\right]$$

$$= -\mathbb{E}\left[\left\langle \begin{bmatrix} \nabla \cdot (\varepsilon \mathbf{E}_h^{n}) \\ \nabla \cdot (\mu \mathbf{H}_h^{n}) \end{bmatrix}, \begin{bmatrix} \psi \\ \phi \end{bmatrix} \right\rangle_{-1}\right].$$

The conclusion of this proposition comes from taking the test functions as $(\phi^\top, 0)^\top$ and $(0, \phi^\top)^\top$, respectively. $\qquad \square$

We have shown in Sect. 3.4 that the law of the exact solution at time T for the following small noise system

$$\begin{cases} du(t) = Mu(t)dt - \sqrt{\lambda}dW(t), & t \in (0, T], \\ u(0) = u_0 \end{cases} \tag{4.79}$$

satisfies a large deviations principle with a good rate function $\mathbb{I}_T^{u_0}$. Here, $\lambda \in \mathbb{R}^+$ and $W(t) = (\varepsilon^{-1} W_1(t)^\top, \mu^{-1} W_2(t)^\top)^\top$. In the sequel, we consider the large deviations principle of the numerical solution of the proposed dG algorithm applied to (4.79), that is

$$u_h^{n+1} = S_{h,\tau} u_h^n - \sqrt{\lambda} T_{h,\tau} \Pi_h \Delta W^{n+1}, \quad n = 0, 1, \ldots, N-1, \qquad (4.80)$$

where $S_{h,\tau} = \left(Id - \frac{\tau}{2} M_h \right)^{-1} \left(Id + \frac{\tau}{2} M_h \right)$ and $T_{h,\tau} = \left(Id - \frac{\tau}{2} M_h \right)^{-1}$. Let

$$W_{M;N,h} := \sum_{j=1}^{N} S_{h,\tau}^{N-j} T_{h,\tau} \Pi_h \Delta W^j.$$

Then it is Gaussian on $\mathbb{H}_{h,r}$ with mean zero and covariance operator

$$Q_{T;N,h} := \mathrm{Cov}(W_{M;N,h}) = \tau \sum_{j=1}^{N} \left(S_{h,\tau}^{N-j} T_{h,\tau} \Pi_h \right) Q (S_{h,\tau}^{N-j} T_{h,\tau} \Pi_h)^*.$$

Proposition 4.28 *Suppose that Assumption 4.1 holds. For $N \in \mathbb{N}_+$ and $u_0 \in \mathbb{H}$, the family of random variables $\{u_h^{N;\, u_0,\lambda}\}_{\lambda > 0}$ satisfies the large deviations principle with the good rate function*

$$\mathbb{I}_{T;N,h}^{u_0}(v) = \begin{cases} \frac{1}{2} \| (Q_{T;N,h})^{-\frac{1}{2}} (v - S_{h,\tau}^N u_h^0) \|_{\mathbb{H}}^2, & \text{if } v - S_{h,\tau}^N u_h^0 \in (Q_{T;N,h})^{\frac{1}{2}}(\mathbb{H}), \\ +\infty, & \text{otherwise}, \end{cases}$$

where $Q_{T;N,h}^{-\frac{1}{2}}$ is the pseudo-inverse of $Q_{T;N,h}^{\frac{1}{2}}$ and $u_h^{N;u_0,\lambda}$ is the mild solution of (4.80) given by

$$u_h^{N;u_0,\lambda} = S_{h,\tau}^N u_h^0 - \sqrt{\lambda} W_{M;N,h}, \quad N \in \mathbb{N}_+.$$

Proof The proof is similar to that of Proposition 3.3 and thus is omitted. □

(b) Stochastic multi-symplecticity

Now we turn to the stochastic multi-symplecticity of (4.71)–(4.72). To this end, we focus on the hybrid numerical flux, the cubic domain D, and its cubic partition. More precisely, the disjoint cubic element on D is defined by

$$K_{ijk} := I_i \times J_j \times G_k := [x_{i-\frac{1}{2}}, x_{i+\frac{1}{2}}] \times [y_{j-\frac{1}{2}}, y_{j+\frac{1}{2}}] \times [z_{k-\frac{1}{2}}, z_{k+\frac{1}{2}}]$$

for $i = 1, 2, \ldots, N_1$, $j = 1, 2, \ldots, N_2$, and $k = 1, 2, \ldots, N_3$. In this situation, the spatial domain $D = \bigcup\limits_{k=1}^{N_3} \bigcup\limits_{j=1}^{N_2} \bigcup\limits_{i=1}^{N_1} K_{ijk}$. Define the mesh sizes in x, y, z-directions as

$$h_{x,i} := x_{i+\frac{1}{2}} - x_{i-\frac{1}{2}}, \quad h_{y,j} := y_{j+\frac{1}{2}} - y_{j-\frac{1}{2}}, \quad h_{z,k} := z_{k+\frac{1}{2}} - z_{k-\frac{1}{2}}.$$

Let $h_x := \max_i h_{x,i}$, $h_y := \max_j h_{y,j}$, $h_z := \max_k h_{z,k}$, and $h := \max\{h_x, h_y, h_z\}$. For all $v_h \in \mathbb{H}_{h,r}$, let $(v_h^+)_{i+\frac{1}{2},y,z}$ and $(v_h^-)_{i+\frac{1}{2},y,z}$ be the right and left limits of v at the interface $\{x = x_{i+\frac{1}{2}}\}$, respectively.

In the above framework, the dG algorithm (4.71)–(4.72) for the stochastic Maxwell equations can be written into a compact form, that is, for $u_h = (\mathbf{E}_h^\top, \mathbf{H}_h^\top)^\top \in \mathbb{H}_{h,r}$, $v_h = (\psi_h^\top, \phi_h^\top)^\top \in \mathbb{H}_{h,r}$, we have

$$\int_{G_k} \int_{J_j} \int_{I_i} \mathbb{F} du_h \cdot v_h dx dy dz$$

$$- \int_{G_k} \int_{J_j} \left[\int_{I_i} \mathbb{K}_1 u_h \cdot (v_h)_x dx - \left(\widehat{\mathbb{K}_1 u_h} \cdot v_h^- \right)_{i+\frac{1}{2},y,z} \right.$$

$$\left. + \left(\widehat{\mathbb{K}_1 u_h} \cdot v_h^+ \right)_{i-\frac{1}{2},y,z} \right] dy dz dt$$

$$- \int_{G_k} \int_{I_i} \left[\int_{J_j} \mathbb{K}_2 u_h \cdot (v_h)_y dy - \left(\widehat{\mathbb{K}_2 u_h} \cdot v_h^- \right)_{x,j+\frac{1}{2},z} \right.$$

$$\left. + \left(\widehat{\mathbb{K}_2 u_h} \cdot v_h^+ \right)_{x,j-\frac{1}{2},z} \right] dx dz dt$$

$$- \int_{J_j} \int_{I_i} \left[\int_{G_k} \mathbb{K}_3 u_h \cdot (v_h)_z dz - \left(\widehat{\mathbb{K}_3 u_h} \cdot v_h^- \right)_{x,y,k+\frac{1}{2}} \right.$$

$$\left. + \left(\widehat{\mathbb{K}_3 u_h} \cdot v_h^+ \right)_{x,y,k-\frac{1}{2}} \right] dx dy dt$$

$$= \int_{G_k} \int_{J_j} \int_{I_i} \nabla_u S(u_h) \cdot v_h dx dy dz \circ dW(t), \tag{4.81}$$

where $S(u) = \lambda_2 \cdot \mathbf{E} - \lambda_1 \cdot \mathbf{H}$ and the skew-symmetric matrices $\mathbb{F}$ and $\mathbb{K}_p$ ($p = 1, 2, 3$) are defined in (1.23). Here, the hybrid numerical fluxes are given by

$$\widehat{\mathbb{K}_p u_h} = \mathbb{K}_p \{\{u_h\}\} + \mathbb{A}_p [\![u_h]\!] \tag{4.82}$$

with $\mathbb{A}_p$ ($p = 1, 2, 3$) being real symmetric matrix.

Lemma 4.1 *For all $u_h, v_h \in \mathbb{H}_{h,r}$ and $p = 1, 2, 3$,*

$$\mathbb{K}_p u_h^- \cdot v_h^- - \widehat{\mathbb{K}_p u_h} \cdot v_h^- + \widehat{\mathbb{K}_p v_h} \cdot u_h^- = \mathscr{R}_{\mathbb{K}_p}(u_h, v_h),$$

$$\mathbb{K}_p u_h^+ \cdot v_h^+ - \widehat{\mathbb{K}_p u_h} \cdot v_h^+ + \widehat{\mathbb{K}_p v_h} \cdot u_h^+ = \mathscr{R}_{\mathbb{K}_p}(u_h, v_h),$$

where

$$\mathscr{R}_{\mathbb{K}_p}(u_h, v_h) := \{\!\{\mathbb{K}_p u_h \cdot v_h\}\!\} - \widehat{\mathbb{K}_p u_h} \cdot \{\!\{v_h\}\!\} + \widehat{\mathbb{K}_p v_h} \cdot \{\!\{u_h\}\!\}.$$

Proof We only give the proof of the first equality, as the second one can be handled similarly. Let

$$\mathscr{K}(u_h, v_h) = \mathbb{K}_p u_h^- \cdot v_h^- - \widehat{\mathbb{K}_p u_h} \cdot v_h^- + \widehat{\mathbb{K}_p v_h} \cdot u_h^- - \mathscr{R}_{\mathbb{K}_p}(u_h, v_h),$$

which implies

$$\mathscr{K}(u_h, v_h) = \frac{1}{2}\Big(-[\![\mathbb{K}_p u_h \cdot v_h]\!] + \widehat{\mathbb{K}_p u_h} \cdot [\![v_h]\!] - \widehat{\mathbb{K}_p v_h} \cdot [\![u_h]\!] \Big).$$

By the definition of the numerical flux given in (4.82), we have

$$\mathscr{K}(u_h, v_h) = \frac{1}{2}\Big(-[\![\mathbb{K}_p u_h \cdot v_h]\!] + \mathbb{K}_p\{\!\{u_h\}\!\} \cdot [\![v_h]\!] - \mathbb{K}_p\{\!\{v_h\}\!\} \cdot [\![u_h]\!] \Big)$$

$$+ \frac{1}{2}\Big(\mathbb{A}_p[\![u_h]\!] \cdot [\![v_h]\!] - \mathbb{A}_p[\![v_h]\!] \cdot [\![u_h]\!] \Big) = 0$$

due to the skew-symmetry of $\mathbb{K}_p$ and the symmetry of $\mathbb{A}_p$.   □

Applying the exterior derivative to (4.81) yields

$$\int_{G_k}\int_{J_j}\int_{I_i} \mathbb{F}d(du_h) \cdot v_h \mathrm{d}x\mathrm{d}y\mathrm{d}z$$

$$- \int_{G_k}\int_{J_j}\Big[\int_{I_i} \mathbb{K}_1 du_h \cdot (v_h)_x \mathrm{d}x\mathrm{d}t - \Big(\widehat{\mathbb{K}_1 du_h} \cdot v_h^-\Big)_{i+\frac{1}{2},y,z}$$

$$+ \Big(\widehat{\mathbb{K}_1 du_h} \cdot v_h^+\Big)_{i-\frac{1}{2},y,z}\Big]\mathrm{d}y\mathrm{d}z\mathrm{d}t$$

$$- \int_{G_k}\int_{I_i}\Big[\int_{J_j} \mathbb{K}_2 du_h \cdot (v_h)_y \mathrm{d}y\mathrm{d}t - \Big(\widehat{\mathbb{K}_2 du_h} \cdot v_h^-\Big)_{x,j+\frac{1}{2},z}$$

$$+ \Big(\widehat{\mathbb{K}_2 du_h} \cdot v_h^+\Big)_{x,j-\frac{1}{2},z}\Big]\mathrm{d}x\mathrm{d}z\mathrm{d}t$$

$$-\int_{J_j}\int_{I_i}\Big[\int_{G_k}\mathbb{K}_3du_h\cdot(v_h)_z\mathrm{d}z\mathrm{d}t-\Big(\widehat{\mathbb{K}_3du_h}\cdot v_h^-\Big)_{x,y,k+\frac{1}{2}}$$

$$+\Big(\widehat{\mathbb{K}_3du_h}\cdot v_h^+\Big)_{x,y,k-\frac{1}{2}}\Big]\mathrm{d}x\mathrm{d}y\mathrm{d}t$$

$$=\int_{G_k}\int_{J_j}\int_{I_i}\nabla^2 S(u_h)du_h\cdot v_h\mathrm{d}x\mathrm{d}y\mathrm{d}z\circ\mathrm{d}W(t). \tag{4.83}$$

This gives the following result.

Theorem 4.3 *Suppose that Assumption 4.1 holds and Y_h, Z_h satisfy the variational equation (4.83), then the dG semi-discretization (4.81) with the fluxes defined in (4.82) possesses the stochastic multi-symplectic conservation law*

$$\mathrm{d}\varpi_{h,i,j,k}-(\kappa_1)_{h,i,y,z}\mathrm{d}t-(\kappa_2)_{h,x,j,z}\mathrm{d}t-(\kappa_3)_{h,x,y,k}\mathrm{d}t=0,\qquad\mathbb{P}\text{-}a.s.,$$

where

$$\varpi_{h,i,j,k}=\int_{G_k}\int_{J_j}\int_{I_i}\mathbb{F}Y_h\cdot Z_h\mathrm{d}x\mathrm{d}y\mathrm{d}z,$$

$$(\kappa_1)_{h,i,y,z}=\int_{G_k}\int_{J_j}\Big(\mathscr{R}_{\mathbb{K}_1}(Y_h,Z_h)_{i+\frac{1}{2},y,z}-\mathscr{R}_{\mathbb{K}_1}(Y_h,Z_h)_{i-\frac{1}{2},y,z}\Big)\mathrm{d}y\mathrm{d}z,$$

$$(\kappa_2)_{h,x,j,z}=\int_{G_k}\int_{I_i}\Big(\mathscr{R}_{\mathbb{K}_2}(Y_h,Z_h)_{x,j+\frac{1}{2},z}-\mathscr{R}_{\mathbb{K}_2}(Y_h,Z_h)_{x,j-\frac{1}{2},z}\Big)\mathrm{d}x\mathrm{d}z,$$

$$(\kappa_3)_{h,x,y,k}=\int_{J_j}\int_{I_i}\Big(\mathscr{R}_{\mathbb{K}_3}(Y_h,Z_h)_{x,y,k+\frac{1}{2}}-\mathscr{R}_{\mathbb{K}_3}(Y_h,Z_h)_{x,y,k-\frac{1}{2}}\Big)\mathrm{d}x\mathrm{d}y.$$

Proof By the skew-symmetry of $\mathbb{F}$ and (4.83), we split $\int_{G_k}\int_{J_j}\int_{I_i}\mathrm{d}(\mathbb{F}Y_h\cdot Z_h)\mathrm{d}x\mathrm{d}y\mathrm{d}z$ into four parts:

$$\int_{G_k}\int_{J_j}\int_{I_i}\mathrm{d}(\mathbb{F}Y_h\cdot Z_h)\mathrm{d}x\mathrm{d}y\mathrm{d}z=\int_{G_k}\int_{J_j}\int_{I_i}(\mathbb{F}\mathrm{d}Y_h\cdot Z_h-\mathbb{F}\mathrm{d}Z_h\cdot Y_h)\mathrm{d}x\mathrm{d}y\mathrm{d}z$$

$$=I+II+III+IV, \tag{4.84}$$

where

$$I := \int_{G_k} \int_{J_j} \left(\int_{I_i} \mathbb{K}_1 Y_h \cdot (Z_h)_x \mathrm{d}x \mathrm{d}t - \left(\widehat{\mathbb{K}_1 Y_h} \cdot Z_h^- \right)_{i+\frac{1}{2}, y, z} \right.$$
$$\left. + \left(\widehat{\mathbb{K}_1 Y_h} \cdot Z_h^+ \right)_{i-\frac{1}{2}, y, z} \right) \mathrm{d}y \mathrm{d}z \mathrm{d}t$$

$$- \int_{G_k} \int_{J_j} \left(\int_{I_i} \mathbb{K}_1 Z_h \cdot (Y_h)_x \mathrm{d}x \mathrm{d}t - \left(\widehat{\mathbb{K}_1 Z_h} \cdot Y_h^- \right)_{i+\frac{1}{2}, y, z} \right.$$
$$\left. + \left(\widehat{\mathbb{K}_1 Z_h} \cdot Y_h^+ \right)_{i-\frac{1}{2}, y, z} \right) \mathrm{d}y \mathrm{d}z \mathrm{d}t,$$

$$II := \int_{G_k} \int_{I_i} \left(\int_{J_j} \mathbb{K}_2 Y_h \cdot (Z_h)_y \mathrm{d}y \mathrm{d}t - \left(\widehat{\mathbb{K}_2 Y_h} \cdot Z_h^- \right)_{x, j+\frac{1}{2}, z} \right.$$
$$\left. + \left(\widehat{\mathbb{K}_2 Y_h} \cdot Z_h^+ \right)_{x, j-\frac{1}{2}, z} \right) \mathrm{d}x \mathrm{d}z \mathrm{d}t$$

$$- \int_{G_k} \int_{I_i} \left(\int_{J_j} \mathbb{K}_2 Z_h \cdot (Y_h)_y \mathrm{d}y \mathrm{d}t - \left(\widehat{K_2 Z_h} \cdot Y_h^- \right)_{x, j+\frac{1}{2}, z} \right.$$
$$\left. + \left(\widehat{\mathbb{K}_2 Z_h} \cdot Y_h^+ \right)_{x, j-\frac{1}{2}, z} \right) \mathrm{d}x \mathrm{d}z \mathrm{d}t,$$

$$III := \int_{J_j} \int_{I_i} \left(\int_{G_k} \mathbb{K}_3 Y_h \cdot (Z_h)_z \mathrm{d}z \mathrm{d}t - \left(\widehat{\mathbb{K}_3 Y_h} \cdot Z_h^- \right)_{x, y, k+\frac{1}{2}} \right.$$
$$\left. + \left(\widehat{\mathbb{K}_3 Y_h} \cdot Z_h^+ \right)_{x, y, k-\frac{1}{2}} \right) \mathrm{d}x \mathrm{d}y \mathrm{d}t$$

$$- \int_{J_j} \int_{I_i} \left(\int_{G_k} \mathbb{K}_3 Z_h \cdot (Y_h)_z \mathrm{d}z \mathrm{d}t - \left(\widehat{K_3 Z_h} \cdot Y_h^- \right)_{x, y, k+\frac{1}{2}} \right.$$
$$\left. + \left(\widehat{\mathbb{K}_3 Z_h} \cdot Y_h^+ \right)_{x, y, k-\frac{1}{2}} \right) \mathrm{d}x \mathrm{d}y \mathrm{d}t,$$

$$IV := \int_{G_k} \int_{J_j} \int_{I_i} \nabla^2 S(u_h) Y_h \cdot Z_h \circ \mathrm{d}W(t) \mathrm{d}x \mathrm{d}y \mathrm{d}z$$
$$- \int_{G_k} \int_{J_j} \int_{I_i} \nabla^2 S(u_h) Z_h \cdot Y_h \circ \mathrm{d}W(t) \mathrm{d}x \mathrm{d}y \mathrm{d}z.$$

By the skew-symmetry of $\mathbb{K}_p$ ($p = 1, 2, 3$) and Lemma 4.1, it yields that

$$I = \int_{G_k} \int_{J_j} \left(\mathbb{K}_1 Y_h^- \cdot Z_h^- - \widehat{\mathbb{K}_1 Y_h} \cdot Z_h^- + \widehat{\mathbb{K}_1 Z_h} \cdot Y_h^- \right)_{i+\frac{1}{2}, y, z} dy dz dt$$

$$- \int_{G_k} \int_{J_j} \left(\mathbb{K}_1 Y_h^+ \cdot Z_h^+ - \widehat{\mathbb{K}_1 Y_h} \cdot Z_h^+ + \widehat{\mathbb{K}_1 Z_h} \cdot Y_h^+ \right)_{i-\frac{1}{2}, y, z} dy dz dt$$

$$= \int_{G_k} \int_{J_j} \left(\mathscr{R}_{\mathbb{K}_1}(Y_h, Z_h)_{i+\frac{1}{2}, y, z} - \mathscr{R}_{\mathbb{K}_1}(Y_h, Z_h)_{i-\frac{1}{2}, y, z} \right) dy dz dt.$$

Similarly, we have

$$II = \int_{G_k} \int_{I_i} \left(\mathscr{R}_{\mathbb{K}_2}(Y_h, Z_h)_{x, j+\frac{1}{2}, z} - \mathscr{R}_{\mathbb{K}_2}(Y_h, Z_h)_{x, j-\frac{1}{2}, z} \right) dx dz dt,$$

$$III = \int_{J_j} \int_{I_i} \left(\mathscr{R}_{\mathbb{K}_3}(Y_h, Z_h)_{x, y, k+\frac{1}{2}} - \mathscr{R}_{\mathbb{K}_3}(Y_h, Z_h)_{x, y, k-\frac{1}{2}} \right) dx dy dt.$$

For the term IV, it follows from the symmetry of the Hessian matrix $\nabla^2 S(\cdot)$ that $IV = 0$.

Plugging these equalities into (4.84) yields the desired result.   □

Below we give a comment on the choice of the numerical flux $\widehat{\mathbb{K}_p u_h}$, $p = 1, 2, 3$.

Remark 4.6 Since $\mathbb{K}_p$ is skew-symmetric, there exists an orthogonal matrix $\mathbb{Q}_p$ such that

$$\mathbb{K}_p = \mathbb{Q}_p^\top \begin{bmatrix} 0 & -\Lambda_p^\top \\ \Lambda_p & 0 \end{bmatrix} \mathbb{Q}_p, \quad p = 1, 2, 3.$$

Assume that $\mathbb{Q}_p u_h = (y_h^\top, z_h^\top)^\top \in \mathbb{H}_{h,r}$. If we choose the matrix $\mathbb{A}_p$ as

$$\mathbb{A}_p = \alpha_p \mathbb{Q}_p^\top \begin{bmatrix} 0 & -\Lambda_p^\top \\ \Lambda_p & 0 \end{bmatrix} \mathbb{Q}_p$$

with $\alpha_p \in [-1/2, 1/2]$, the numerical flux $\widehat{\mathbb{K}_p u_h}$ reduces to

$$\widehat{\mathbb{K}_p u_h} = \mathbb{Q}_p^\top \begin{bmatrix} -\Lambda_p^\top \left(\{\!\{ z_h \}\!\} - \alpha_p [\![z_h]\!] \right) \\ \Lambda_p^\top \left(\{\!\{ y_h \}\!\} + \alpha_p [\![y_h]\!] \right) \end{bmatrix},$$

which retrieves alternating fluxes with $\alpha_p = \pm 1/2$, and the central flux with $\alpha_p = 0$.

We refer to [29, 160, 161] for more investigations on dG methods for the stochastic Maxwell equations.

4.2.4 Stochastic Conformal Multi-Symplectic and Ergodic Algorithm

This section is devoted to constructing a full discretization of the stochastic Maxwell equations (3.34) with damping, i.e.,

$$
\begin{cases}
du(t) = [Mu(t) - \sigma u(t)]dt + \lambda \mathbb{J}^{-1} u(t) \circ dW_1(t) + \theta dW_2(t), & t > 0, \\
u(0) = u_0
\end{cases}
$$

(4.85)

on a cubic domain $D = [a_1^-, a_1^+] \times [a_2^-, a_2^+] \times [a_3^-, a_3^+]$, where $\lambda \in \mathbb{R}$ and $\theta = (\theta_1^\top, \theta_2^\top) \in \mathbb{R}^6$. As shown in Sect. 3.5, (4.85) possesses the ergodicity and the stochastic conformal multi-symplectic conservation law. Below we aim to propose a fully discrete algorithm to inherit these two properties.

More precisely, we apply the stochastic midpoint method to (4.34) in the spatial direction and obtain

$$
\delta_t^{\sigma_{i,j,k}} u_{i+\frac{1}{2},j+\frac{1}{2},k+\frac{1}{2}}^n = \widetilde{M}(A_t^{\sigma_{i,j,k}} u_{i,j,k}^n) + \theta (\gamma_2)_{i+\frac{1}{2},j+\frac{1}{2},k+\frac{1}{2}}^n
$$

$$
+ \lambda \mathbb{J}^{-1}(A_t^{\sigma_{i,j,k}} u_{i+\frac{1}{2},j+\frac{1}{2},k+\frac{1}{2}}^n)(\overline{\gamma}_1)_{i+\frac{1}{2},j+\frac{1}{2},k+\frac{1}{2}}^n,
$$

(4.86)

where $\sigma_{i,j,k} := \sigma(x_i, y_j, z_k)$,

$$
\delta_t^{\sigma_{i,j,k}} u_{i+\frac{1}{2},j+\frac{1}{2},k+\frac{1}{2}}^n = \frac{u_{i+\frac{1}{2},j+\frac{1}{2},k+\frac{1}{2}}^{n+1} - e^{-\tau \sigma_{i,j,k}} u_{i+\frac{1}{2},j+\frac{1}{2},k+\frac{1}{2}}^n}{\tau},
$$

$$
A_t^{\sigma_{i,j,k}} u_{i,j,k}^n = \frac{u_{i,j,k}^{n+1} + e^{-\tau \sigma_{i,j,k}} u_{i,j,k}^n}{2},
$$

$$
(\overline{\gamma}_1)_{i,j,k}^{n+1} = \frac{(\Delta \overline{W}_1)_{i,j,k}^{n+1}}{\tau} = \frac{\overline{W}_1(t_{n+1}, x_i, y_j, z_k) - \overline{W}_1(t_n, x_i, y_j, z_k)}{\tau},
$$

$$
(\gamma_2)_{i,j,k}^{n+1} = \frac{(\Delta W_2)_{i,j,k}^{n+1}}{\tau} = \frac{W_2(t_{n+1}, x_i, y_j, z_k) - W_2(t_n, x_i, y_j, z_k)}{\tau},
$$

$\overline{W}_1$ is the truncation of W_1 (see (4.20)), and the discrete Maxwell operator $\widetilde{M}$ is defined by

$$
\widetilde{M} := \begin{bmatrix} 0 & \widetilde{\nabla\times} \\ -\widetilde{\nabla\times} & 0 \end{bmatrix} \quad \text{with} \quad \widetilde{\nabla\times} := \begin{bmatrix} 0 & -\delta_z A_x A_y & \delta_y A_x A_z \\ \delta_z A_x A_y & 0 & -\delta_x A_y A_z \\ -\delta_y A_x A_z & \delta_x A_y A_z & 0 \end{bmatrix}.
$$

Set $S_1(u) := \frac{1}{2}\lambda|u|^2$ and $S_2(u) := (\mathbb{J}\boldsymbol{\theta}) \cdot u$. Note that (4.86) can be written compactly as

$$
\mathbb{F}\delta_t^{\sigma_{i,j,k}} u^n_{i+\frac{1}{2},j+\frac{1}{2},k+\frac{1}{2}} + \mathbb{K}_1\delta_x(A_t^{\sigma_{i,j,k}} u^n_{i,j+\frac{1}{2},k+\frac{1}{2}})
$$

$$
+ \mathbb{K}_2\delta_y(A_t^{\sigma_{i,j,k}} u^n_{i+\frac{1}{2},j,k+\frac{1}{2}}) + \mathbb{K}_3\delta_z(A_t^{\sigma_{i,j,k}} u^n_{i+\frac{1}{2},j+\frac{1}{2},k})
$$

$$
= \nabla_u S_1(A_t^{\sigma_{i,j,k}} u^n_{i+\frac{1}{2},j+\frac{1}{2},k+\frac{1}{2}})(\overline{\gamma}_1)^{n+1}_{i+\frac{1}{2},j+\frac{1}{2},k+\frac{1}{2}}
$$

$$
+ \nabla_u S_2(A_t^{\sigma_{i,j,k}} u^n_{i+\frac{1}{2},j+\frac{1}{2},k+\frac{1}{2}})(\gamma_2)^{n+1}_{i+\frac{1}{2},j+\frac{1}{2},k+\frac{1}{2}},
$$

where $\mathbb{F}, \mathbb{K}_1, \mathbb{K}_2$, and $\mathbb{K}_3$ are defined in (3.12). It can be verified that (4.86) possesses the following stochastic conformal multi-symplectic conservation law.

Proposition 4.29 *The full discretization* (4.86) *possesses the discrete stochastic conformal multi-symplectic conservation law*

$$
\delta_t^{2\sigma_{i,j,k}} \varpi^n_{i+\frac{1}{2},j+\frac{1}{2},k+\frac{1}{2}} + \delta_x(\kappa_1)^{n,\sigma_{i,j,k}}_{i,j+\frac{1}{2},k+\frac{1}{2}}
$$

$$
+ \delta_y(\kappa_2)^{n,\sigma_{i,j,k}}_{i+\frac{1}{2},j,k+\frac{1}{2}} + \delta_z(\kappa_3)^{n,\sigma_{i,j,k}}_{i+\frac{1}{2},j+\frac{1}{2},k} = 0, \quad \mathbb{P}\text{-}a.s.,
$$

where $\varpi^n_{i,j,k} = \frac{1}{2}du^n_{i,j,k} \wedge \mathbb{F}du^n_{i,j,k}$, *and* $(\kappa_p)^{n,\sigma_{i,j,k}}_{i,j,k} = \frac{1}{2}d(A_t^{\sigma_{i,j,k}} u^n_{i,j,k}) \wedge \mathbb{K}_p d(A_t^{\sigma_{i,j,k}} u^n_{i,j,k})$, $p = 1, 2, 3$.

Proof The proof is analogous to that of Proposition 4.11 and is omitted. □

Denote the discrete energy of (4.86) by

$$
\Phi(t_n) := \Delta x \Delta y \Delta z \sum_{i=0}^{I-1} \sum_{j=0}^{J-1} \sum_{k=0}^{K-1} \left|u^n_{i+\frac{1}{2},j+\frac{1}{2},k+\frac{1}{2}}\right|^2.
$$

It can be shown that $\mathbb{E}[\Phi(t_n)]$ is uniformly bounded, which plays a key role in the proof of the ergodicity of (4.86).

Proposition 4.30 *Assume that*

(i) *the orthonormal basis $\{e_m\}_{m \in \mathbb{N}}$ in $U = L^2(D)$ is smooth and its first order derivative is bounded ;*
(ii) *the initial datum $u_0 \in L^2(\Omega, \mathbb{H})$, and the damped coefficient $\sigma \geq \sigma_0 > 0$ with a constant σ_0;*
(iii) *the operators $Q_i^{\frac{1}{2}} \in HS(U, H^{\gamma_i}(D))$, $i = 1, 2$ with $\gamma_1 > 3/2$ and $\gamma_2 > 5/2$.*

Then, under the periodic boundary condition, for sufficiently small τ, it holds

$$\mathbb{E}\big[\Phi(t_n)\big] \leq e^{-\sigma_0 n \tau} \mathbb{E}\big[\Phi(t_0)\big] + C, \tag{4.87}$$

where the positive constant $C = C(\sigma_0, \lambda, \theta, |D|, Q_1, Q_2)$ is independent of τ and n.

Proof Similar to the proof of Proposition 4.15, one can check that the discrete energy Φ satisfies the following evolution relation:

$$\Phi(t_{n+1}) = \Delta x \Delta y \Delta z \sum_{i=0}^{I-1} \sum_{j=0}^{J-1} \sum_{k=0}^{K-1} e^{-2\tau \sigma_{i,j,k}} \Big| u^n_{i+\frac{1}{2}, j+\frac{1}{2}, k+\frac{1}{2}} \Big|^2$$

$$+ 2\Delta x \Delta y \Delta z \sum_{i=0}^{I-1} \sum_{j=0}^{J-1} \sum_{k=0}^{K-1} \Big(\Upsilon^{n, \sigma_{i,j,k}}_{i+\frac{1}{2}, j+\frac{1}{2}, k+\frac{1}{2}} (\Delta W_2)^{n+1}_{i+\frac{1}{2}, j+\frac{1}{2}, k+\frac{1}{2}} \Big)$$

$$\tag{4.88}$$

for all $n \in \mathbb{N}$, where $\Upsilon^{n, \sigma_{i,j,k}}_{i+\frac{1}{2}, j+\frac{1}{2}, k+\frac{1}{2}} := \theta \cdot A_t^{\sigma_{i,j,k}} u^n_{i+\frac{1}{2}, j+\frac{1}{2}, k+\frac{1}{2}}$.
For the first term on the right-hand side of (4.88), we have

$$\Delta x \Delta y \Delta z \sum_{i=0}^{I-1} \sum_{j=0}^{J-1} \sum_{k=0}^{K-1} e^{-2\tau \sigma_{i,j,k}} \Big| u^n_{i+\frac{1}{2}, j+\frac{1}{2}, k+\frac{1}{2}} \Big|^2 \leq e^{-2\sigma_0 \tau} \Phi(t_n) \tag{4.89}$$

according to the assumption $\sigma \geq \sigma_0 > 0$. For the second term on the right-hand side of (4.88), the estimate is more complicated. For the simplicity of notations, we denote $\bar{s} := s + \frac{1}{2}$ for $s = i, j, k$ in the following proof.

Now let us first consider the sub-term $A_t^{\sigma_{i,j,k}} (E_1)^n_{\bar{i}, \bar{j}, \bar{k}} (\Delta W_2)^{n+1}_{\bar{i}, \bar{j}, \bar{k}}$. Using the fact that ΔW_2^{n+1} is independent of $\mathscr{F}_{t_n}$, we have

$$\mathbb{E}\Big[A_t^{\sigma_{i,j,k}} (E_1)^n_{\bar{i}, \bar{j}, \bar{k}} (\Delta W_2)^{n+1}_{\bar{i}, \bar{j}, \bar{k}} \Big]$$

$$= \frac{1}{2} \mathbb{E}\Big[\Big((E_1)^{n+1}_{\bar{i}, \bar{j}, \bar{k}} - e^{-\sigma_{i,j,k} \tau} (E_1)^n_{\bar{i}, \bar{j}, \bar{k}} \Big) (\Delta W_2)^{n+1}_{\bar{i}, \bar{j}, \bar{k}} \Big] \tag{4.90}$$

$$= \frac{1}{2} \mathbb{E}\Big[\Big(\tau \delta_t^{\sigma_{i,j,k}} (E_1)^n_{\bar{i}, \bar{j}, \bar{k}} \Big) (\Delta W_2)^{n+1}_{\bar{i}, \bar{j}, \bar{k}} \Big].$$

It follows from (4.86) that the first component $(E_1)^n_{i,j,k}$ of $u^n_{i,j,k}$ satisfies

$$\delta_t^{\sigma_{i,j,k}}(E_1)^n_{\bar{i},\bar{j},\bar{k}} = \delta_y A_t^{\sigma_{i,j,k}}(H_3)^n_{i,\bar{j},\bar{k}} - \delta_z A_t^{\sigma_{i,j,k}}(H_2)^n_{\bar{i},\bar{j},k}$$

$$- \lambda A_t^{\sigma_{i,j,k}}(H_1)^n_{\bar{i},j,k}(\overline{\gamma}_1)^n_{\bar{i},\bar{j},\bar{k}} + \theta_1^{(1)}(\gamma_2)^n_{\bar{i},\bar{j},\bar{k}},$$

where $\theta_1^{(\ell)}$ is the ℓ-th component of $\boldsymbol{\theta}_1$, $\ell = 1,2,3$. Plugging the above identity into (4.90) and adding up all indices yield

$$\mathbb{E}\Big[\sum_{i=0}^{I-1}\sum_{j=0}^{J-1}\sum_{k=0}^{K-1} A_t^{\sigma_{i,j,k}}(E_1)^n_{\bar{i},\bar{j},\bar{k}}(\Delta W_2)^{n+1}_{\bar{i},\bar{j},\bar{k}}\Big]$$

$$= \frac{1}{2}\mathbb{E}\Big[\sum_{i=0}^{I-1}\sum_{j=0}^{J-1}\sum_{k=0}^{K-1}\Big(\tau\delta_y A_t^{\sigma_{i,j,k}}(H_3)^n_{i,\bar{j},\bar{k}} - \tau\delta_z A_t^{\sigma_{i,j,k}}(H_2)^n_{\bar{i},\bar{j},k}$$

$$- \lambda A_t^{\sigma_{i,j,k}}(H_1)^n_{\bar{i},j,k}(\Delta\overline{W}_1)^{n+1}_{\bar{i},\bar{j},\bar{k}} + \theta_1^{(1)}(\Delta W_2)^{n+1}_{\bar{i},\bar{j},\bar{k}}\Big)(\Delta W_2)^{n+1}_{\bar{i},\bar{j},\bar{k}}\Big]$$

$$= \frac{1}{2}\mathbb{E}\Big[\sum_{i=0}^{I-1}\sum_{j=0}^{J-1}\sum_{k=0}^{K-1}\Big(-\tau A_t^{\sigma_{i,j,k}}(H_3)^n_{i,\bar{j},\bar{k}}\delta_y(\Delta W_2)^{n+1}_{i,\bar{j},\bar{k}}$$

$$+ \tau A_t^{\sigma_{i,j,k}}(H_2)^n_{\bar{i},\bar{j},k}\delta_z(\Delta W_2)^{n+1}_{\bar{i},\bar{j},k}$$

$$- \lambda A_t^{\sigma_{i,j,k}}(H_1)^n_{\bar{i},j,k}(\Delta\overline{W}_1)^{n+1}_{\bar{i},\bar{j},\bar{k}}(\Delta W_2)^{n+1}_{\bar{i},\bar{j},\bar{k}} + \theta_1^{(1)}\big[(\Delta W_2)^{n+1}_{\bar{i},\bar{j},\bar{k}}\big]^2\Big)\Big], \tag{4.91}$$

where we used the fact that

$$\sum_{j=0}^{J-1}\delta_y(A_t^{\sigma_{i,j,k}}(H_3)^n_{i,\bar{j},\bar{k}})(\Delta W_2)^{n+1}_{\bar{i},\bar{j},\bar{k}} = -\sum_{j=0}^{J-1}(A_t^{\sigma_{i,j,k}}(H_3)^n_{i,\bar{j},\bar{k}})\delta_y(\Delta W_2)^{n+1}_{i,\bar{j},\bar{k}},$$

$$\sum_{k=0}^{K-1}\delta_z(A_t^{\sigma_{i,j,k}}(H_2)^n_{\bar{i},\bar{j},k})(\Delta W_2)^{n+1}_{\bar{i},\bar{j},\bar{k}} = -\sum_{k=0}^{K-1}(A_t^{\sigma_{i,j,k}}(H_2)^n_{\bar{i},\bar{j},k})\delta_z(\Delta W_2)^{n+1}_{\bar{i},\bar{j},k}$$

$$\tag{4.92}$$

due to the periodic boundary condition. Hence, the Hölder inequality and the Young inequality imply that

$$\theta_1^{(1)}\mathbb{E}\Big[\sum_{i=0}^{I-1}\sum_{j=0}^{J-1}\sum_{k=0}^{K-1} A_t^{\sigma_{i,j,k}}(E_1)^n_{\bar{i},\bar{j},\bar{k}}(\Delta W_2)^{n+1}_{\bar{i},\bar{j},\bar{k}}\Big]$$

$$\leq \frac{\sigma_0}{8}\tau\mathbb{E}\Big[\sum_{i=0}^{I-1}\sum_{j=0}^{J-1}\sum_{k=0}^{K-1}\big|A_t^{\sigma_{i,j,k}}\mathbf{H}^n_{\bar{i},\bar{j},\bar{k}}\big|^2\Big]$$

$$+ C\tau \mathbb{E}\Big[\sum_{i=0}^{I-1}\sum_{j=0}^{J-1}\sum_{k=0}^{K-1} \Big(|\delta_y (\Delta W_2)_{i,j,k}^{n+1}|^2 + |\delta_z (\Delta W_2)_{i,j,k}^{n+1}|^2 \Big) \Big]$$

$$+ C\frac{1}{\tau} \sum_{i=0}^{I-1}\sum_{j=0}^{J-1}\sum_{k=0}^{K-1} \Big(\mathbb{E}\big[|(\Delta \overline{W}_1)_{i,j,k}^{n+1}|^2 \big]\mathbb{E}\big[|(\Delta W_2)_{i,j,k}^{n+1}|^2 \big] + \tau \mathbb{E}\big[|(\Delta W_2)_{i,j,k}^{n+1}|^2 \big] \Big).$$

Notice that

$$\Delta x \Delta y \Delta z \sum_{i=0}^{I-1}\sum_{j=0}^{J-1}\sum_{k=0}^{K-1} \mathbb{E}\Big[|(\Delta W_2)_{i,\bar{j},\bar{k}}^{n}|^2 \Big]$$

$$= \Delta x \Delta y \Delta z \sum_{i=0}^{I-1}\sum_{j=0}^{J-1}\sum_{k=0}^{K-1} \mathbb{E}\Big[\Big| \sum_{m\in\mathbb{N}} \sqrt{\eta_m^{(2)}}\, e_m(x_{\bar{i}}, y_{\bar{j}}, z_{\bar{k}})(\beta_m^{(2)}(t_{n+1}) - \beta_m^{(2)}(t_n)) \Big|^2 \Big]$$

$$= \tau \Delta x \Delta y \Delta z \sum_{i=0}^{I-1}\sum_{j=0}^{J-1}\sum_{k=0}^{K-1}\sum_{m\in\mathbb{N}} \eta_m^{(2)} |e_m(x_{\bar{i}}, y_{\bar{j}}, z_{\bar{k}})|^2$$

$$\leq \tau \Delta x \Delta y \Delta z \sum_{i=0}^{I-1}\sum_{j=0}^{J-1}\sum_{k=0}^{K-1}\sum_{m\in\mathbb{N}} \eta_m^{(2)} \|e_m\|_{L^\infty(D)}^2$$

$$\leq C\Big(|D|, \|Q_2^{\frac{1}{2}}\|_{HS(U,H^{\gamma_1}(D))} \Big)\tau$$

and

$$\Delta x \Delta y \Delta z \sum_{i=0}^{I-1}\sum_{j=0}^{J-1}\sum_{k=0}^{K-1} \Big(\mathbb{E}\big[|(\Delta \overline{W}_1)_{i,j,\bar{k}}^{n}|^2 \big]\mathbb{E}\big[|(\Delta W_2)_{i,\bar{j},\bar{k}}^{n}|^2 \big] \Big)$$

$$\leq |D|\Big[\Big(\sum_{m\in\mathbb{N}} \eta_m |e_m(x_{\bar{i}}, y_{\bar{j}}, z_{\bar{k}})|^2 \tau \Big)\Big(\sum_{m\in\mathbb{N}} \eta_m |e_m(x_{\bar{i}}, y_{\bar{j}}, z_{\bar{k}})|^2 \tau \Big) \Big]$$

$$\leq C\Big(|D|, \|Q_1^{\frac{1}{2}}\|_{HS(U,H^{\gamma_1}(D))}, \|Q_2^{\frac{1}{2}}\|_{HS(U,H^{\gamma_1}(D))} \Big)\tau^2,$$

where in the last steps of the above two inequalities we used the Sobolev embedding $H^\gamma \hookrightarrow L^\infty(D)$ with $\gamma > 3/2$. Combining the above three estimates, we obtain

$$\Delta x \Delta y \Delta z \theta_1^{(1)}\mathbb{E}\Big[\sum_{i=0}^{I-1}\sum_{j=0}^{J-1}\sum_{k=0}^{K-1} A_t^{\sigma_{i,j,k}}(E_1)_{i,\bar{j},\bar{k}}^{n}(\Delta W_2)_{i,\bar{j},\bar{k}}^{n+1} \Big]$$

$$\leq \frac{\sigma_0}{16}\tau \mathbb{E}\Big[\sum_{i=0}^{I-1}\sum_{j=0}^{J-1}\sum_{k=0}^{K-1} \Big(|\mathbf{H}_{i,j,k}^{n+1}|^2 + e^{-2\tau\sigma_{i,j,k}}|\mathbf{H}_{i,\bar{j},\bar{k}}^{n}|^2 \Big) \Big]$$

$$+ C\tau \mathbb{E}\Big[\sum_{i=0}^{I-1}\sum_{j=0}^{J-1}\sum_{k=0}^{K-1} \Big(|\delta_y (\Delta W_2)_{i,j,k}^{n+1}|^2 + |\delta_z (\Delta W_2)_{i,j,k}^{n+1}|^2 \Big) \Big] + C\tau.$$

The other five sub-terms of the second term on the right-hand side of (4.88) can be handled in a similar way. As a consequence, we obtain

$$\Delta x \Delta y \Delta z \sum_{i=0}^{I-1} \sum_{j=0}^{J-1} \sum_{k=0}^{K-1} \mathbb{E}\Big[\Upsilon^n_{i+\frac{1}{2},j+\frac{1}{2},k+\frac{1}{2}} (\Delta W_2)^{n+1}_{i+\frac{1}{2},j+\frac{1}{2},k+\frac{1}{2}} \Big]$$

$$\leq \frac{3\sigma_0}{16} \tau \, \mathbb{E}\big[\Phi(t_{n+1})\big] + \frac{3\sigma_0}{16} \tau e^{-2\sigma_0 \tau} \mathbb{E}\big[\Phi(t_n)\big] + C\tau$$

$$+ C\tau \Delta x \Delta y \Delta z \, \mathbb{E}\Big[\sum_{i=0}^{I-1} \sum_{j=0}^{J-1} \sum_{k=0}^{K-1} \Big(|\delta_x (\Delta W_2)^{n+1}_{i,j,k}|^2 $$

$$+ |\delta_y(\Delta W_2)^{n+1}_{i,j,k}|^2 + |\delta_z(\Delta W_2)^{n+1}_{i,j,k}|^2 \Big)\Big].$$

Moreover, by the Sobolev embedding $H^\gamma \hookrightarrow L^\infty(D)$ with $\gamma > 3/2$, it yields that

$$\Delta x \Delta y \Delta z \, \mathbb{E}\Big[\sum_{i=0}^{I-1} \sum_{j=0}^{J-1} \sum_{k=0}^{K-1} |\delta_x(\Delta W_2)^n_{i,\bar{j},\bar{k}}|^2 \Big]$$

$$= \tau \Delta x \Delta y \Delta z \sum_{i=0}^{I-1} \sum_{j=0}^{J-1} \sum_{k=0}^{K-1} \sum_{m\in\mathbb{N}} \eta_m^{(2)} |\partial_x e_m(x_i + \xi_i \Delta x, y_{\bar{j}}, z_{\bar{k}})|^2$$

$$\leq \tau \sum_{i=0}^{I-1} \sum_{j=0}^{J-1} \sum_{k=0}^{K-1} \sum_{m\in\mathbb{N}} \eta_m^{(2)} \|e_m\|^2_{W^{1,\infty}(D)} \leq C\Big(|D|, \|Q_2^{\frac{1}{2}}\|_{HS(U,H^{\gamma_2}(D))}\Big)\tau,$$

where $\xi_i \in [0, 1]$, $i = 0, 1, \ldots, I - 1$. Similarly, we can derive

$$\Delta x \Delta y \Delta z \, \mathbb{E}\Big[\sum_{i=0}^{I-1} \sum_{j=0}^{J-1} \sum_{k=0}^{K-1} |\delta_y(\Delta W_2)^n_{i,j,\bar{k}}|^2 \Big] \leq C\Big(|D|, \|Q_2^{\frac{1}{2}}\|_{HS(U,H^{\gamma_2}(D))}\Big)\tau,$$

$$\Delta x \Delta y \Delta z \, \mathbb{E}\Big[\sum_{i=0}^{I-1} \sum_{j=0}^{J-1} \sum_{k=0}^{K-1} |\delta_z(\Delta W_2)^n_{i,\bar{j},k}|^2 \Big] \leq C\Big(|D|, \|Q_2^{\frac{1}{2}}\|_{HS(U,H^{\gamma_2}(D))}\Big)\tau.$$

Hence,

$$\Delta x \Delta y \Delta z \sum_{i=0}^{I-1} \sum_{j=0}^{J-1} \sum_{k=0}^{K-1} \mathbb{E}\Big[\Upsilon^n_{i+\frac{1}{2},j+\frac{1}{2},k+\frac{1}{2}} (\Delta W_2)^{n+1}_{i+\frac{1}{2},j+\frac{1}{2},k+\frac{1}{2}} \Big]$$

$$\leq \frac{3\sigma_0}{16} \tau \mathbb{E}\big[\Phi(t_{n+1})\big] + \frac{3\sigma_0}{16} \tau e^{-2\sigma_0 \tau} \mathbb{E}\big[\Phi(t_n)\big] + C\tau,$$

(4.93)

which, together with (4.88) and (4.89), yields that

$$\mathbb{E}\big[\Phi(t_{n+1})\big] \le e^{-2\sigma_0\tau}\mathbb{E}\big[\Phi(t_n)\big] + \frac{3\sigma_0}{16}\tau\mathbb{E}\big[\Phi(t_{n+1})\big] + \frac{3\sigma_0}{16}\tau e^{-2\sigma_0\tau}\mathbb{E}\big[\Phi(t_n)\big] + C\tau.$$

By the Grönwall inequality, it can be shown that for sufficiently small τ,

$$\mathbb{E}\big[\Phi(t_n)\big] \le e^{-\sigma_0 n\tau}\mathbb{E}\big[\Phi(t_0)\big] + C.$$

Thus, the proof is finished. □

Let

$$U^n = \Big((E_1)^n_{\frac{1}{2},\frac{1}{2},\frac{1}{2}}, (E_1)^n_{\frac{3}{2},\frac{1}{2},\frac{1}{2}}, \dots, (E_1)^n_{I-\frac{1}{2},\frac{1}{2},\frac{1}{2}}, (E_1)^n_{\frac{1}{2},\frac{3}{2},\frac{1}{2}}, \dots, (E_1)^n_{I-\frac{1}{2},\frac{3}{2},\frac{1}{2}},$$

$$\dots, (E_1)^n_{I-\frac{1}{2},J-\frac{1}{2},K-\frac{1}{2}}, (E_2)^n_{\frac{1}{2},\frac{1}{2},\frac{1}{2}}, \dots, (E_2)^n_{I-\frac{1}{2},J-\frac{1}{2},K-\frac{1}{2}}, (E_3)^n_{\frac{1}{2},\frac{1}{2},\frac{1}{2}},$$

$$\dots, (E_3)^n_{I-\frac{1}{2},J-\frac{1}{2},K-\frac{1}{2}}, (H_1)^n_{\frac{1}{2},\frac{1}{2},\frac{1}{2}}, \dots, (H_3)^n_{I-\frac{1}{2},J-\frac{1}{2},K-\frac{1}{2}}\Big)^{\top}.$$

Then the discrete energy $\Phi(t_n)$ can be rewritten as

$$\Phi(t_n) = \Delta x \Delta y \Delta z |U^n|^2, \quad n \in \mathbb{N}. \tag{4.94}$$

By taking (4.94) as a Lyapunov function, the proof of the ergodicity of (4.86) is similar to that of Theorem 3.6, and hence is skipped.

Proposition 4.31 *Under conditions in Proposition 4.30 and the periodic boundary condition, the following statements hold.*

(i) *Let $\{U^n; n \in \mathbb{N}\}$ and $\{\widetilde{U}^n; n \in \mathbb{N}\}$ be solutions of (4.86) with initial values U^0 and $\widetilde{U}^0$, respectively. Then*

$$\mathbb{E}\Big[|U^n - \widetilde{U}^n|^2\Big] \le e^{-2\sigma_0 t_n}\mathbb{E}\Big[|U^0 - \widetilde{U}^0|^2\Big].$$

(ii) *For sufficiently small τ, the numerical solution $\{U^n; n \in \mathbb{N}\}$ of (4.86) has a unique invariant measure $\pi^{\tau,h} \in \mathscr{P}_2(\mathbb{H})$. Thus, $\{U^n; n \in \mathbb{N}\}$ is ergodic. Moreover, $\{U^n; n \in \mathbb{N}\}$ is exponentially mixing.*

(iii) *For arbitrary two distributions $\pi_1, \pi_2 \in \mathscr{P}_2(\mathbb{H})$, $n \in \mathbb{N}$, it holds that*

$$\mathscr{W}_2((P_n^{\tau,h})^*\pi_1, (P_n^{\tau,h})^*\pi_2) \le e^{-\sigma_0 t_n}\mathscr{W}_2(\pi_1, \pi_2),$$

where $(P_n^{\tau,h})^*\pi, n \in \mathbb{N}$ denotes the probability distribution of U^n with initial probability distribution π. Moreover,

$$\mathscr{W}_2((P_n^{\tau,h})^*\pi, \pi^{\tau,h}) \le e^{-\sigma_0 t_n}\mathscr{W}_2(\pi, \pi^{\tau,h}) \quad \forall \pi \in \mathscr{P}_2(\mathbb{H}), n \in \mathbb{N}.$$

4.3 Splitting Techniques for Stochastic Maxwell Equations

As is well known, the numerical simulation of the stochastic Maxwell equations is consuming, which requires the investigation of highly efficient numerical algorithms. In order to reduce the computational cost and improve efficiency, this section presents the splitting technique for the stochastic Maxwell equations with additive noise

$$
\begin{cases}
\varepsilon d\mathbf{E}(t) = \nabla \times \mathbf{H}(t)dt + \lambda_1 dW(t), & t \in (0, T], \\
\mu d\mathbf{H}(t) = -\nabla \times \mathbf{E}(t)dt + \lambda_2 dW(t), & t \in (0, T], \\
\mathbf{E}(0) = \mathbf{E}_0, \quad \mathbf{H}(t) = \mathbf{H}_0
\end{cases}
\tag{4.95}
$$

on a cuboid $D = (a_1^-, a_1^+) \times (a_2^-, a_2^+) \times (a_3^-, a_3^+)$ with a Lipschitz boundary Γ. For convenience, we denote

$$
\Gamma_1^+ := \{\mathbf{x} \in \overline{D} : x = a_1^+\}, \quad \Gamma_1^- := \{\mathbf{x} \in \overline{D} : x = a_1^-\}.
$$

And $\Gamma_2^+, \Gamma_2^-, \Gamma_3^+$, and Γ_3^- are defined similarly. Let $\Gamma_j := \Gamma_j^- \cup \Gamma_j^+$ for $j = 1, 2, 3$.

The approach studied here is to split the Maxwell operator

$$
M = \begin{bmatrix} 0 & \varepsilon^{-1}\nabla\times \\ -\mu^{-1}\nabla\times & 0 \end{bmatrix}, \quad \mathscr{D}(M) = H_0(\mathrm{curl}, D) \times H(\mathrm{curl}, D)
$$

with the aim of the feasible implementation and economic memory for the numerical algorithms. For this, we classify the splitting technique into two classes: local one-dimensional splitting and alternating direction implicit splitting.

(a) The local one-dimensional splitting

The local one-dimensional technique, which can allow the multi-dimensional problems to be solved by treating the spatial variables individually in a cyclic fashion, is popular in saving memory and CPU time when applied to discretize partial differential equations (see e.g., [25, 36, 72, 157]). For the Maxwell operator M, it can be decomposed into

$$
M_\alpha := \begin{bmatrix} 0 & \varepsilon^{-1}\mathrm{curl}_\alpha \\ -\mu^{-1}\mathrm{curl}_\alpha & 0 \end{bmatrix}, \quad \alpha = x, y, z
\tag{4.96}
$$

with one-dimensional differential operator-valued matrices

$$\text{curl}_x = \begin{bmatrix} 0 & 0 & 0 \\ 0 & 0 & -\partial_x \\ 0 & \partial_x & 0 \end{bmatrix}, \quad \text{curl}_y = \begin{bmatrix} 0 & 0 & \partial_y \\ 0 & 0 & 0 \\ -\partial_y & 0 & 0 \end{bmatrix}, \quad \text{curl}_z = \begin{bmatrix} 0 & -\partial_z & 0 \\ \partial_z & 0 & 0 \\ 0 & 0 & 0 \end{bmatrix}.$$

The domains of M_x, M_y, and M_z are given by

$$\mathscr{D}(M_x) = \{u \in \mathbb{H}: \ M_x u \in \mathbb{H}, \ u_2 = u_3 = 0 \ \text{on} \ \Gamma_1\},$$

$$\mathscr{D}(M_y) = \{u \in \mathbb{H}: \ M_y u \in \mathbb{H}, \ u_1 = u_3 = 0 \ \text{on} \ \Gamma_2\},$$

$$\mathscr{D}(M_z) = \{u \in \mathbb{H}: \ M_z u \in \mathbb{H}, \ u_1 = u_2 = 0 \ \text{on} \ \Gamma_3\},$$

respectively.

It can be verified that $\mathscr{D}(M_x) \cap \mathscr{D}(M_y) \cap \mathscr{D}(M_z) \subset \mathscr{D}(M)$ and $M_x u + M_y u + M_z u = M u$ for $u \in \mathscr{D}(M_x) \cap \mathscr{D}(M_y) \cap \mathscr{D}(M_z)$. Furthermore, operators $(M_x, \mathscr{D}(M_x))$, $(M_y, \mathscr{D}(M_y))$, and $(M_z, \mathscr{D}(M_z))$ generate unitary C_0-semigroups. The proof is similar to that of Theorem 2.1 and is omitted.

Lemma 4.2 *If ε, μ satisfy Assumption 2.1, then the operator $M_\alpha : \mathscr{D}(M_\alpha) \subset \mathbb{H} \rightarrow \mathbb{H}$ is skew-adjoint, and generates a unitary C_0-semigroup $\{S_\alpha(t) := e^{t M_\alpha}, t \geq 0\}$ on $\mathbb{H}$ for $\alpha = x, y, z$.*

(b) The alternating direction implicit splitting

The main idea of the alternating direction implicit splitting is, roughly speaking, to decompose the operator into two parts and to propagate the associated sub-flows in such a way that the implicitness is reduced to one-dimensional problems. This invention has attracted a lot of interest, and a large number of follow-up papers can be found in the literature, see, e.g., [75, 85, 98, 141] and references therein. In this setting, the Maxwell operator is decomposed as $M = M_1 + M_2$ with

$$M_1 = \begin{bmatrix} 0 & \varepsilon^{-1}\text{curl}_1 \\ -\mu^{-1}\text{curl}_2 & 0 \end{bmatrix} \quad \text{and} \quad M_2 = \begin{bmatrix} 0 & \varepsilon^{-1}\text{curl}_2 \\ -\mu^{-1}\text{curl}_1 & 0 \end{bmatrix}. \tag{4.97}$$

Here,

$$\text{curl}_1 = \begin{bmatrix} 0 & 0 & \partial_y \\ \partial_z & 0 & 0 \\ 0 & \partial_x & 0 \end{bmatrix}, \quad \text{curl}_2 = \begin{bmatrix} 0 & -\partial_z & 0 \\ 0 & 0 & -\partial_x \\ -\partial_y & 0 & 0 \end{bmatrix},$$

and the domains of M_1 and M_2 are

$$\mathscr{D}(M_1) = \{u \in \mathbb{H} : M_1 u \in \mathbb{H}, \ u_1 = 0 \text{ on } \Gamma_2, \ u_2 = 0 \text{ on } \Gamma_3, \ u_3 = 0 \text{ on } \Gamma_1\},$$

$$\mathscr{D}(M_2) = \{u \in \mathbb{H} : M_2 u \in \mathbb{H}, \ u_1 = 0 \text{ on } \Gamma_3, \ u_2 = 0 \text{ on } \Gamma_1, \ u_3 = 0 \text{ on } \Gamma_2\},$$

respectively.

Similarly, we have $\mathscr{D}(M_1) \cap \mathscr{D}(M_2) \subset \mathscr{D}(M)$ and $M_1 u + M_2 u = Mu$ for $u \in \mathscr{D}(M_1) \cap \mathscr{D}(M_2)$. Moreover, we can derive the following lemma.

Lemma 4.3 *If ε, μ satisfy Assumption 2.1, then the operator $M_j : \mathscr{D}(M_j) \subset \mathbb{H} \to \mathbb{H}$ is skew-adjoint, and generates a unitary C_0-semigroup $\{S_j(t) := e^{tM_j}, t \geq 0\}$ on $\mathbb{H}$ for $j = 1, 2$.*

Hence, based on the above claims, the stochastic Maxwell equations (4.95) can be split into several lower dimensional subsystems. Without loss of generality, we just focus on the local one-dimensional splitting case and the alternating direction implicit splitting can be presented in the same procedure. For simplicity, below we assume that $\varepsilon = \mu \equiv 1$.

Denote $\boldsymbol{\lambda}^{[1]} := (\lambda_1^{(1)}, 0, 0, \lambda_2^{(1)}, 0, 0)^\top$, $\boldsymbol{\lambda}^{[2]} := (0, \lambda_1^{(2)}, 0, 0, \lambda_2^{(2)}, 0)^\top$, and $\boldsymbol{\lambda}^{[3]} := (0, 0, \lambda_1^{(3)}, 0, 0, \lambda_2^{(3)})^\top$. Then we have the following three one-dimensional subsystems when applying the local one-dimensional splitting:

$$\begin{cases} du^{[j]}(t) = M_\alpha u^{[j]}(t)dt + \boldsymbol{\lambda}^{[j]}dW(t), & t \in (0, T], \\ u^{[j]}(0) = u_0^{[j]} \end{cases} \tag{4.98}$$

for $(\alpha, j) \in \{(x, 1), (y, 2), (z, 3)\}$.

Remark 4.7 It can be observed from (4.98) that each of them can be implemented easily. To show this clearly, we take $\alpha = x$, $j = 1$ as an example, namely,

$$\begin{cases} dE_2^{[1]}(t) = -\partial_x H_3^{[1]}(t)dt, \\ dH_3^{[1]}(t) = -\partial_x E_2^{[1]}(t)dt, \end{cases} \quad \begin{cases} dE_3^{[1]}(t) = \partial_x H_2^{[1]}(t)dt, \\ dH_2^{[1]}(t) = \partial_x E_3^{[1]}(t)dt, \end{cases} \tag{4.99}$$

$$dE_1^{[1]}(t) = \lambda_1^{(1)}dW(t), \quad dH_1^{[1]}(t) = \lambda_2^{(1)}dW(t). \tag{4.100}$$

It can be observed that one only needs to solve some small deterministic linear systems in each time step. In addition, we can obtain the analytical expression of the solution of (4.100). These characteristics of the splitting technique lead to a dramatic reduction of computational costs in solving the stochastic Maxwell equations.

It follows from Theorem 2.2 and Lemma 4.2 that (4.98) has a unique mild solution given by

$$u^{[j]}(t) = S_\alpha(t)u_0^{[j]} + \int_0^t S_\alpha(t-s)\lambda^{[j]}dW(s), \quad \mathbb{P}\text{-}a.s. \tag{4.101}$$

for all $t \in [0, T]$ and $(\alpha, j) \in \{(x, 1), (y, 2), (z, 3)\}$.

Similar to Remark 3.2 (a), we have the following averaged energy evolution law.

Proposition 4.32 *Let $u_0^{[j]}$ be $\mathscr{F}_0$-measurable $\mathbb{H}$-valued random variables satisfying $\|u_0^{[j]}\|_{L^2(\Omega,\mathbb{H})} < \infty$ for $j = 1, 2, 3$. Then*

$$\mathbb{E}\big[\|u^{[j]}(t)\|_{\mathbb{H}}^2\big] = \mathbb{E}\big[\|u_0^{[j]}\|_{\mathbb{H}}^2\big] + t|\lambda^{[j]}|^2\text{Tr}(Q)$$

for all $t \in [0, T]$ and $j = 1, 2, 3$.

Below, we focus on investigating the stochastic symplectic and multi-symplectic structures of (4.98). To present the formulation of the stochastic Hamiltonian system, we define

$$\mathscr{H}_1^{[j]}(\mathbf{E}^{[j]}, \mathbf{H}^{[j]}) = \frac{1}{2}\int_D \Big(|\mathbf{E}^{[j]}|^2 + |\mathbf{H}^{[j]}|^2\Big)dx,$$

$$\mathscr{H}_2^{[j]}(\mathbf{E}^{[j]}, \mathbf{H}^{[j]}) = \int_D \Big(\lambda_2^{(j)}E_j^{[j]} - \lambda_1^{(j)}H_j^{[j]}\Big)dx,$$

from which we can reformulate (4.98) as the following non-canonical stochastic Hamiltonian system

$$d\begin{bmatrix}\mathbf{E}^{[j]} \\ \mathbf{H}^{[j]}\end{bmatrix} = \begin{bmatrix} 0 & \text{curl}_\alpha \\ -\text{curl}_\alpha & 0 \end{bmatrix}\begin{bmatrix} \frac{\delta\mathscr{H}_1^{[j]}}{\delta\mathbf{E}^{[j]}} \\ \frac{\delta\mathscr{H}_1^{[j]}}{\delta\mathbf{H}^{[j]}} \end{bmatrix}dt + \mathbb{J}^{-1}\begin{bmatrix} \frac{\delta\mathscr{H}_2^{[j]}}{\delta\mathbf{E}^{[j]}} \\ \frac{\delta\mathscr{H}_2^{[j]}}{\delta\mathbf{H}^{[j]}} \end{bmatrix} \circ dW(t) \tag{4.102}$$

for $(\alpha, j) \in \{(x, 1), (y, 2), (z, 3)\}$. Taking the exterior derivative on both sides of (4.102) and utilizing the skew-adjointness of M_α, $\alpha = x, y, z$, yield the following assertion.

Lemma 4.4 *The phase flow of (4.98) preserves the stochastic symplectic structures, that is,*

$$\varpi^{[j]}(t) := \int_D d\mathbf{E}^{[j]}(t, \mathbf{x}) \wedge \text{curl}_\alpha(d\mathbf{H}^{[j]}(t, \mathbf{x}))dx = \varpi^{[j]}(0), \quad \mathbb{P}\text{-}a.s.$$

for all $t \in [0, T]$ and $(\alpha, j) \in \{(x, 1), (y, 2), (z, 3)\}$.

Using the Hamiltonian

$$\mathscr{H}_1^{[\alpha]}(\mathbf{E}^{[j]}, \mathbf{H}^{[j]}) = -\frac{1}{2}\int_D \left(\mathbf{E}^{[j]}\cdot\mathrm{curl}_\alpha\mathbf{E}^{[j]} + \mathbf{H}^{[j]}\cdot\mathrm{curl}_\alpha\mathbf{H}^{[j]}\right)\mathrm{dx},$$

we can derive the canonical stochastic Hamiltonian system of (4.98):

$$\mathrm{d}\begin{bmatrix}\mathbf{E}^{[j]}\\\mathbf{H}^{[j]}\end{bmatrix} = \mathbb{J}^{-1}\begin{bmatrix}\frac{\delta\mathscr{H}_1^{[\alpha]}}{\delta\mathbf{E}^{[j]}}\\\frac{\delta\mathscr{H}_1^{[\alpha]}}{\delta\mathbf{H}^{[j]}}\end{bmatrix}\mathrm{d}t + \mathbb{J}^{-1}\begin{bmatrix}\frac{\delta\mathscr{H}_2^{[j]}}{\delta\mathbf{E}^{[j]}}\\\frac{\delta\mathscr{H}_2^{[j]}}{\delta\mathbf{H}^{[j]}}\end{bmatrix}\circ\mathrm{d}W(t) \tag{4.103}$$

for $(\alpha, j) \in \{(x, 1), (y, 2), (z, 3)\}$. In this situation, the stochastic symplectic structure

$$\varpi^{[j]}(t) := \int_D d\mathbf{E}^{[j]}(t, \mathbf{x}) \wedge d\mathbf{H}^{[j]}(t, \mathbf{x})\mathrm{dx}$$

is preserved by the phase flow of (4.98), if the homogeneous boundary condition is enforced (cf. Theorem 3.1).

Now we turn to the stochastic multi-symplecticity of (4.98). Let

$$S(u^{[j]}) = \lambda_2^{(j)}E_j^{[j]} - \lambda_1^{(j)}H_j^{[j]}, \quad j = 1, 2, 3.$$

Then (4.98) can be rewritten as

$$\mathbb{F}du^{[j]} + \mathbb{K}_j\partial_\alpha u^{[j]}\mathrm{d}t = \nabla_{u^{[j]}}S(u^{[j]})\circ\mathrm{d}W(t) \tag{4.104}$$

for $(\alpha, j) \in \{(x, 1), (y, 2), (z, 3)\}$. Here, $\mathbb{F}$ and $\mathbb{K}_j$, $j = 1, 2, 3$ are defined in (3.12). Similar to the proof of Theorem 3.2, we obtain the following result.

Lemma 4.5 *The system* (4.98) *preserves the stochastic multi-symplectic conservative law, that is, for* $(\alpha, j) \in \{(x, 1), (y, 2), (z, 3)\}$,

$$\mathrm{d}\varpi^{[j]} + \partial_\alpha\kappa^{[j]}\mathrm{d}t = 0, \quad \mathbb{P}\text{-}a.s.,$$

which means

$$\int_{\alpha_0}^{\alpha_1}\varpi^{[j]}(t_1, \alpha)\mathrm{d}\alpha + \int_{t_0}^{t_1}\kappa^{[j]}(t, \alpha_1)\mathrm{d}t = \int_{\alpha_0}^{\alpha_1}\varpi^{[j]}(t_0, \alpha)\mathrm{d}\alpha + \int_{t_0}^{t_1}\kappa^{[j]}(t, \alpha_0)\mathrm{d}t,$$

where $\varpi^{[j]}(t, \mathbf{x}) = \frac{1}{2}du^{[j]} \wedge \mathbb{F}du^{[j]}$ *and* $\kappa^{[j]}(t, \mathbf{x}) = \frac{1}{2}du^{[j]} \wedge \mathbb{K}_jdu^{[j]}$ *are differential 2-forms associated with skew-symmetric matrices* $\mathbb{F}$ *and* $\mathbb{K}_j$, $j = 1, 2, 3$.

Summary and Outlook

This chapter is devoted to the construction and analysis of structure-preserving algorithms, which can inherit the intrinsic properties of the stochastic Maxwell equations with Stratonovich noise.

For the temporal semi-discretizations for the stochastic Maxwell equations, we investigate stochastic symplectic Runge–Kutta methods and exponential-type methods. The well-posedness and regularity of exponential-type methods are presented for both the additive and multiplicative cases. While we restrict the analysis to the additive noise case for stochastic Runge–Kutta methods. For the multiplicative noise case, we only establish the well-posedness and regularity of the stochastic midpoint method. It is worth studying these problems for the general stochastic Runge–Kutta methods in the multiplicative noise case.

Then we present the construction of full discretizations by discretizing the temporal semi-discretizations further in the spatial direction via the finite difference method, the wavelet method, and the dG method, respectively. The *a priori* estimates and intrinsic discrete structures of these obtained full discretization are analyzed, which will play an important role in the error analysis in Chap. 5. There are some other approaches to constructing structure-preserving algorithms for stochastic partial differential equations. For instance, we refer to [180] for the stochastic multi-symplectic Runge–Kutta method, and to [104] for the mesh-less local radial basis function collocation method, the splitting multi-symplectic Runge–Kutta method and the multi-symplectic partitioned Runge–Kutta method for the stochastic Hamiltonian partial differential equations. Moreover, we propose and analyze the stochastic conformal multi-symplectic and ergodic algorithm for the stochastic Maxwell equations with damping; see [45] for more details. We also refer to [156] for the study of stochastic conformal schemes for the damped stochastic Klein–Gordon equation, and to [10] for the approach based on the Wiener chaos expansion for the stochastic wave equation.

In the implementation of numerical algorithms for the three-dimensional stochastic Maxwell equations, we remark that it is required to solve at least a 10^6-scale algebraic equation at every time step provided that the considered spatial domain is divided into $100 \times 100 \times 100$ cells. Moreover, the computational cost will be at least a multiple of P, where P is the number of samples. Thus, the numerical implementation of the three-dimensional stochastic Maxwell equations is a very difficult issue due to the limitation of memory and the performance of the CPU in a common computing environment. The splitting technique proposed in this chapter is a good tool to reduce computational costs. It is also a challenging and meaningful topic to study whether the time parallel method, the domain decomposition method, and the multilevel Monte–Carlo method can be utilized to construct both highly efficient and structure-preserving algorithms for the stochastic Maxwell equations.

Chapter 5
Convergence Analysis of Structure-Preserving Algorithms for Stochastic Maxwell Equations

This chapter is concerned with the convergence analysis of structure-preserving algorithms proposed in Chap. 4 for the stochastic Maxwell equations.

Section 5.1 concentrates on the convergence analysis of the temporally semi-discrete algorithms. The mean-square convergence orders of stochastic midpoint methods for the considered system with either additive noise or multiplicative noise are presented in Sect. 5.1.1. Moreover, we introduce error estimates of the general stochastic symplectic Runge–Kutta methods for the additive noise case in Sect. 5.1.2. Finally, the convergence analyses of exponential-type methods for the considered system with either additive noise or multiplicative noise are shown in Sect. 5.1.3.

Section 5.2 is devoted to studying the mean-square convergence order of the fully discrete algorithms. We first present the convergence analysis of the dG method in space in Sect. 5.2.1. Then for the stochastic symplectic dG full discretization, its mean-square convergence order is obtained in Sect. 5.2.2.

As discussed in the previous chapter, the splitting technique for the stochastic Maxwell equations is proposed to reduce the computational cost. In order to clarify the influence of the splitting technique on the convergence order of the numerical algorithm, we analyze the splitting error and derive the mean-square convergence order of the splitting midpoint method in Sect. 5.3.

5.1 Convergence Analysis for Temporally Semi-Discrete Algorithms

In this section, we focus on the convergence analysis of temporally semi-discrete structure-preserving algorithms, including the stochastic midpoint method, stochastic symplectic Runge–Kutta methods, and exponential-type methods, for the

C. Chen et al., *Numerical Approximations of Stochastic Maxwell Equations*,
Lecture Notes in Mathematics 2341, https://doi.org/10.1007/978-981-99-6686-8_5

stochastic Maxwell equations. Let $D \subset \mathbb{R}^3$ be an open, bounded, and Lipschitz domain with boundary ∂D.

5.1.1 Stochastic Midpoint Method

Let us first consider the stochastic Maxwell equations with additive noise

$$
\begin{cases}
du(t) = \Big[Mu(t) + F(t, u(t)) \Big] dt + B(t) dW(t), & t \in (0, T], \\
u(0) = u_0
\end{cases}
\tag{5.1}
$$

and the corresponding stochastic midpoint method

$$
u^{n+1} = u^n + \tau M u^{n+\frac{1}{2}} + \tau F^{n+\frac{1}{2}}(u^{n+\frac{1}{2}}) + B^{n+\frac{1}{2}} \Delta W^{n+1}, \quad n = 0, 1, \ldots, N-1
\tag{5.2}
$$

with $u^0 = u_0$, $u^{n+\frac{1}{2}} = \frac{1}{2}(u^n + u^{n+1})$, $F^{n+\frac{1}{2}}(u^{n+\frac{1}{2}}) = F(t_{n+\frac{1}{2}}, u^{n+\frac{1}{2}})$, and $B^{n+\frac{1}{2}} = B(t_{n+\frac{1}{2}})$. We have the following mean-square convergence result.

Theorem 5.1 *Let conditions with $k \in \{1, 2\}$ in Proposition 4.3 hold. Assume that $F(t, \cdot)$ is twice Fréchet differentiable with bounded derivatives for $t \in [0, T]$. Then the solution of (5.2) satisfies*

$$
\max_{1 \leq n \leq N} \Big(\mathbb{E}\Big[\| u(t_n) - u^n \|_{\mathbb{H}}^2 \Big] \Big)^{1/2} \leq C \tau^{k/2},
$$

where the positive constant C depends on T, F, B, and u_0.

Proof For $n = 1, 2, \ldots, N$, note that (5.1) has a unique mild solution given by

$$
u(t_n) = S(t_n) u_0 + \int_0^{t_n} S(t_n - r) F(r, u(r)) dr + \int_0^{t_n} S(t_n - r) B(r) dW(r),
\tag{5.3}
$$

and that the mild form of (5.2) is

$$
u^n = (S_\tau)^n u_0 + \tau \sum_{j=0}^{n-1} (S_\tau)^{n-j-1} T_\tau F^{j+\frac{1}{2}}(u^{j+\frac{1}{2}}) + \sum_{j=0}^{n-1} (S_\tau)^{n-j-1} T_\tau B^{j+\frac{1}{2}} \Delta W^{j+1},
\tag{5.4}
$$

where $S_\tau := \left(Id - \frac{\tau}{2}M\right)^{-1}\left(Id + \frac{\tau}{2}M\right)$ and $T_\tau := \left(Id - \frac{\tau}{2}M\right)^{-1}$. Denoting $e^n := u(t_n) - u^n$, we have

$$
\begin{aligned}
e^n ={} & \left(S(t_n) - (S_\tau)^n\right)u_0 \\
&+ \sum_{j=0}^{n-1}\int_{t_j}^{t_{j+1}}\left[S(t_n - r)F(r, u(r)) - (S_\tau)^{n-j-1}T_\tau F^{j+\frac{1}{2}}(u^{j+\frac{1}{2}})\right]dr \\
&+ \sum_{j=0}^{n-1}\int_{t_j}^{t_{j+1}}\left[S(t_n - r)B(r) - (S_\tau)^{n-j-1}T_\tau B^{j+\frac{1}{2}}\right]dW(r) \\
=:{} & \left(S(t_n) - (S_\tau)^n\right)u_0 + I_n + J_n.
\end{aligned} \tag{5.5}
$$

It follows from Lemma C.3 (iii) that

$$
\mathbb{E}\left[\left\|\left(S(t_n) - (S_\tau)^n\right)u_0\right\|_{\mathbb{H}}^2\right] \le C\mathbb{E}\left[\|u_0\|_{\mathscr{D}(M^k)}^2\right]\tau^k.
$$

Step 1. Estimate of the term I_n. We have

$$
\begin{aligned}
I_n ={} & \sum_{j=0}^{n-1}\int_{t_j}^{t_{j+1}}\left[S(t_n - r)\Big(F(r, u(r)) - F(t_j, u(t_j))\Big)\right]dr \\
&+ \sum_{j=0}^{n-1}\int_{t_j}^{t_{j+1}}\left[S(t_n - r)\Big(F(t_j, u(t_j)) - F\big(t_{j+\frac{1}{2}}, \frac{u(t_j) + u(t_{j+1})}{2}\big)\Big)\right]dr
\end{aligned}
$$

$$ \tag{5.6} $$

$$
\begin{aligned}
&+ \sum_{j=0}^{n-1}\int_{t_j}^{t_{j+1}}\left[\Big(S(t_n - r) - (S_\tau)^{n-j-1}T_\tau\Big)F\big(t_{j+\frac{1}{2}}, \frac{u(t_j) + u(t_{j+1})}{2}\big)\right]dr \\
&+ \sum_{j=0}^{n-1}\int_{t_j}^{t_{j+1}}\left[(S_\tau)^{n-j-1}T_\tau\Big(F\big(t_{j+\frac{1}{2}}, \frac{u(t_j) + u(t_{j+1})}{2}\big) - F\big(t_{j+\frac{1}{2}}, u^{j+\frac{1}{2}}\big)\Big)\right]dr \\
=:{} & I_n^1 + I_n^2 + I_n^3 + I_n^4.
\end{aligned}
$$

Using the Taylor expansion gives

$$
\begin{aligned}
F(r, u(r)) - F(t_j, u(t_j)) ={} & F(r, u(r)) - F(t_j, u(r)) + F_u(t_j, u(t_j))\big(u(r) - u(t_j)\big) \\
&+ \int_0^1 \theta F_{uu}(t_j, u_\theta)\big(u(r) - u(t_j), u(r) - u(t_j)\big)d\theta,
\end{aligned}
$$

where $u_\theta = \theta u(t_j) + (1 - \theta)u(r)$ for $\theta \in [0, 1]$. Combining this with (5.3), the term I_n^1 can be rewritten as

$$I_n^1 = \sum_{j=0}^{n-1} \int_{t_j}^{t_{j+1}} \left[S(t_n - r) F_u(t_j, u(t_j)) \Big(S(r - t_j) - Id \Big) u(t_j) \right] dr$$

$$+ \sum_{j=0}^{n-1} \int_{t_j}^{t_{j+1}} \left[S(t_n - r) F_u(t_j, u(t_j)) \int_{t_j}^{r} S(r - \xi) F(\xi, u(\xi)) d\xi \right] dr$$

$$+ \sum_{j=0}^{n-1} \int_{t_j}^{t_{j+1}} \left[S(t_n - r) F_u(t_j, u(t_j)) \int_{t_j}^{r} S(r - \xi) B(\xi) dW(\xi) \right] dr \qquad (5.7)$$

$$+ \sum_{j=0}^{n-1} \int_{t_j}^{t_{j+1}} \left[S(t_n - r) \int_0^1 F_{uu}(t_j, u_\theta) \Big(u(r) - u(t_j), u(r) - u(t_j) \Big) d\theta \right] dr$$

$$+ \sum_{j=0}^{n-1} \int_{t_j}^{t_{j+1}} \left[S(t_n - r) \Big(F(r, u(r)) - F(t_j, u(r)) \Big) \right] dr$$

$$=: I_n^{1,1} + I_n^{1,2} + I_n^{1,3} + I_n^{1,4} + I_n^{1,5}.$$

For the term $I_n^{1,1}$, by the unitarity of the semigroup $\{S(t), t \in [0, T]\}$, Lemma C.1, and Theorem 2.4, we obtain

$$\mathbb{E}\big[\|I_n^{1,1}\|_{\mathbb{H}}^2\big] \leq C\,\mathbb{E} \sum_{j=0}^{n-1} \int_{t_j}^{t_{j+1}} \left\| F_u(t_j, u(t_j)) \Big(S(r - t_j) - Id \Big) u(t_j) \right\|_{\mathbb{H}}^2 dr$$

$$\leq C \sum_{j=0}^{n-1} \int_{t_j}^{t_{j+1}} (r - t_j)^2 \mathbb{E}\big[\|u(t_j)\|_{\mathscr{D}(M)}^2\big] dr$$

$$\leq C\tau^2.$$

For the term $I_n^{1,2}$, by the linear growth of F, it holds that

$$\mathbb{E}\big[\|I_n^{1,2}\|_{\mathbb{H}}^2\big] \leq C \sum_{j=0}^{n-1} \int_{t_j}^{t_{j+1}} (r - t_j) \int_{t_j}^{r} \mathbb{E}\big[\|F(\xi, u(\xi))\|_{\mathbb{H}}^2\big] d\xi\, dr$$

$$\leq C \sum_{j=0}^{n-1} \int_{t_j}^{t_{j+1}} (r - t_k) \int_{t_j}^{r} (1 + \mathbb{E}\big[\|u(\xi)\|_{\mathbb{H}}^2\big]) d\xi\, dr$$

$$\leq C\tau^2.$$

For the third term $I_n^{1,3}$, the stochastic Fubini theorem and the Itô isometry lead to

$$\mathbb{E}\big[\|I_n^{1,3}\|_{\mathbb{H}}^2\big]$$

$$= \mathbb{E}\Big[\big\| \sum_{j=0}^{n-1} \int_{t_j}^{t_{j+1}} \int_{\xi}^{t_{j+1}} \big[S(t_{n+1}-r)F_u(t_j,u(t_j))S(r-\xi)B(\xi)\big]dr\,dW(\xi) \big\|_{\mathbb{H}}^2\Big]$$

$$= \sum_{j=0}^{n-1} \int_{t_j}^{t_{j+1}} \mathbb{E}\Big[\big\| \int_{\xi}^{t_{j+1}} \big[S(t_{n+1}-r)F_u(t_j,u(t_j))S(r-\xi)B(\xi)\big]dr \big\|_{HS(U_0,\mathbb{H})}^2\Big]d\xi$$

$$\leq C\tau^2.$$

It follows from Assumption 2.4 and Theorem 2.5 that

$$\mathbb{E}\Big[\|I_n^{1,4}\|_{\mathbb{H}}^2 + \|I_n^{1,5}\|_{\mathbb{H}}^2\Big] \leq C\tau^2.$$

Combining the above estimates, we have

$$\mathbb{E}\Big[\|I_n^1\|_{\mathbb{H}}^2\Big] \leq C\tau^2.$$

Similar to the estimate of I_n^1, the term I_n^2 is estimated as

$$\mathbb{E}\Big[\|I_n^2\|_{\mathbb{H}}^2\Big] \leq C\tau^2.$$

For the term I_n^3, one has

$$\mathbb{E}\Big[\|I_n^3\|_{\mathbb{H}}^2\Big] \leq C\sum_{j=0}^{n-1} \int_{t_j}^{t_{j+1}} \mathbb{E}\Big[\big\|\big(S(t_n-r)-(S_\tau)^{n-j-1}T_\tau\big)$$

$$\times F\big(t_{j+\frac{1}{2}}, \frac{u(t_j)+u(t_{j+1})}{2}\big)\big\|_{\mathbb{H}}^2\Big]dr$$

$$\leq C\tau^k \sum_{j=0}^{n-1} \int_{t_j}^{t_{j+1}} \mathbb{E}\Big[1 + \|u(t_j)\|_{\mathscr{D}(M^k)}^2 + \|u(t_{j+1})\|_{\mathscr{D}(M^k)}^2\Big]dr$$

$$\leq C\tau^k,$$

where we used Lemma C.3 (iv) and Theorem 2.4. For the last term I_n^4, one has

$$\mathbb{E}\Big[\|I_n^4\|_{\mathbb{H}}^2\Big] \leq C\tau \sum_{j=0}^{n-1} \mathbb{E}\Big[\|e^j\|_{\mathbb{H}}^2 + \|e^{j+1}\|_{\mathbb{H}}^2\Big].$$

Therefore, we obtain

$$\mathbb{E}\Big[\|I_n\|_{\mathbb{H}}^2\Big] \le C\tau^k + C\tau \sum_{j=0}^{n} \mathbb{E}\Big[\|e^j\|_{\mathbb{H}}^2 + \|e^{j+1}\|_{\mathbb{H}}^2\Big]. \tag{5.8}$$

Step 2. Estimate of the term J_n. We decompose the term J_n as

$$J_n = \sum_{j=0}^{n-1} \int_{t_j}^{t_{j+1}} \Big[S(t_n - r) - (S_\tau)^{n-j-1}T_\tau\Big]B(r)\mathrm{d}W(r)$$

$$+ \sum_{j=0}^{n-1} \int_{t_j}^{t_{j+1}} (S_\tau)^{n-j-1}T_\tau\Big[B(r) - B(t_{j+\frac{1}{2}})\Big]\mathrm{d}W(r)$$

$$=: J_n^1 + J_n^2.$$

Notice that

$$\mathbb{E}\Big[\|J_n^1\|_{\mathbb{H}}^2\Big] = \sum_{j=0}^{n-1} \int_{t_j}^{t_{j+1}} \mathbb{E}\Big[\Big\|\Big(S(t_n - r) - (S_\tau)^{n-j-1}T_\tau\Big)B(r)\Big\|_{HS(U_0,\mathbb{H})}^2\Big]\mathrm{d}r$$

$$\le \sum_{j=0}^{n-1} \int_{t_j}^{t_{j+1}} \Big\|S(t_n - r)-(S_\tau)^{n-j-1}\Big\|_{\mathscr{L}(\mathscr{D}(M^k),\mathbb{H})}^2 \|B(r)\|_{HS(U_0,\mathscr{D}(M^k))}^2\mathrm{d}r$$

$$\le C\tau^k$$

due to Lemma C.3 (iii). Similarly, we can derive $\mathbb{E}\Big[\|J_n^2\|_{\mathbb{H}}^2\Big] \le C\tau^2$.

Combining *Step 1* and *Step 2* yields

$$\mathbb{E}\Big[\|e^n\|_{\mathbb{H}}^2\Big] \le C\tau^k + C\tau \sum_{j=0}^{n-1} \mathbb{E}\Big[\|e^j\|_{\mathbb{H}}^2 + \|e^{j+1}\|_{\mathbb{H}}^2\Big].$$

The conclusion of Theorem 5.1 follows from the discrete Grönwall inequality given in Proposition A.5. □

Now we are in the position to study the stochastic Maxwell equations driven by linear multiplicative noise

$$\begin{cases} \mathrm{d}u(t) = \Big[Mu(t) + F(t, u(t))\Big]\mathrm{d}t + \lambda\mathbb{J}^{-1}u(t) \circ \mathrm{d}W(t), \quad t \in (0, T], \\ u(0) = u_0, \end{cases}$$

$$\tag{5.9}$$

and the convergence analysis of the corresponding stochastic midpoint method

$$u^{n+1} = u^n + \tau M u^{n+\frac{1}{2}} + \tau F^{n+\frac{1}{2}}(u^{n+\frac{1}{2}}) + \lambda \mathbb{J}^{-1} u^{n+\frac{1}{2}} \zeta^{n+1}, \qquad (5.10)$$

where $u^0 = u_0$, and $\zeta^{n+1} := \Delta \overline{W}^{n+1}$ is defined in (4.20). Recall that $F_Q(\mathbf{x}) := \sum_{j \in \mathbb{N}} (Q^{\frac{1}{2}} e_j(\mathbf{x}))^2$.

Proposition 5.1 *Let conditions in Proposition 4.5 hold. Assume that $F_Q \in W^{1,\infty}(D)$ and*

$$\mathbb{E}\left[\|M u_0\|_{\mathbb{H}}^4 + \|\nabla \cdot \mathbf{E}_0\|_U^4 + \|\nabla \cdot \mathbf{H}_0\|_U^4 \right] < \infty.$$

Then there exists a positive constant $C = C(T, \lambda, u_0, F, Q, F_Q)$ such that the solution of (5.10) satisfies

$$\max_{1 \leq n \leq N} \left(\mathbb{E}\left[\|u(t_n) - u^n\|_{\mathbb{H}}^2 \right] \right)^{\frac{1}{2}} \leq C \tau^{\frac{1}{2}}.$$

Proof Notice that (5.9) is equivalent to

$$\begin{cases} du(t) = \left[M u(t) + F(t, u(t)) - \frac{\lambda^2}{2} F_Q u(t) \right] dt + \lambda \mathbb{J}^{-1} u(t) dW(t), \quad t \in (0, T], \\ u(0) = u_0. \end{cases}$$

From the definition of the mild solution,

$$u(t_n) = S(t_n) u_0 + \int_0^{t_n} S(t_n - r) F(r, u(r)) dr$$

$$- \frac{\lambda^2}{2} \int_0^{t_n} S(t_n - r) F_Q u(r) dr + \lambda \int_0^{t_n} S(t_n - r) \mathbb{J}^{-1} u(r) dW(r) \tag{5.11}$$

for all $t_n \in [0, T]$. Equation (5.10) implies that for $n = 1, 2, \ldots, N$,

$$u^n = (S_\tau)^n u_0 + \tau \sum_{j=0}^{n-1} (S_\tau)^{n-j-1} T_\tau F^{j+\frac{1}{2}}(u^{j+\frac{1}{2}}) + \lambda \sum_{j=0}^{n-1} (S_\tau)^{n-j-1} T_\tau \mathbb{J}^{-1} u^{j+\frac{1}{2}} \zeta^{j+1}.$$

$$\tag{5.12}$$

Denoting $e^n := u(t_n) - u^n$ and subtracting (5.12) from (5.11), we arrive at

$$
\begin{aligned}
e^n = & \left(S(t_n) - (S_\tau)^n\right)u_0 \\
& + \sum_{j=0}^{n-1} \int_{t_j}^{t_{j+1}} \left(S(t_n - r)F(r, u(r)) - (S_\tau)^{n-j-1}T_\tau F^{j+\frac{1}{2}}(u^{j+\frac{1}{2}})\right)dr \\
& + \sum_{j=0}^{n-1} \left[\lambda \int_{t_j}^{t_{j+1}} S(t_n - r)\mathbb{J}^{-1}u(r)dW(r) - \frac{\lambda^2}{2}\int_{t_j}^{t_{j+1}} S(t_n - r)F_Q u(r)dr \right. \\
& \left. \quad - \lambda(S_\tau)^{n-j-1}T_\tau \mathbb{J}^{-1}u^{j+\frac{1}{2}}\zeta^{j+1}\right] \\
=: & \left(S(t_n) - (S_\tau)^n\right)u_0 + I_n + J_n.
\end{aligned}
$$

For the term I_n, we proceed similarly as (5.6), but use the unitarity of the semigroup $\{S(t), t \in [0, T]\}$ and the Lipschitz continuity of F to estimate the term I_n^1,

$$
\begin{aligned}
\mathbb{E}\left[\|I_n^1\|_{\mathbb{H}}^2\right] &= \mathbb{E}\left[\left\|\sum_{j=0}^{n-1}\int_{t_j}^{t_{j+1}}\left[S(t_n - r)\left(F(r, u(r)) - F(t_j, u(t_j))\right)\right]dr\right\|_{\mathbb{H}}^2\right] \\
&\le C\sum_{j=0}^{n-1}\int_{t_j}^{t_{j+1}}\mathbb{E}\left[\left\|S(t_n - r)\left(F(r, u(r)) - F(t_j, u(t_j))\right)\right\|_{\mathbb{H}}^2\right]dr \\
&\le C\sum_{j=0}^{n-1}\int_{t_j}^{t_{j+1}}\left((r - t_j)^2 + \mathbb{E}\left[\|u(r) - u(t_j)\|_{\mathbb{H}}^2\right]\right)dr \le C\tau
\end{aligned}
$$

due to the Hölder continuity of u in Theorem 2.5. Combining the estimates of terms I_n^2, I_n^3, and I_n^4 in *Step 1* of Theorem 5.1, we have

$$
\mathbb{E}\left[\|I_n\|_{\mathbb{H}}^2\right] \le C\tau + C\tau \sum_{j=0}^{n-1}\mathbb{E}\left[\|e^j\|_{\mathbb{H}}^2 + \|e^{j+1}\|_{\mathbb{H}}^2\right]. \tag{5.13}
$$

To estimate the term J_n, we decompose it into

$$
J_n = \sum_{j=0}^{n-1}\left(J_{n,j}^1 + J_{n,j}^2 + J_{n,j}^3\right),
$$

where

$$J_{n,j}^1 := \frac{\lambda^2}{2}(S_\tau)^{n-j-1}T_\tau u^j(\Delta W^{j+1})^2 - \frac{\lambda^2}{2}\int_{t_j}^{t_{j+1}} S(t_n - r)F_Q u(r)dr,$$

$$J_{n,j}^2 := \lambda(S_\tau)^{n-j-1}T_\tau \mathbb{J}^{-1}u^j(\Delta W^{j+1} - \zeta^{j+1}) - \frac{\lambda\tau}{2}(S_\tau)^{n-j-1}T_\tau \mathbb{J}^{-1}Mu^{j+\frac{1}{2}}\zeta^{j+1}$$

$$- \frac{\lambda\tau}{2}(S_\tau)^{n-j-1}T_\tau \mathbb{J}^{-1}F^{j+\frac{1}{2}}(u^{j+\frac{1}{2}})\zeta^{j+1}$$

$$+ \frac{\lambda^2}{4}(S_\tau)^{n-j-1}T_\tau(u^{j+1} - u^j)(\zeta^{j+1})^2$$

$$- \frac{\lambda^2}{2}(S_\tau)^{n-j-1}T_\tau u^j((\Delta W^{j+1})^2 - (\zeta^{j+1})^2),$$

$$J_{n,j}^3 := \lambda\int_{t_j}^{t_{j+1}}\left(S(t_n - r)\mathbb{J}^{-1}u(r) - (S_\tau)^{n-j-1}T_\tau \mathbb{J}^{-1}u^j\right)dW(r).$$

(i) *Estimate of the term* $\sum_{j=0}^{n-1} J_{n,j}^1$. Note that

$$J_{n,j}^1 = \lambda^2\int_{t_j}^{t_{j+1}}\int_{t_j}^{r}(S_\tau)^{n-j-1}T_\tau u^j dW(\rho)dW(r)$$

$$+ \frac{\lambda^2}{2}\int_{t_j}^{t_{j+1}}\left[(S_\tau)^{n-j-1}T_\tau F_Q u^j - S(t_n - r)F_Q u(r)\right]dr$$

$$= \lambda^2\int_{t_j}^{t_{j+1}}\int_{t_j}^{r}(S_\tau)^{n-j-1}T_\tau u^j dW(\rho)dW(r)$$

$$+ \frac{\lambda^2}{2}\int_{t_j}^{t_{j+1}}\left[(S_\tau)^{n-j-1}T_\tau - S(t_n - r)\right]F_Q u^j dr$$

$$- \frac{\lambda^2}{2}\int_{t_j}^{t_{j+1}} S(t_n - r)F_Q e^j dr$$

$$- \frac{\lambda^2}{2}\int_{t_j}^{t_{j+1}} S(t_n - r)F_Q(u(r) - u(t_j))dr$$

$$=: J_{n,j,1}^1 + J_{n,j,2}^1 + J_{n,j,3}^1 + J_{n,j,4}^1.$$

For the term $J^1_{n,j,1}$, it follows from the Itô isometry, the Sobolev embedding $H^\gamma(D) \hookrightarrow L^\infty(D)$ with $\gamma > 3/2$, Lemma C.3 (i)–(ii), and Proposition 4.4 that

$$
\mathbb{E}\big[\|J^1_{n,j,1}\|^2_{\mathbb{H}}\big] = \lambda^2 \int_{t_j}^{t_{j+1}} \mathbb{E}\Big[\big\| \int_{t_j}^{r} (S_\tau)^{n-j-1} T_\tau u^j \, dW(\rho) Q^{\frac{1}{2}} \big\|^2_{HS(U,\mathbb{H})} \Big] dr
$$

$$
\leq C \|Q^{\frac{1}{2}}\|^2_{HS(U,H^\gamma(D))} \int_{t_j}^{t_{j+1}} \int_{t_j}^{r}
$$

$$
\mathbb{E}\Big[\big\| (S_\tau)^{n-j-1} T_\tau u^j Q^{\frac{1}{2}} \big\|^2_{HS(U,\mathbb{H})} \Big] d\rho \, dr
$$

$$
\leq C\tau^2 \|Q^{\frac{1}{2}}\|^4_{HS(U,H^\gamma(D))} \mathbb{E}\big[\|u^j\|^2_{\mathbb{H}}\big] \leq C\tau^2.
$$

Hence,

$$
\mathbb{E}\Big[\big\| \sum_{j=0}^{n-1} J^1_{n,j,1} \big\|^2_{\mathbb{H}}\Big] = \sum_{j=0}^{n-1} \mathbb{E}\big[\|J^1_{n,j,1}\|^2_{\mathbb{H}}\big] \leq C\tau.
$$

For the term $J^1_{n,j,2}$, we utilize Lemma C.3 (iv) and Propositions 4.4–4.5 to obtain

$$
\mathbb{E}\big[\|J^1_{n,j,2}\|^2_{\mathbb{H}}\big] \leq C\tau \int_{t_j}^{t_{j+1}} \mathbb{E}\Big[\big\| [(S_\tau)^{n-j-1} T_\tau - S(t_n - r)] F_Q u^j \big\|^2_{\mathbb{H}} \Big] dr
$$

$$
\leq C\tau \int_{t_j}^{t_{j+1}} \big\| (S_\tau)^{n-j-1} T_\tau - S(t_n - r) \big\|^2_{\mathscr{L}(\mathscr{D}(M),\mathbb{H})}
$$

$$
\times \mathbb{E}\big[\|F_Q u^j\|^2_{\mathscr{D}(M)}\big] dr
$$

$$
\leq C\tau^3,
$$

which leads to

$$
\mathbb{E}\Big[\big\| \sum_{j=0}^{n-1} J^1_{n,j,2} \big\|^2_{\mathbb{H}}\Big] \leq n \sum_{j=0}^{n-1} \mathbb{E}\big[\|J^1_{n,j,2}\|^2_{\mathbb{H}}\big] \leq C\tau.
$$

For the term $J^1_{n,j,3}$, one has

$$
\mathbb{E}\big[\|J^1_{n,j,3}\|^2_{\mathbb{H}}\big] \leq C\tau \int_{t_j}^{t_{j+1}} \mathbb{E}\big[\|S(t_n - r) F_Q e^j\|^2_{\mathbb{H}}\big] dr \leq C\tau^2 \mathbb{E}\big[\|e^j\|^2_{\mathbb{H}}\big],
$$

which leads to

$$\mathbb{E}\Big[\Big\|\sum_{j=0}^{n-1} J^1_{n,j,3}\Big\|^2_{\mathbb{H}}\Big] \le n \sum_{j=0}^{n-1} \mathbb{E}\big[\|J^1_{n,j,3}\|^2_{\mathbb{H}}\big] \le C\tau \sum_{j=0}^{n-1} \mathbb{E}\big[\|e^j\|^2_{\mathbb{H}}\big].$$

For the term $J^1_{n,j,4}$, by similar arguments, we have

$$\mathbb{E}\Big[\Big\|\sum_{j=0}^{n-1} J^1_{n,j,4}\Big\|^2_{\mathbb{H}}\Big] \le n \sum_{j=0}^{n-1} \mathbb{E}\big[\|J^1_{n,j,4}\|^2_{\mathbb{H}}\big] \le C\tau.$$

Combining estimates of $\sum_{j=0}^{n-1} J^1_{n,j,1}, \ldots, \sum_{j=0}^{n-1} J^1_{n,j,4}$ yields

$$\mathbb{E}\Big[\Big\|\sum_{j=0}^{n-1} J^1_{n,j}\Big\|^2_{\mathbb{H}}\Big] \le C\tau + C\tau \sum_{j=0}^{n-1} \mathbb{E}\big[\|e^j\|^2_{\mathbb{H}}\big].$$

(ii) *Estimate of the term $\sum_{j=0}^{n-1} J^2_{n,j}$.* It follows from Lemma C.3, the Lipschitz continuity of F, and the Sobolev embedding $H^\gamma(D) \hookrightarrow L^\infty(D)$ with $\gamma > 3/2$ that

$$\mathbb{E}\Big[\|J^2_{n,j}\|^2_{\mathbb{H}}\Big] \le C\mathbb{E}\big[\|u^j\|^2_{\mathbb{H}}\|\Delta W^{j+1} - \zeta^{j+1}\|^2_{H^\gamma(D)}\big]$$

$$+ C\tau^2 \mathbb{E}\Big[\|Mu^{j+\frac{1}{2}}\|^2_{\mathbb{H}}\|\zeta^{j+1}\|^2_{H^\gamma(D)}\Big]$$

$$+ C\tau^2 \mathbb{E}\Big[\big(1 + \|u^{j+\frac{1}{2}}\|^2_{\mathbb{H}}\big)\|\zeta^{j+1}\|^2_{H^\gamma(D)}\Big]$$

$$+ C\mathbb{E}\Big[\|u^{j+\frac{1}{2}}(\zeta^{j+1})^3\|^2_{\mathbb{H}}\Big]$$

$$+ C\mathbb{E}\Big[\|u^j\|^2_{\mathbb{H}}\|(\Delta W^{j+1})^2 - (\zeta^{j+1})^2\|^2_{H^\gamma(D)}\Big].$$

Note that Remark 4.2 yields $\sup_{1 \le n \le N} \mathbb{E}\Big[\|u^n\|^4_{H^1(D)^6}\Big] \le C$, which together with (4.21) and Proposition 4.4 leads to

$$\mathbb{E}\Big[\Big\|\sum_{j=0}^{n-1} J^2_{n,j}\Big\|^2_{\mathbb{H}}\Big] \le n \sum_{j=0}^{n-1} \mathbb{E}\Big[\|J^2_{n,j}\|^2_{\mathbb{H}}\Big] \le C\tau.$$

(iii) *Estimate of the term* $\sum_{j=0}^{n-1} J_{n,j}^3$. Based on the Itô isometry and the Sobolev embedding $H^\gamma(D) \hookrightarrow L^\infty(D)$ with $\gamma > 3/2$, we obtain

$$\mathbb{E}\Big[\big\|J_{n,j}^3\big\|_{\mathbb{H}}^2\Big]$$

$$= \lambda^2 \int_{t_j}^{t_{j+1}} \mathbb{E}\Big[\big\|\big(S(t_n - r)\mathbb{J}^{-1}u(r) - (S_\tau)^{n-j-1}T_\tau \mathbb{J}^{-1}u^j\big)Q^{\frac{1}{2}}\big\|_{HS(U,\mathbb{H})}^2\Big]dr$$

$$\leq C \int_{t_j}^{t_{j+1}} \mathbb{E}\Big[\big\|S(t_n - r)\mathbb{J}^{-1}u(r) - (S_\tau)^{n-j-1}T_\tau \mathbb{J}^{-1}u^j\big\|_{HS(U,\mathbb{H})}^2\Big]dr$$

$$\leq C \int_{t_j}^{t_{j+1}} \big\|S(t_n - r) - (S_\tau)^{n-j-1}T_\tau\big\|_{\mathscr{L}(\mathscr{D}(M),\mathbb{H})}^2 \mathbb{E}\Big[\big\|\mathbb{J}^{-1}u(r)\big\|_{\mathscr{D}(M)}^2\Big]dr$$

$$+ C \int_{t_j}^{t_{j+1}} \big\|(S_\tau)^{n-j-1}T_\tau\big\|_{\mathscr{L}(\mathbb{H},\mathbb{H})}^2 \mathbb{E}\Big[\big\|\mathbb{J}^{-1}(u(r) - u(t_j))\big\|_{\mathbb{H}}^2 + \big\|\mathbb{J}^{-1}e^j\big\|_{\mathbb{H}}^2\Big]dr$$

$$\leq C\tau^2 + C\tau\mathbb{E}\Big[\big\|e^j\big\|_{\mathbb{H}}^2\Big],$$

which yields

$$\mathbb{E}\Big[\big\|\sum_{j=0}^{n-1} J_{n,j}^3\big\|_{\mathbb{H}}^2\Big] = \sum_{j=0}^{n-1}\mathbb{E}\Big[\big\|J_{n,j}^3\big\|_{\mathbb{H}}^2\Big] \leq C\tau + C\tau\sum_{j=0}^{n-1}\mathbb{E}\Big[\big\|e^j\big\|_{\mathbb{H}}^2\Big].$$

Combining (i)–(iii), we obtain

$$\mathbb{E}\Big[\big\|J_n\big\|_{\mathbb{H}}^2\Big] \leq C\mathbb{E}\Big[\big\|\sum_{j=0}^{n-1} J_{n,j}^1\big\|_{\mathbb{H}}^2 + \big\|\sum_{j=0}^{n-1} J_{n,j}^2\big\|_{\mathbb{H}}^2 + \big\|\sum_{j=0}^{n-1} J_{n,j}^3\big\|_{\mathbb{H}}^2\Big]$$

$$\leq C\tau + C\tau\sum_{j=0}^{n-1}\mathbb{E}\Big[\big\|e^j\big\|_{\mathbb{H}}^2\Big].$$

Altogether, we arrive at

$$\mathbb{E}\big[\|e^n\|_{\mathbb{H}}^2\big] \leq C\tau + C\tau\sum_{j=0}^{n-1}\mathbb{E}\big[\|e^j\|_{\mathbb{H}}^2 + \|e^{j+1}\|_{\mathbb{H}}^2\big].$$

Then, the discrete Grönwall inequality given in Proposition A.5 implies the assertion.

$\square$

Remark 5.1 Similarly, we can obtain the error estimate of the modified stochastic midpoint method (4.34), that is, there exists a positive constant C such that the solution u^n of (4.34) satisfies

$$\max_{n \in \mathbb{N}} \left(\mathbb{E} \left[\| u(t_n) - u^n \|_{\mathbb{H}}^2 \right] \right)^{\frac{1}{2}} \leq C \tau^{\frac{1}{2}}, \tag{5.14}$$

where $u(t_n)$ is the exact solution of the damped stochastic Maxwell equations (3.34) at time t_n. See [45] for more details.

Furthermore, by Theorem 3.6 (iii) and (5.14), the error between the invariant measure π^* of the exact solution and the invariant measure π^τ of the numerical solution in the L^2-Wasserstein distance can be estimated. More precisely, we have

$$\mathscr{W}_2(\pi^*, \pi^\tau) \leq \mathscr{W}_2((P_n^\tau)^* \pi^\tau, P_{t_n}^* \pi^\tau) + \mathscr{W}_2(P_{t_n}^* \pi^\tau, P_{t_n}^* \pi^*)$$

$$\leq C \tau^{\frac{1}{2}} + e^{-\sigma_0 t_n} \mathscr{W}_2(\pi^*, \pi^\tau).$$

Letting $n \to \infty$ leads to

$$\mathscr{W}_2(\pi^*, \pi^\tau) \leq C \tau^{\frac{1}{2}}.$$

5.1.2 Stochastic Symplectic Runge–Kutta Methods

This subsection is devoted to discussing the generalization of Theorem 5.1 to stochastic symplectic Runge–Kutta methods of the stochastic Maxwell equations with additive noise. More precisely, we study

$$U_i^n = u^n + \tau \sum_{j=1}^{s} a_{ij} \left(M U_j^n + F^{nj}(U_j^n) \right) + \sum_{j=1}^{s} \widetilde{a}_{ij} B^{nj} \Delta W^{n+1},$$

$$u^{n+1} = u^n + \tau \sum_{i=1}^{s} b_i \left(M U_i^n + F^{ni}(U_i^n) \right) + \sum_{i=1}^{s} \widetilde{b}_i B^{ni} \Delta W^{n+1} \tag{5.15}$$

with the symplectic condition $b_i a_{ij} + b_j a_{ji} - b_i b_j = 0$ for all $i, j = 1, 2, \ldots, s$. For convenience of notations, we denote $A := (a_{ij})_{i,j=1}^{s}$, $\widetilde{A} := (\widetilde{a}_{ij})_{i,j=1}^{s}$, $b := (b_1, \ldots, b_s)^\top$ and $\widetilde{b} := (\widetilde{b}_1, \ldots, \widetilde{b}_s)^\top$, and rewrite (5.15) as

$$U_n = \mathbf{1}_s \otimes u^n + \tau (A \otimes M) U_n + \tau (A \otimes Id) F^n(U_n) + (\widetilde{A} \otimes Id) B^n \Delta W^{n+1},$$

$$u^{n+1} = u^n + \tau (b^\top \otimes M) U_n + \tau (b^\top \otimes Id) F^n(U_n) + (\widetilde{b}^\top \otimes Id) B^n \Delta W^{n+1}. \tag{5.16}$$

Now we formulate the main result of this subsection.

Theorem 5.2 *Let conditions with* $k = 2$ *in Proposition 4.6 hold. Suppose that* $\sum_{i=1}^{s} b_i = \sum_{i=1}^{s} \widetilde{b}_i = 1$. *Then there exists a positive constant* $C = C(T, u_0, F, B)$ *such that*

$$\max_{1 \le n \le N} \left(\mathbb{E}\left[\|u(t_n) - u^n\|_{\mathbb{H}}^2 \right] \right)^{\frac{1}{2}} \le C\tau. \tag{5.17}$$

Proof Eliminating the intermediate variable U_n from (5.16) yields that

$$
\begin{aligned}
u^{n+1} &= u^n + \tau\left(b^\top \otimes M\right)\left[Id - \tau\left(A \otimes M\right)\right]^{-1}\left(\mathbf{1}_s \otimes u^n\right) + \tau\left(b^\top \otimes Id\right)F^n(U_n) \\
&\quad + \tau^2\left(b^\top \otimes M\right)\left[Id - \tau\left(A \otimes M\right)\right]^{-1}\left(A \otimes Id\right)F^n(U_n) \\
&\quad + \tau\left(b^\top \otimes M\right)\left[Id - \tau\left(A \otimes M\right)\right]^{-1}\left(\widetilde{A} \otimes Id\right)B^n \Delta W^{n+1} \\
&\quad + \left(\widetilde{b}^\top \otimes Id\right)B^n \Delta W^{n+1}.
\end{aligned} \tag{5.18}
$$

Let $e^n := u(t_n) - u^n$. From the definition of the strong solution of (5.1),

$$u(t_{n+1}) = u(t_n) + \int_{t_n}^{t_{n+1}} \left[Mu(s) + F(s, u(s))\right]ds + \int_{t_n}^{t_{n+1}} B(s)dW(s). \tag{5.19}$$

Subtracting (5.18) from (5.19), we obtain

$$
\begin{aligned}
e^{n+1} - e^n &= \int_{t_n}^{t_{n+1}} Mu(s)ds - \tau\left(b^\top \otimes M\right)\left[Id - \tau\left(A \otimes M\right)\right]^{-1}\left(\mathbf{1}_s \otimes u^n\right) \\
&\quad + \int_{t_n}^{t_{n+1}} F(s, u(s))ds - \tau\left(b^\top \otimes Id\right)F^n(U_n) \\
&\quad + \tau^2\left(b^\top \otimes M\right)\left[Id - \tau\left(A \otimes M\right)\right]^{-1}\left(A \otimes Id\right)F^n(U_n) \tag{5.20} \\
&\quad + \int_{t_n}^{t_{n+1}} B(s)dW(s) - \left(\widetilde{b}^\top \otimes Id\right)B^n \Delta W^{n+1} \\
&\quad + \tau\left(b^\top \otimes M\right)\left[Id - \tau\left(A \otimes M\right)\right]^{-1}\left(\left(\widetilde{A} \otimes Id\right)B^n \Delta W^{n+1}\right) \\
&=: I + II_a - II_b + III_a - III_b.
\end{aligned}
$$

Taking the $\mathbb{H}$-norm on both sides of (5.20) leads to

$$
\begin{aligned}
\|e^{n+1}\|_{\mathbb{H}}^2 &\le \|e^n\|_{\mathbb{H}}^2 + 3\|I\|_{\mathbb{H}}^2 + 2\langle e^n, I\rangle_{\mathbb{H}} + 3\|II\|_{\mathbb{H}}^2 + 2\langle e^n, II\rangle_{\mathbb{H}} \\
&\quad + 3\|III\|_{\mathbb{H}}^2 + 2\langle e^n, III\rangle_{\mathbb{H}}
\end{aligned}
$$

with $II = II_a - II_b$ and $III = III_a - III_b$. Now we estimate terms $\|I\|_{\mathbb{H}}^2$, $\langle e^n, I \rangle_{\mathbb{H}}$, $\|II\|_{\mathbb{H}}^2$, $\langle e^n, II \rangle_{\mathbb{H}}$, $\|III\|_{\mathbb{H}}^2$, and $\langle e^n, III \rangle_{\mathbb{H}}$ separately.

Step 1. *Estimates of terms $\|I\|_{\mathbb{H}}^2$ and $\langle e^n, I \rangle_{\mathbb{H}}$*. Notice that

$$I = \tau M e^n + \int_{t_n}^{t_{n+1}} M\big[u(s) - u(t_n)\big]ds$$

$$+ \Big(\tau M u^n - \tau\big(b^\top \otimes M\big)\big[Id - \tau(A \otimes M)\big]^{-1}\big(\mathbf{1}_s \otimes u^n\big)\Big)$$

$$=: \tau M e^n + I_a + I_b.$$

Theorem 2.4 and Proposition 4.6 with $p = 2$ and $k = 2$ imply that

$$\mathbb{E}\big[\|\tau M e^n\|_{\mathbb{H}}^2\big] = -\tau^2 \mathbb{E}\big[\langle e^n, M^2 e^n \rangle_{\mathbb{H}}\big]$$

$$\leq \tau \mathbb{E}\big[\|e^n\|_{\mathbb{H}}^2\big] + C\tau^3 \mathbb{E}\Big[\|M^2 u(t_n)\|_{\mathbb{H}}^2 + \|M^2 u^n\|_{\mathbb{H}}^2\Big]$$

$$\leq \tau \mathbb{E}\big[\|e^n\|_{\mathbb{H}}^2\big] + C\tau^3.$$

It follows from Theorem 2.5 that

$$\mathbb{E}\big[\|I_a\|_{\mathbb{H}}^2\big] \leq \tau \int_{t_n}^{t_{n+1}} \mathbb{E}\big[\|u(s) - u(t_n)\|_{\mathscr{D}(M)}^2\big]ds \leq C\tau^3$$

and

$$\mathbb{E}\Big[\big\|\mathbb{E}[I_a | \mathscr{F}_{t_n}]\big\|_{\mathbb{H}}^2\Big] \leq \tau \int_{t_n}^{t_{n+1}} \big\|\mathbb{E}[(u(s) - u(t_n)) | \mathscr{F}_{t_n}]\big\|_{\mathscr{D}(M)}^2 ds \leq C\tau^4.$$

For the term I_b, using the assumption $\sum_{i=1}^{s} b_i = 1$, we have

$$\big(b^\top \otimes Id\big)\big(\mathbf{1}_s \otimes M u^n\big) = (b^\top \mathbf{1}_s) \otimes (M u^n) = \sum_{i=1}^{s} b_i \otimes (M u^n) = M u^n. \quad (5.21)$$

Additionally, it holds that

$$\big(b^\top \otimes M\big)\big[Id - \tau(A \otimes M)\big]^{-1}\big(\mathbf{1}_s \otimes u^n\big)$$

$$= (b^\top \otimes Id)\big[Id - \tau(A \otimes M)\big]^{-1}\big(Id \otimes M\big)\big(\mathbf{1}_s \otimes u^n\big) \quad (5.22)$$

$$= (b^\top \otimes Id)\big[Id - \tau(A \otimes M)\big]^{-1}\big(\mathbf{1}_s \otimes M u^n\big).$$

Hence, plugging (5.21)–(5.22) into the term I_b yields

$$I_b = \tau(b^\top \otimes Id)(\mathbf{1}_s \otimes Mu^n) - \tau(b^\top \otimes Id)\big[Id - \tau(A \otimes M)\big]^{-1}(\mathbf{1}_s \otimes Mu^n)$$

$$= \tau(b^\top \otimes Id)\Big(Id - \big[Id - \tau(A \otimes M)\big]^{-1}\Big)(\mathbf{1}_s \otimes Mu^n),$$

$$(5.23)$$

which combining Lemma C.2 (ii) leads to

$$\mathbb{E}\big[\|I_b\|_{\mathbb{H}}^2\big] \le C\tau \mathbb{E}\Big[\Big\|\Big(Id - \big[Id - \tau(A \otimes M)\big]^{-1}\Big)(\mathbf{1}_s \otimes Mu^n)\Big\|_{\mathbb{H}^{\otimes s}}^2\Big] \le C\tau^4.$$

Combining the above estimates, we see that

$$\mathbb{E}\big[\|I\|_{\mathbb{H}}^2\big] \le C\tau \mathbb{E}\big[\|e^n\|_{\mathbb{H}}^2\big] + C\tau^3$$

and

$$\mathbb{E}\big[\langle e^n, I\rangle_{\mathbb{H}}\big] = \mathbb{E}\big[\langle e^n, \mathbb{E}[I_a|\mathscr{F}_{t_n}]\rangle_{\mathbb{H}}\big] + \mathbb{E}\big[\langle e^n, I_b\rangle_{\mathbb{H}}\big]$$

$$\le \tau \mathbb{E}\big[\|e^n\|_{\mathbb{H}}^2\big] + C\tau^{-1}\mathbb{E}\big[\|\mathbb{E}[I_a|\mathscr{F}_{t_n}]\|_{\mathbb{H}}^2\big] + C\tau^{-1}\mathbb{E}\big[\|I_b\|_{\mathbb{H}}^2\big]$$

$$\le \tau \mathbb{E}\big[\|e^n\|_{\mathbb{H}}^2\big] + C\tau^3.$$

Step 2.   *Estimates of terms $\|II\|_{\mathbb{H}}$ and $\langle e^n, II\rangle_{\mathbb{H}}$.* We have

$$II_a = \tau\Big(F(t_n, u(t_n)) - F(t_n, u^n)\Big) + \int_{t_n}^{t_{n+1}}\Big(F(r, u(r)) - F(t_n, u(t_n))\Big)dr$$

$$+ \tau\sum_{i=1}^{s} b_i\Big(F(t_n, u^n) - F(t_n + c_i\tau, U_i^n)\Big)$$

due to the assumption $\sum_{i=1}^{s} b_i = 1$. From the Lipschitz continuity of F, we obtain

$$\|II_a\|_{\mathbb{H}}^2 \le C\tau^2\|e^n\|_{\mathbb{H}}^2 + C\tau^4 + C\tau\int_{t_n}^{t_{n+1}}\|u(r) - u(t_n)\|_{\mathbb{H}}^2 dr$$

$$+ C\tau^2\|U_j^n - u^n\|_{\mathbb{H}}^2.$$

Based on Theorem 2.5 and Proposition 4.6 (iii), we derive that

$$\mathbb{E}\big[\|II_a\|_{\mathbb{H}}^2\big] \le C\tau^2\mathbb{E}\big[\|e^n\|_{\mathbb{H}}^2\big] + C\tau^3.$$

For the term $\langle e^n, II_a \rangle_{\mathbb{H}}$, we need to estimate $\mathbb{E}\big[\|\mathbb{E}[II_a|\mathscr{F}_{t_n}]\|_{\mathbb{H}}^2\big]$, whose estimate is technical. We take the term

$$\int_{t_n}^{t_{n+1}} \Big(F(u(r)) - F(u(t_n))\Big) dr$$

in II_a as an example, where F is supposed to be independent of t for the simplicity of notations. The dependence on t causes no substantial problems in the analysis but just leads to longer formulas.

Thanks to the Taylor formula, we have

$$\int_{t_n}^{t_{n+1}} \Big(F(u(r)) - F(u(t_n))\Big) dr$$

$$= \int_{t_n}^{t_{n+1}} F_u(u(t_n))\big(u(r) - u(t_n)\big) dr$$

$$+ \int_{t_n}^{t_{n+1}} \int_0^1 \theta F_{uu}(\theta u(t_n) + (1-\theta)u(r))(u(r) - u(t_n), u(r) - u(t_n)) d\theta dr.$$

$$(5.24)$$

The second moment of the second term on the right-hand side of (5.24) is bounded by $C\tau^4$ due to Theorem 2.5. For the first term on the right-hand side of (5.24), we apply the conditional expectation first to obtain

$$\mathbb{E}\bigg[\int_{t_n}^{t_{n+1}} F_u(u(t_n))\big(u(r) - u(t_n)\big) dr \,\bigg|\, \mathscr{F}_{t_n}\bigg]$$

$$(5.25)$$

$$= \int_{t_n}^{t_{n+1}} F_u(u(t_n)) \mathbb{E}\big[u(r) - u(t_n)\big|\mathscr{F}_{t_n}\big] dr,$$

where the adaptedness of $\{u(t)\}_{t\in[0,T]}$ and properties of the conditional expectation are used. Then using again Theorem 2.5, we can show that

$$\mathbb{E}\big[\|\mathbb{E}[II_a|\mathscr{F}_{t_n}]\|_{\mathbb{H}}^2\big] \leq C\tau^2 \mathbb{E}\big[\|e^n\|_{\mathbb{H}}^2\big] + C\tau^4.$$

For the term II_b, we have

$$II_b = \tau^2(b^\top \otimes Id)(Id \otimes M)\big[Id - \tau(A \otimes M)\big]^{-1}(A \otimes Id)F^n(U_n)$$

$$= \tau^2(b^\top \otimes Id)\big[Id - \tau(A \otimes M)\big]^{-1}(Id \otimes M)(A \otimes Id)F^n(U_n)$$

$$= \tau^2(b^\top \otimes Id)\big[Id - \tau(A \otimes M)\big]^{-1}(A \otimes Id)(Id \otimes M)F^n(U_n).$$

$$(5.26)$$

Hence, it follows from Lemma C.2 and the linear growth of F that

$$\|II_b\|_{\mathbb{H}} \le C\tau^2 \left\| [Id - \tau(A \otimes M)]^{-1} (A \otimes Id)(Id \otimes M) F^n(U_n) \right\|_{\mathbb{H}^{\otimes s}}$$

$$\le C\tau^2 \left\| (A \otimes Id)(Id \otimes M) F^n(U_n) \right\|_{\mathbb{H}^{\otimes s}}$$

$$\le C\tau^2 (1 + \|U_n\|_{\mathscr{D}(M)^{\otimes s}}),$$

which leads to $\mathbb{E}\left[\|II_b\|_{\mathbb{H}}^2\right] \le C\tau^4$.
Therefore, we deduce that

$$\mathbb{E}\left[\|II\|_{\mathbb{H}}^2\right] \le C\tau^2 \mathbb{E}\left[\|e^n\|_{\mathbb{H}}^2\right] + C\tau^3$$

and

$$\mathbb{E}\left[\langle e^n, II\rangle_{\mathbb{H}}\right] = \mathbb{E}\left[\langle e^n, \mathbb{E}(II_a|\mathscr{F}_{t_n})\rangle_{\mathbb{H}}\right] - \mathbb{E}\left[\langle e^n, II_b\rangle_{\mathbb{H}}\right]$$

$$\le C\tau \mathbb{E}\left[\|e^n\|_{\mathbb{H}}^2\right] + \frac{C}{\tau}\mathbb{E}\left[\|\mathbb{E}[II_a|\mathscr{F}_{t_n}]\|_{\mathbb{H}}^2\right] + \frac{C}{\tau}\mathbb{E}\left[\|II_b\|_{\mathbb{H}}^2\right]$$

$$\le C\tau \mathbb{E}\left[\|e^n\|_{\mathbb{H}}^2\right] + C\tau^3.$$

Step 3.   Estimates of terms $\|III\|_{\mathbb{H}}$ *and* $\langle e^n, III\rangle_{\mathbb{H}}$. For the term III_a, it follows from $\sum_{i=1}^{s} \widetilde{b}_i = 1$ that

$$III_a = \int_{t_n}^{t_{n+1}} \left(B(s) - \sum_{i=1}^{s} \widetilde{b}_i B^{ni} \right) dW(s) = \int_{t_n}^{t_{n+1}} \sum_{i=1}^{s} \widetilde{b}_i \left(B(s) - B^{ni} \right) dW(s).$$

Hence,

$$\mathbb{E}\left[\|III_a\|_{\mathbb{H}}^2\right] = \int_{t_n}^{t_{n+1}} \left\| \sum_{i=1}^{s} \widetilde{b}_i \left(B(s) - B^{ni} \right) \right\|_{HS(U_0, \mathbb{H})}^2 ds \le C\tau^3.$$

Similar to II_b, for the term III_b, we obtain

$$III_b = \tau (b^\top \otimes Id)(Id \otimes M)[Id - \tau(A \otimes M)]^{-1} (\widetilde{A} \otimes Id) B^n \Delta W^{n+1}$$

$$= \tau (b^\top \otimes Id)[Id - \tau(A \otimes M)]^{-1} (Id \otimes M)(\widetilde{A} \otimes Id) B^n \Delta W^{n+1}$$

$$= \tau (b^\top \otimes Id)[Id - \tau(A \otimes M)]^{-1} (\widetilde{A} \otimes Id)(Id \otimes M) B^n \Delta W^{n+1},$$

which yields

$$\mathbb{E}\big[\|III_b\|_{\mathbb{H}}^2\big]$$

$$\leq C\tau^2\,\mathbb{E}\Big[\Big\|[Id - \tau(A \otimes M)]^{-1}(\widetilde{A} \otimes Id)(Id \otimes M)(B^n \Delta W^{n+1})\Big\|_{\mathbb{H}^{\otimes s}}^2\Big]$$

$$\leq C\tau^2\,\mathbb{E}\Big[\big\|(\widetilde{A} \otimes Id)(Id \otimes M)(B^n \Delta W^{n+1})\big\|_{\mathbb{H}^{\otimes s}}^2\Big]$$

$$\leq C\tau^3$$

due to Lemma C.2. Therefore,

$$\mathbb{E}\big[\|III\|_{\mathbb{H}}^2\big] \leq C\tau^3, \quad \mathbb{E}\big[\langle e^n, III\rangle_{\mathbb{H}}\big] = 0.$$

Combining *Steps 1–3* above, we conclude that

$$\mathbb{E}\big[\|e^{n+1}\|_{\mathbb{H}}^2\big] \leq (1 + C\tau)\mathbb{E}\big[\|e^n\|_{\mathbb{H}}^2\big] + C\tau^3.$$

Then the Grönwall inequality gives the desired result. The proof of Theorem 5.2 is finished. □

5.1.3   Exponential-Type Methods

This part presents the convergence analysis of the exponential Euler method and the accelerated exponential Euler method for the stochastic Maxwell equations

$$\begin{cases} du(t) = \Big[Mu(t) + \widetilde{F}(t, u(t))\Big]dt + B(t, u(t))dW(t), \quad t \in (0, T], \\ u(0) = u_0, \end{cases}$$

$$(5.27)$$

where

$$\widetilde{F}(t, u(t)) := F(t, u(t)) + \frac{1}{2}B_u(t, u(t))B(t, u(t))F_Q, \quad F_Q(\mathbf{x}) = \sum_{j \in \mathbb{N}}\big(Q^{\frac{1}{2}}e_j(\mathbf{x})\big)^2.$$

Recall that for $n = 0, 1, \ldots, N - 1$, the exponential Euler method for (5.27) reads as

$$u^{n+1} = S(\tau)u^n + \tau S(\tau)\widetilde{F}(t_n, u^n) + S(\tau)B(t_n, u^n)\Delta W^{n+1}, \quad (5.28)$$

and the accelerated exponential Euler method is

$$u^{n+1} = S(\tau)u^n + \int_{t_n}^{t_{n+1}} S(t_{n+1} - s)\widetilde{F}(t_n, u^n)\mathrm{d}s$$

$$+ \int_{t_n}^{t_{n+1}} S(t_{n+1} - s)B(t_n, u^n)\mathrm{d}W(s). \tag{5.29}$$

First, we present the convergence result for the additive noise case.

Theorem 5.3 *Let conditions in Proposition 4.7 hold. Assume that $\widetilde{F}$ is twice Fréchet differentiable with bounded derivatives. Then for $p \geq 1$ there exists a positive constant $C = C(p, T, Q, u_0)$ such that solutions of (5.28) and (5.29) in the additive noise case satisfy*

$$\mathbb{E}\left[\max_{1 \leq n \leq N} \|u(t_n) - u^n\|_{\mathbb{H}}^{2p} \right] \leq C\tau^{2p}.$$

Proof We only present the proof for the exponential Euler method. The proof for the accelerated exponential Euler method is similar and is omitted. Denoting $e^n := u(t_n) - u^n$, it follows from mild solutions of (5.27) and (5.28) that

$$e^{n+1} = \sum_{k=0}^{n} \int_{t_k}^{t_{k+1}} \left[S(t_{n+1} - r)\widetilde{F}(r, u(r)) - S(t_{n+1} - t_k)\widetilde{F}(t_k, u^k) \right]\mathrm{d}r$$

$$+ \sum_{k=0}^{n} \int_{t_k}^{t_{k+1}} \left[S(t_{n+1} - r)B(r) - S(t_{n+1} - t_k)B(t_k) \right]\mathrm{d}W(r) \tag{5.30}$$

$$=: I_n + J_n$$

for all $n = 0, 1, \ldots, N - 1$.

Step 1. Estimate of the term I_n. We decompose I_n as

$$I_n = \sum_{k=0}^{n} \int_{t_k}^{t_{k+1}} \left[S(t_{n+1} - r)\left(\widetilde{F}(r, u(r)) - \widetilde{F}(t_k, u(t_k))\right) \right]\mathrm{d}r$$

$$+ \sum_{k=0}^{n} \int_{t_k}^{t_{k+1}} \left[\left(S(t_{n+1} - r) - S(t_{n+1} - t_k)\right)\widetilde{F}(t_k, u(t_k)) \right]\mathrm{d}r \tag{5.31}$$

$$+ \sum_{k=0}^{n} \int_{t_k}^{t_{k+1}} \left[S(t_{n+1} - t_k)\left(\widetilde{F}(t_k, u(t_k)) - \widetilde{F}(t_k, u^k)\right) \right]\mathrm{d}r$$

$$=: I_n^1 + I_n^2 + I_n^3.$$

Similar to (5.7), the term I_n^1 can be rewritten as

$$
I_n^1 = \sum_{k=0}^{n} \int_{t_k}^{t_{k+1}} \left[S(t_{n+1} - r) \widetilde{F}_u(t_k, u(t_k)) \big(S(r - t_k) - I \big) u(t_k) \right] dr
$$

$$
+ \sum_{k=0}^{n} \int_{t_k}^{t_{k+1}} \left[S(t_{n+1} - r) \widetilde{F}_u(t_k, u(t_k)) \int_{t_k}^{r} \Big(S(r - \xi) \widetilde{F}(\xi, u(\xi)) \Big) d\xi \right] dr
$$

$$
+ \sum_{k=0}^{n} \int_{t_k}^{t_{k+1}} \left[S(t_{n+1} - r) \widetilde{F}_u(t_k, u(t_k)) \int_{t_k}^{r} \Big(S(r - \xi) B(\xi) \Big) dW(\xi) \right] dr
$$

$$
+ \sum_{k=0}^{n} \int_{t_k}^{t_{k+1}} \left[S(t_{n+1} - r) \int_{0}^{1} \Big(\theta \widetilde{F}_{uu}(t_k, u_\theta) \big(u(r) \right.
$$

$$
\left. - u(t_k), u(r) - u(t_k) \big) \Big) d\theta \right] dr
$$

$$
+ \sum_{k=0}^{n} \int_{t_k}^{t_{k+1}} \left[S(t_{n+1} - r) \big(\widetilde{F}(r, u(r)) - \widetilde{F}(t_k, u(r)) \big) \right] dr
$$

$$
=: I_n^{1,1} + I_n^{1,2} + I_n^{1,3} + I_n^{1,4} + I_n^{1,5},
$$

where $u_\theta := \theta u(t_k) + (1 - \theta) u(r)$ for $\theta \in [0, 1]$. By the assumption on $\widetilde{F}$ and the Hölder continuity of u in Theorem 2.5, we deduce that

$$
\mathbb{E} \left[\max_{0 \le n \le N-1} \Big(\| I_n^{1,4} \|_{\mathbb{H}}^{2p} + \| I_n^{1,5} \|_{\mathbb{H}}^{2p} \Big) \right] \le C \tau^{2p}.
$$

For the term $I_n^{1,1}$, in virtue of the unitarity of the semigroup $\{ S(t), t \in [0, T] \}$ and Lemma C.1, we have

$$
\| I_n^{1,1} \|_{\mathbb{H}} \le \sum_{k=0}^{n} \int_{t_k}^{t_{k+1}} \big\| \widetilde{F}_u(t_k, u(t_k)) \big(S(r - t_k) - Id \big) u(t_k) \big\|_{\mathbb{H}} dr
$$

$$
\le C \sum_{k=0}^{n} \int_{t_k}^{t_{k+1}} (r - t_k) \| u(t_k) \|_{\mathscr{D}(M)} dr
$$

$$
\le C \tau \Big(\max_{0 \le k \le n} \| u(t_k) \|_{\mathscr{D}(M)}^{2p} \Big)^{\frac{1}{2p}},
$$

then

$$
\mathbb{E} \left[\max_{0 \le n \le N-1} \| I_n^{1,1} \|_{\mathbb{H}}^{2p} \right] \le C \tau^{2p}
$$

due to the $\mathscr{D}(M)$-regularity of u given in Theorem 2.4.

For the term $I_n^{1,2}$, using the linear growth of $\widetilde{F}$ gives

$$
\|I_n^{1,2}\|_{\mathbb{H}} \le C \sum_{k=0}^{n} \int_{t_k}^{t_{k+1}} \left[\int_{t_k}^{r} \|\widetilde{F}(\xi, u(\xi))\|_{\mathbb{H}} d\xi \right] dr
$$

$$
\le C \sum_{k=0}^{n} \int_{t_k}^{t_{k+1}} \left[\int_{t_k}^{r} (1 + \|u(\xi)\|_{\mathbb{H}}) d\xi \right] dr
$$

$$
\le C\tau + C\tau \left(\sup_{0 \le t \le T} \|u(t)\|_{\mathbb{H}}^{2p} \right)^{\frac{1}{2p}}.
$$

Applying Theorem 2.4 again yields

$$
\mathbb{E}\left[\max_{0 \le n \le N-1} \|I_n^{1,2}\|_{\mathbb{H}}^{2p} \right] \le C\tau^{2p}.
$$

For the term $I_n^{1,3}$, the stochastic Fubini theorem leads to

$$
I_n^{1,3} = \sum_{k=0}^{n} \int_{t_k}^{t_{k+1}} \int_{\xi}^{t_{k+1}} \left[S(t_{n+1} - r)\widetilde{F}_u(t_k, u(t_k))S(r - \xi)B(\xi) \right] dr dW(\xi)
$$

$$
= \int_{0}^{t_{n+1}} \int_{\xi}^{(\lfloor \frac{\xi}{\tau} \rfloor + 1)\tau} \left[S(t_{n+1} - r)\widetilde{F}_u(t_{\lfloor \frac{r}{\tau} \rfloor \tau}, u(t_{\lfloor \frac{r}{\tau} \rfloor \tau}))S(r - \xi)B(\xi) \right] dr dW(\xi).
$$

Using Proposition D.5, we obtain

$$
\mathbb{E}\left[\max_{0 \le n \le N-1} \|I_n^{1,3}\|_{\mathbb{H}}^{2p} \right]
$$

$$
\le C\mathbb{E}\left[\left(\int_{0}^{T} \left\| \int_{\xi}^{(\lfloor \frac{\xi}{\tau} \rfloor + 1)\tau} \left(S(\xi - r)\widetilde{F}_u(t_{\lfloor \frac{r}{\tau} \rfloor \tau}, u(t_{\lfloor \frac{r}{\tau} \rfloor \tau}))S(r - \xi)B(\xi) \right) dr \right\|_{HS(U_0, \mathbb{H})}^{2} d\xi \right)^{p} \right]
$$

$$
\le C\mathbb{E}\left[\left(\sum_{k=0}^{N-1} \int_{t_k}^{t_{k+1}} \left(\int_{\xi}^{t_{k+1}} \|S(-r)\widetilde{F}_u(t_k, u(t_k))S(r - \xi)B(\xi)\|_{HS(U_0, \mathbb{H})} dr \right)^{2} d\xi \right)^{p} \right]
$$

$$
\le C\mathbb{E}\left[\left(\sum_{k=0}^{N-1} \int_{t_k}^{t_{k+1}} \left(\int_{\xi}^{t_{k+1}} \|B(\xi)\|_{HS(U_0, \mathbb{H})} dr \right)^{2} d\xi \right)^{p} \right] \le C\tau^{2p}.
$$

Combining the above estimates, one has

$$
\mathbb{E}\left[\max_{0\leq n\leq N-1}\|I_n^1\|_{\mathbb{H}}^{2p}\right] \leq C\tau^{2p}.
$$

For the term I_n^2, the unitarity of the semigroup $\{S(t), t \in [0, T]\}$, the linear growth of $\widetilde{F}$, and Theorem 2.4 imply

$$
\mathbb{E}\left[\max_{0\leq n\leq N-1}\|I_n^2\|_{\mathbb{H}}^{2p}\right]
$$

$$
\leq \mathbb{E}\left[\max_{0\leq n\leq N-1}\left(\sum_{k=0}^{n}\int_{t_k}^{t_{k+1}}\left\|\Big(S(t_{n+1}-r)-S(t_{n+1}-t_k)\Big)\widetilde{F}(t_k, u(t_k))\right\|_{\mathbb{H}}\mathrm{d}r\right)^{2p}\right]
$$

$$
= \mathbb{E}\left[\max_{0\leq n\leq N-1}\left(\sum_{k=0}^{n}\int_{t_k}^{t_{k+1}}\left\|\Big(S(r-t_k)-Id\Big)\widetilde{F}(t_k, u(t_k))\right\|_{\mathbb{H}}\mathrm{d}r\right)^{2p}\right]
$$

$$
\leq C\tau^{2p} + C\tau^{2p}\mathbb{E}\left[\max_{0\leq t\leq T}\|u(t)\|_{\mathscr{D}(M)}^{2p}\right] \leq C\tau^{2p}.
$$

By a similar argument, one has

$$
\mathbb{E}\left[\max_{0\leq n\leq N-1}\|I_n^3\|_{\mathbb{H}}^{2p}\right] \leq C\tau\sum_{n=0}^{N-1}\mathbb{E}\left[\max_{0\leq k\leq n}\|e^k\|_{\mathbb{H}}^{2p}\right].
$$

Therefore, we obtain

$$
\mathbb{E}\left[\max_{0\leq n\leq N-1}\|I_n\|_{\mathbb{H}}^{2p}\right] \leq C\tau^{2p} + C\tau\sum_{n=0}^{N-1}\mathbb{E}\left[\max_{0\leq k\leq n}\|e^k\|_{\mathbb{H}}^{2p}\right]. \tag{5.32}
$$

Step 2.   *Estimate of the term J_n.* We decompose the term J_n as

$$
J_n = \sum_{k=0}^{n}\int_{t_k}^{t_{k+1}}\Big[S(t_{n+1}-r)-S(t_{n+1}-t_k)\Big]B(r)\mathrm{d}W(r)
$$

$$
+ \sum_{k=0}^{n}\int_{t_k}^{t_{k+1}}S(t_{n+1}-t_k)\Big[B(r)-B(t_k)\Big]\mathrm{d}W(r)
$$

$$
=: J_n^1 + J_n^2.
$$

Notice that

$$\mathbb{E}\left[\max_{0\leq n\leq N-1}\|J_n^1\|_{\mathbb{H}}^{2p}\right]$$

$$=\mathbb{E}\left[\max_{0\leq n\leq N-1}\left\|\int_0^{t_{n+1}}\left(S\left(t_{n+1}-\left\lfloor\frac{r}{\tau}\right\rfloor\tau\right)-S(t_{n+1}-r)\right)B(r)\mathrm{d}W(r)\right\|_{\mathbb{H}}^{2p}\right]$$

$$\leq\mathbb{E}\left[\sup_{0\leq t\leq T}\left\|\int_0^t S(t-r)\left(S\left(r-\left\lfloor\frac{r}{\tau}\right\rfloor\tau\right)-Id\right)B(r)\mathrm{d}W(r)\right\|_{\mathbb{H}}^{2p}\right].$$

It follows from Proposition D.5 and Lemma C.1 that

$$\mathbb{E}\left[\max_{0\leq n\leq N-1}\|J_n^1\|_{\mathbb{H}}^{2p}\right]$$

$$\leq C\,\mathbb{E}\left[\left(\int_0^t\left\|\left(S\left(r-\left\lfloor\frac{r}{\tau}\right\rfloor\tau\right)-Id\right)B(r)\right\|_{HS(U_0,\mathbb{H})}^2\mathrm{d}r\right)^p\right]$$

$$=C\,\mathbb{E}\left[\left(\sum_{k=0}^{N-1}\int_{t_k}^{t_{k+1}}\left\|(S(r-t_k)-Id)B(r)\right\|_{HS(U_0,\mathbb{H})}^2\mathrm{d}r\right)^p\right]$$

$$\leq C\,\mathbb{E}\left[\left(\sum_{k=0}^{N-1}\int_{t_k}^{t_{k+1}}(r-t_k)^2\|B(r)\|_{HS(U_0,\mathscr{D}(M))}^2\mathrm{d}r\right)^p\right]$$

$$\leq C\tau^{2p}.$$

The estimate of the term J_n^2 is similar to that of J_n^1, which gives

$$\mathbb{E}\left[\max_{0\leq n\leq N-1}\|J_n^2\|_{\mathbb{H}}^{2p}\right]\leq C\tau^{2p}.$$

Therefore,

$$\mathbb{E}\left[\max_{0\leq n\leq N-1}\|J_n\|_{\mathbb{H}}^{2p}\right]\leq C\tau^{2p}. \tag{5.33}$$

Combining (5.32) and (5.33) yields

$$\mathbb{E}\left[\max_{0\leq n\leq N-1}\|e^{n+1}\|_{\mathbb{H}}^{2p}\right]\leq C\tau^{2p}+C\tau\sum_{n=0}^{N-1}\mathbb{E}\left[\max_{0\leq k\leq n}\|e^k\|_{\mathbb{H}}^{2p}\right].$$

The conclusion of Theorem 5.3 follows from the Grönwall inequality.

$$\square$$

Next, we consider the mean-square errors of the exponential Euler method and the accelerated exponential Euler method for the multiplicative noise case.

Theorem 5.4 *Let conditions in Proposition 4.7 hold. Assume that $\widetilde{F}$ is twice Fréchet differentiable with bounded derivatives. Then for $p \geq 1$ there exists a positive constant $C = C(p, T, Q, u_0)$ such that solutions of (5.28) and (5.29) in the multiplicative noise case satisfy*

$$\mathbb{E}\left[\max_{0 \leq n \leq N} \|u(t_n) - u^n\|_{\mathbb{H}}^{2p} \right] \leq C\tau^p.$$

Proof We only present the proof for the exponential Euler method. The proof for the accelerated exponential Euler method is similar and is omitted. We use the same notations as those in Theorem 5.3. Note that the main difference lies in terms I_n^1 and J_n.

For the term I_n^1 in (5.31), we use the Lipschitz continuity of $\widetilde{F}$ and Theorem 2.5 to obtain

$$\mathbb{E}\left[\max_{0 \leq n \leq N-1} \|I_n^1\|_{\mathbb{H}}^{2p} \right]$$

$$\leq \mathbb{E}\left[\max_{0 \leq n \leq N-1} \left(\sum_{k=0}^{n} \int_{t_k}^{t_{k+1}} \left\| S(t_{n+1} - r)\left(\widetilde{F}(r, u(r)) - \widetilde{F}(t_k, u(t_k)) \right) \right\|_{\mathbb{H}} dr \right)^{2p} \right]$$

$$\leq C\mathbb{E}\left[\max_{0 \leq n \leq N-1} \sum_{k=0}^{n} \int_{t_k}^{t_{k+1}} \left((r - t_k)^{2p} + \|u(r) - u(t_k)\|_{\mathbb{H}}^{2p} \right) dr \right]$$

$$\leq C\mathbb{E}\left[\sum_{k=0}^{N-1} \int_{t_k}^{t_{k+1}} (r - t_k)^p dr \right] \leq C\tau^p.$$

By Theorem 5.3, we know that

$$\mathbb{E}\left[\max_{0 \leq n \leq N-1} \|I_n^2\|_{\mathbb{H}}^{2p} \right] \leq C\tau^{2p},$$

$$\mathbb{E}\left[\max_{0 \leq n \leq N-1} \|I_n^3\|_{\mathbb{H}}^{2p} \right] \leq C\tau \sum_{n=0}^{N-1} \mathbb{E}\left[\max_{0 \leq k \leq n} \|e^k\|_{\mathbb{H}}^{2p} \right].$$

Therefore,

$$\mathbb{E}\left[\max_{0 \leq n \leq N-1} \|I_n\|_{\mathbb{H}}^{2p} \right] \leq C\tau^p + C\tau \sum_{n=0}^{N-1} \mathbb{E}\left[\max_{0 \leq k \leq n} \|e^k\|_{\mathbb{H}}^{2p} \right]. \tag{5.34}$$

We are now in the position to give the estimate of J_n. Here J_n reads as

$$J_n = \sum_{k=0}^{n} \int_{t_k}^{t_{k+1}} \left[S(t_{n+1} - r)\Big(B(r, u(r)) - B(t_k, u(t_k))\Big) \right] \mathrm{d}W(r)$$

$$+ \sum_{k=0}^{n} \int_{t_k}^{t_{k+1}} \left[\Big(S(t_{n+1} - r) - S(t_{n+1} - t_k)\Big) B(t_k, u(t_k)) \right] \mathrm{d}W(r)$$

$$+ \sum_{k=0}^{n} \int_{t_k}^{t_{k+1}} \left[S(t_{n+1} - t_k)\Big(B(t_k, u(t_k)) - B(t_k, u^k)\Big) \right] \mathrm{d}W(r)$$

$$=: J_n^1 + J_n^2 + J_n^3.$$

Thanks to Proposition D.5 and the Lipschitz continuity of B, we have

$$\mathbb{E}\left[\max_{0 \leq n \leq N-1} \| J_n^1 \|_{\mathbb{H}}^{2p} \right]$$

$$\leq C\mathbb{E}\left[\left(\int_0^T \left\| B(r, u(r)) - B\Big(\Big\lfloor \frac{r}{\tau} \Big\rfloor \tau, u\Big(\Big\lfloor \frac{r}{\tau} \Big\rfloor \tau\Big)\Big) \right\|_{HS(U_0, \mathbb{H})}^2 \mathrm{d}r \right)^p \right]$$

$$\leq C\mathbb{E}\left[\left(\int_0^T \left(\Big(r - \Big\lfloor \frac{r}{\tau} \Big\rfloor \tau\Big)^2 + \Big\| u(r) - u\Big(\Big\lfloor \frac{r}{\tau} \Big\rfloor \tau\Big) \Big\|_{\mathbb{H}}^2 \right) \mathrm{d}r \right)^p \right].$$

Thus,

$$\mathbb{E}\left[\max_{0 \leq n \leq N-1} \| J_n^1 \|_{\mathbb{H}}^{2p} \right] \leq C \sum_{k=0}^{N-1} \int_{t_k}^{t_{k+1}} (r - t_k)^p \mathrm{d}r \leq C\tau^p$$

due to the Hölder continuity of u in Theorem 2.5. Similarly, for the term J_n^2, we have

$$\mathbb{E}\left[\max_{0 \leq n \leq N-1} \| J_n^2 \|_{\mathbb{H}}^{2p} \right]$$

$$\leq \mathbb{E}\left[\sup_{0 \leq t \leq T} \left\| \int_0^t \Big(S(t - r) - S\Big(t - \Big\lfloor \frac{r}{\tau} \Big\rfloor \tau\Big)\Big) B\Big(\Big\lfloor \frac{r}{\tau} \Big\rfloor \tau, u\Big(\Big\lfloor \frac{r}{\tau} \Big\rfloor \tau\Big)\Big) \mathrm{d}W(r) \right\|_{\mathbb{H}}^{2p} \right]$$

$$\leq C\mathbb{E}\left[\left(\int_0^T \left\| \Big(Id - S\Big(r - \Big\lfloor \frac{r}{\tau} \Big\rfloor \tau\Big)\Big) B\Big(\Big\lfloor \frac{r}{\tau} \Big\rfloor \tau, u\Big(\Big\lfloor \frac{r}{\tau} \Big\rfloor \tau\Big)\Big) \right\|_{HS(U_0, \mathbb{H})}^2 \mathrm{d}r \right)^p \right]$$

$$\leq C \sum_{k=0}^{N-1} \int_{t_k}^{t_{k+1}} \Big(r - \Big\lfloor \frac{r}{\tau} \Big\rfloor \tau\Big)^{2p} \mathbb{E}\left[\left\| B\Big(\Big\lfloor \frac{r}{\tau} \Big\rfloor \tau, u\Big(\Big\lfloor \frac{r}{\tau} \Big\rfloor \tau\Big)\Big) \right\|_{HS(U_0, \mathscr{D}(M))}^{2p} \mathrm{d}r \right]$$

$$\leq C\tau^{2p}.$$

For the last term J_n^3, the Lipschitz continuity of B yields

$$\mathbb{E}\left[\max_{0\leq n\leq N-1}\|J_n^3\|_{\mathbb{H}}^{2p}\right] \leq C\tau \sum_{n=0}^{N-1}\mathbb{E}\left[\max_{0\leq k\leq n}\|e^k\|_{\mathbb{H}}^{2p}\right].$$

Altogether, we obtain

$$\mathbb{E}\left[\max_{0\leq n\leq N-1}\|J_n\|_{\mathbb{H}}^{2p}\right] \leq C\tau^p + C\tau \sum_{n=0}^{N-1}\mathbb{E}\left[\max_{0\leq k\leq n}\|e^k\|_{\mathbb{H}}^{2p}\right]. \tag{5.35}$$

Combining (5.34) and (5.35), and using the Grönwall inequality give the desired result.   □

5.2   Convergence Analysis for Fully Discrete Algorithms

In Sect. 4.2.3, we have constructed several fully discrete algorithms for

$$\begin{cases} du(t) = Mu(t)dt + \lambda dW(t), & t \in (0, T], \\ u(0) = u_0, \end{cases} \tag{5.36}$$

where $\lambda = (\lambda_1^\top, \lambda_2^\top)^\top \in \mathbb{R}^6$ and

$$M = \begin{bmatrix} 0 & \varepsilon^{-1}\nabla\times \\ -\mu^{-1}\nabla\times & 0 \end{bmatrix} \quad \text{with } \mathscr{D}(M) = H_0(\text{curl}, D) \times H(\text{curl}, D).$$

Those fully discrete algorithms are constructed via dG methods in space and the stochastic midpoint method in time. And the well-posedness, regularities of the numerical solution, and some intrinsic properties of the fully discrete system were analyzed. In this section, we continue to give their mean-square convergence analysis. We take the error analysis of the following stochastic symplectic dG algorithm as an example

$$\begin{cases} u_h^{n+1} = u_h^n + \tau M_h^{\text{upw}} u_h^{n+\frac{1}{2}} + \Pi_h\lambda\Delta W^{n+1}, & n = 0, 1, \ldots, N-1, \\ u_h^0 = \Pi_h u_0, \end{cases}$$

$$\tag{5.37}$$

where $\Delta W^{n+1} = W(t_{n+1}) - W(t_n)$ and M_h^{upw} is given in Definition 4.4. And the finite element space is taken as

$$\mathbb{H}_{h,1} = \left\{ v_h \in L^2(D) : v_h|_K \in P^1(K) \text{ for all } K \in \mathscr{T}_h \right\}^6.$$

5.2.1 Error Estimate of the Spatial Semi-Discretization

To give the convergence analysis of (5.37), we first study the mean-square convergence of the spatial semi-discretization (4.75), that is,

$$\begin{cases} du_h(t) = M_h^{\text{upw}} u_h(t)dt + \Pi_h \lambda dW(t), & t \in (0, T], \\ u_h(0) = \Pi_h u_0. \end{cases} \tag{5.38}$$

Proposition 5.2 *Suppose that Assumption 4.1 holds. The spatial semi-discretization (5.38) is well-posed, i.e., there exists a unique mild solution $u_h \in L^2(\Omega, C([0, T], \mathbb{H}_{h,1}))$ given by*

$$u_h(t) = e^{tM_h^{\text{upw}}} u_h(0) + \int_0^t e^{(t-s)M_h^{\text{upw}}} \Pi_h \lambda dW(s), \quad \mathbb{P}\text{-}a.s. \tag{5.39}$$

for all $t \in [0, T]$. Moreover,

$$\mathbb{E}\left[\sup_{0 \le t \le T} \|u_h(t)\|_{\mathbb{H}}^2 \right] \le C\left(1 + \mathbb{E}\left[\|u_0\|_{\mathbb{H}}^2\right]\right),$$

where the positive constant C depends on T, ε, μ, λ, and $\text{Tr}(Q)$.

Proof Note that

$$Id - M_h^{\text{upw}} : \mathbb{H}_{h,1} \to \mathbb{H}_{h,1}$$

is injective and surjective, and thus $\text{ran}(Id - M_h^{\text{upw}}) = \mathbb{H}_{h,1}$. It follows from Proposition 4.25 (ii) that M_h^{upw} is dissipative on $\mathbb{H}_{h,1}$, and thus it generates a contraction semigroup $\{S_h(t) := e^{tM_h^{\text{upw}}}, t \ge 0\}$. Similar to Theorem 2.2, for any $t \in [0, T]$, there exists a unique mild solution u_h of (5.38) given by (5.39).

Combining (5.39) and Proposition D.5 yields

$$\mathbb{E}\left[\sup_{0 \le t \le T} \|u_h(t)\|_{\mathbb{H}}^2 \right]$$

$$\le 2\mathbb{E}\left[\sup_{0 \le t \le T} \left\| S_h(t) u_h(0) \right\|_{\mathbb{H}}^2 \right] + 2\mathbb{E}\left[\sup_{0 \le t \le T} \left\| \int_0^t S_h(t-s) \Pi_h \lambda dW(s) \right\|_{\mathbb{H}}^2 \right]$$

$$\leq 2\mathbb{E}\big[\|u_h(0)\|_{\mathbb{H}}^2\big] + 2T\,\mathbb{E}\big[\|\Pi_h \lambda Q^{\frac{1}{2}}\|_{HS(U,\mathbb{H})}^2\big]$$

$$\leq 2\mathbb{E}\big[\|u_h(0)\|_{\mathbb{H}}^2\big] + 2T\max\{\|\varepsilon\|_{L^\infty(D)}, \|\mu\|_{L^\infty(D)}\}|\lambda|^2 \mathrm{Tr}(Q)$$

$$\leq C\big(1 + \mathbb{E}\big[\|u_0\|_{\mathbb{H}}^2\big]\big)$$

due to the property $\|\Pi_h v\|_{\mathbb{H}} \leq \|v\|_{\mathbb{H}}$ for $v \in \mathbb{H}$ of the projection operator. $\qquad\square$

Now, we apply the projection operator Π_h to (5.36) and use the property $M_h^{\mathrm{upw}} = \Pi_h M u$ in Proposition 4.25 (i) to obtain

$$\begin{cases} \mathrm{d}\Pi_h u(t) = M_h^{\mathrm{upw}} u(t)\mathrm{d}t + \Pi_h \lambda \mathrm{d}W(t), & t \in (0, T], \\ \Pi_h u(0) = \Pi_h u_0. \end{cases} \tag{5.40}$$

We need the following result of projection operator Π_h (see e.g., [143] for more details).

Lemma 5.1 *For all $v \in H^k(\mathscr{T}_h)$ with $k \in \{1, 2\}$,*

$$|v - \Pi_h v|_{H^s(\mathscr{T}_h)} \leq Ch^{k-s}|v|_{H^k(\mathscr{T}_h)} \quad \forall s \in \{0, 1, \ldots, k\}$$

and

$$\sum_{F \in \mathscr{G}_h} \|v - \Pi_h v\|_{L^2(F)^6}^2 \leq Ch^{2k-1}|v|_{H^k(\mathscr{T}_h)}^2,$$

where the positive constant C is independent of h.

After these preparations, the mean-square convergence order of (5.38) is stated as follows.

Theorem 5.5 *Let $u \in C\big([0, T], L^2(\Omega, H^k(D)^6)\big)$ with $k \in \{1, 2\}$ be the solution of (5.36), and $u_h \in C\big([0, T], L^2(\Omega, \mathbb{H}_{h,1})\big)$ be the solution of (5.38). Then there exists a positive constant C independent of h such that*

$$\sup_{0 \leq t \leq T} \mathbb{E}\big[\|u_h(t) - u(t)\|_{\mathbb{H}}^2\big] \leq Ch^{2k-1}.$$

Proof Define the error $e(t) := u_h(t) - u(t)$ for all $t \in [0, T]$, which can be decomposed as

$$e(t) = \big(u_h(t) - \Pi_h u(t)\big) - \big(u(t) - \Pi_h u(t)\big) =: e_h(t) - e_\pi(t).$$

For the error $e_\pi(t)$, it follows immediately from Lemma 5.1 that

$$\mathbb{E}\big[\|e_\pi(t)\|_{\mathbb{H}}^2\big] = \mathbb{E}\big[\|u(t) - \Pi_h u(t)\|_{\mathbb{H}}^2\big] \leq Ch^{2k}\mathbb{E}\big[\|u(t)\|_{H^k(D)^6}^2\big]. \tag{5.41}$$

For the error $e_h(t)$, subtracting (5.40) from (5.38) obtains

$$de_h(t) = M_h^{\text{upw}} e_h(t)dt - M_h^{\text{upw}} e_\pi(t)dt, \quad t \in (0, T]$$

with $e_h(0) = 0$. Then for any $t \in [0, T]$, we have

$$\frac{1}{2}\|e_h(t)\|_{\mathbb{H}}^2 - \int_0^t \left\langle M_h^{\text{upw}} e_h(s), e_h(s) \right\rangle_{\mathbb{H}} ds = - \int_0^t \left\langle M_h^{\text{upw}} e_\pi(s), e_h(s) \right\rangle_{\mathbb{H}} ds. \tag{5.42}$$

Let $e_\pi =: \left((e_\pi^{\mathbf{E}})^\top, (e_\pi^{\mathbf{H}})^\top\right)^\top$ and $e_h =: \left((e_h^{\mathbf{E}})^\top, (e_h^{\mathbf{H}})^\top\right)^\top$. Notice that $e_h(t) \in \mathbb{H}_{h,1}$ and $e_\pi(t) \in \mathbb{H}_{h,1} + (\mathcal{D}(M) \cap H^1(D)^6)$ for $t \in [0, T]$. Proposition 4.25 (iii) implies

$$\left\langle M_h^{\text{upw}} e_\pi, e_h \right\rangle_{\mathbb{H}}$$

$$= \sum_{K \in \mathcal{T}_h} \left[\left\langle e_\pi^{\mathbf{H}}, \nabla \times e_h^{\mathbf{E}} \right\rangle_{L^2(K)^3} - \left\langle e_\pi^{\mathbf{E}}, \nabla \times e_h^{\mathbf{H}} \right\rangle_{L^2(K)^3} \right]$$

$$+ \sum_{G \in G_h^I} \left[\left\langle \beta_K e_{\pi,K_G}^{\mathbf{H}} + \beta_{K_G} e_{\pi,K}^{\mathbf{H}} - \gamma_G \mathbf{n}_G \times [[e_\pi^{\mathbf{E}}]]_G, \mathbf{n}_G \times [[e_h^{\mathbf{E}}]]_G \right\rangle_{L^2(G)^3} \right.$$

$$\left. - \left\langle \alpha_K e_{\pi,K_G}^{\mathbf{E}} + \alpha_{K_G} e_{\pi,K}^{\mathbf{E}} + \delta_G \mathbf{n}_G \times [[e_\pi^{\mathbf{H}}]]_G, \mathbf{n}_G \times [[e_h^{\mathbf{H}}]]_G \right\rangle_{L^2(G)^3} \right]$$

$$- \sum_{G \in G_h^B} \left[\left\langle e_\pi^{\mathbf{H}}, \mathbf{n}_G \times e_h^{\mathbf{E}} \right\rangle_{L^2(G)^3} + 2\gamma_G \left\langle \mathbf{n}_G \times e_\pi^{\mathbf{E}}, \mathbf{n}_G \times e_h^{\mathbf{E}} \right\rangle_{L^2(G)^3} \right].$$

Using the fact that $\langle v - \Pi_h v, w_h \rangle_{\mathbb{H}} = 0$ for all $v \in \mathbb{H}$ and $w_h \in \mathbb{H}_{h,1}$ leads to

$$\left\langle e_\pi^{\mathbf{H}}, \nabla \times e_h^{\mathbf{E}} \right\rangle_{L^2(K)^3} = \left\langle e_\pi^{\mathbf{E}}, \nabla \times e_h^{\mathbf{H}} \right\rangle_{L^2(K)^3} = 0.$$

Then the Young inequality yields

$$\left| \left\langle M_h^{\text{upw}} e_\pi, e_h \right\rangle_{\mathbb{H}} \right|$$

$$\leq \sum_{G \in G_h^B} \gamma_G \|\mathbf{n}_G \times e_h^{\mathbf{E}}\|_{L^2(G)^3}^2$$

$$+ \frac{1}{2} \sum_{G \in G_h^I} \left[\gamma_G \|\mathbf{n}_G \times [[e_h^{\mathbf{E}}]]_G\|_{L^2(G)^3}^2 + \delta_G \|\mathbf{n}_G \times [[e_h^{\mathbf{H}}]]_G\|_{L^2(G)^3}^2 \right]$$

$$+ \frac{1}{2} \sum_{G \in G_h^I} \left[\frac{1}{\gamma_G} \|\beta_K e_{\pi,K_G}^{\mathbf{H}} + \beta_{K_G} e_{\pi,K}^{\mathbf{H}} - \gamma_G \mathbf{n}_G \times [[e_\pi^{\mathbf{E}}]]_G\|_{L^2(G)^3}^2 \right. \tag{5.43}$$

$$+ \frac{1}{\delta_G} \left\| \alpha_K e^{\mathbf{E}}_{\pi,K_G} + \alpha_{K_G} e^{\mathbf{E}}_{\pi,K} + \delta_G \mathbf{n}_G \times [[e^{\mathbf{H}}_\pi]]_G \right\|^2_{L^2(G)^3} \right]$$

$$+ \sum_{G \in G^B_h} \left[\frac{1}{2\gamma_G} \left\| e^{\mathbf{H}}_\pi \right\|^2_{L^2(G)^3} + 2\gamma_G \left\| \mathbf{n}_G \times e^{\mathbf{E}}_\pi \right\|^2_{L^2(G)^3} \right]$$

$$\leq -\frac{1}{2} \left\langle M^{\mathrm{upw}}_h e_h, e_h \right\rangle_{\mathbb{H}} + Ch^{2k-1} \| u(s) \|^2_{H^k(D)^6},$$

where in the last step we used Proposition 4.25 (ii) and Lemma 5.1. Plugging (5.43) into (5.42), we obtain

$$\frac{1}{2} \| e_h(t) \|^2_{\mathbb{H}} - \frac{1}{2} \int_0^t \left\langle M^{\mathrm{upw}}_h e_h(s), e_h(s) \right\rangle_{\mathbb{H}} ds \leq Ch^{2k-1} \int_0^t \| u(s) \|^2_{H^k(D)^6} ds.$$

By $\left\langle M^{\mathrm{upw}}_h v_h, v_h \right\rangle_{\mathbb{H}} \leq 0$ for all $v_h \in \mathbb{H}_{h,1}$ given in Proposition 4.25 (ii), and the H^k-regularity of the solution u (see Theorems 2.6 and 2.7), one has

$$\sup_{0 \leq t \leq T} \mathbb{E} \left[\| e_h(t) \|^2_{\mathbb{H}} \right] \leq Ch^{2k-1} \int_0^T \mathbb{E} \left[\| u(s) \|^2_{H^k(D)^6} \right] ds,$$

which combining with (5.41) completes the proof.   □

5.2.2 Error Estimate of the Full Discretization

Now, we focus on the convergence analysis of the full discretization (5.37) for the stochastic Maxwell equations (5.36). Denote

$$S_{h,\tau} := \left(Id - \frac{\tau}{2} M^{\mathrm{upw}}_h \right)^{-1} \left(Id + \frac{\tau}{2} M^{\mathrm{upw}}_h \right) \quad \text{and} \quad T_{h,\tau} := \left(Id - \frac{\tau}{2} M^{\mathrm{upw}}_h \right)^{-1},$$

then (5.37) can be rewritten in a compact form

$$u^{n+1}_n = S_{h,\tau} u^n_h + T_{h,\tau} \Pi_h \lambda \Delta W^{n+1} \tag{5.44}$$

for all $n = 0, 1, \ldots, N - 1$.

Proposition 5.3 *Suppose that Assumption 4.1 holds. Let u^n_h be the solution of (5.37). Then there exists a positive constant $C = C(T, \varepsilon, \mu, \lambda, Q)$ such that*

$$\max_{1 \leq n \leq N} \mathbb{E} \left[\| u^n_h \|^2_{\mathbb{H}} \right] \leq C \left(1 + \mathbb{E} \left[\| u_0 \|^2_{\mathbb{H}} \right] \right).$$

Proof It follows from (5.44) that

$$u_h^n = S_{h,\tau}^n \Pi_h u_0 + \sum_{j=1}^{n} S_{h,\tau}^{n-j} T_{h,\tau} \Pi_h \lambda \Delta W^j, \quad n = 1, 2, \ldots, N.$$

Taking the $\mathbb{H}$-norm on both sides of the above equation, one obtains

$$\mathbb{E}\big[\|u_h^n\|_{\mathbb{H}}^2\big] \le 2\mathbb{E}\Big[\|S_{h,\tau}^n \Pi_h u_0\|_{\mathbb{H}}^2\Big] + 2\mathbb{E}\Big[\big\|\sum_{j=1}^{n} S_{h,\tau}^{n-j} T_{h,\tau} \Pi_h \lambda \Delta W^j\big\|_{\mathbb{H}}^2\Big].$$

This combining with Lemma C.4 and the property $\|\Pi_h v\|_{\mathbb{H}} \le \|v\|_{\mathbb{H}}$ for all $v \in \mathbb{H}$ leads to

$$\mathbb{E}\big[\|u_h^n\|_{\mathbb{H}}^2\big] \le 2\mathbb{E}\big[\|\Pi_h u_0\|_{\mathbb{H}}^2\big] + 2\sum_{j=1}^{n} \mathbb{E}\big[\|\Pi_h \lambda \Delta W^j\|_{\mathbb{H}}^2\big]$$

$$\le 2\mathbb{E}\big[\|\Pi_h u_0\|_{\mathbb{H}}^2\big] + 2n\tau \max\{\|\varepsilon\|_{L^\infty(D)}, \|\mu\|_{L^\infty(D)}\}|\lambda|^2 \mathrm{Tr}(Q)$$

$$\le C\Big(1 + \mathbb{E}\big[\|u_0\|_{\mathbb{H}}^2\big]\Big),$$

which completes the proof.   □

In order to present the mean-square convergence result of the full discretization (5.37), we need $H^k(D)^6$-regularity ($k = 1, 2$) of the solution u^n of the stochastic midpoint method

$$u^{n+1} = u^n + \tau M u^{n+\frac{1}{2}} + \lambda \Delta W^{n+1}, \quad n = 0, 1, \ldots, N-1. \tag{5.45}$$

Similar to Theorems 2.6 and 2.7, we have the following proposition.

Proposition 5.4 *Under conditions in Theorem 2.7 and the PEC boundary conditions* $\mathbf{n} \times \mathbf{E}|_{\partial D} = 0$, $\mathbf{n} \cdot \mathbf{H}|_{\partial D} = 0$, *the solution* u^n *of* (5.45) *satisfies*

$$\max_{1 \le n \le N} \mathbb{E}\big[\|u^n\|_{H^k(D)^6}^2\big] \le C\Big(1 + \mathbb{E}\big[\|u_0\|_{H^k(D)^6}^2\big]\Big), \quad k = 1, 2,$$

where the positive constant C depends on T, λ, and $\|Q^{\frac{1}{2}}\|_{HS(U, H^k(D))}$.

Based on the above analyses, we finally derive the following mean-square error for the full discretization (5.37).

Theorem 5.6 *Under conditions in Theorem 5.1 and Proposition 5.4, there exists a positive constant C such that*

$$\max_{1 \leq n \leq N} \left(\mathbb{E}\big[\|u(t_n) - u_h^n\|_{\mathbb{H}}^2 \big] \right)^{\frac{1}{2}} \leq C\tau^{\frac{k}{2}} + Ch^{k-\frac{1}{2}}, \quad k = 1, 2.$$

Proof We divide the error $u_h^n - u(t_n)$ into two parts:

$$u_h^n - u(t_n) = \underbrace{\left(u_h^n - u^n \right)}_{\text{Spatial error}} + \underbrace{\left(u^n - u(t_n) \right)}_{\text{Temporal error}}.$$

The convergence analysis of the temporal error has been established in Theorem 5.1, namely,

$$\max_{0 \leq n \leq N} \left(\mathbb{E}\big[\|u(t_n) - u^n\|_{\mathbb{H}}^2 \big] \right)^{\frac{1}{2}} \leq C\tau^{\frac{k}{2}}.$$

Now we give the spatial error estimate. Note that

$$u_h^n - u^n = \left(u_h^n - \Pi_h u^n \right) + \left(\Pi_h u^n - u^n \right) =: e_h^n + e_\pi^n.$$

It follows from Lemma 5.1 that

$$\mathbb{E}\big[\|e_\pi^n\|_{\mathbb{H}}^2 \big] \leq Ch^{2k} \mathbb{E}\big[\|u^n\|_{H^k(D)^6}^2 \big],$$

which combining Proposition 5.4 gives

$$\mathbb{E}\big[\|e_\pi^n\|_{\mathbb{H}}^2 \big] \leq Ch^{2k} \left(1 + \mathbb{E}\big[\|u_0\|_{H^k(D)^6}^2 \big] \right). \tag{5.46}$$

The estimate of e_h^n is stated below. We apply the projection operator Π_h to (5.45) and use Proposition 4.25 (i) to obtain

$$\Pi_h u^{n+1} = \Pi_h u^n + \frac{\tau}{2} \left(M_h^{\text{upw}} u^n + M_h^{\text{upw}} u^{n+1} \right) + \Pi_h \lambda \Delta W^{n+1}. \tag{5.47}$$

Subtracting (5.37) from (5.47) yields

$$e_n^{n+1} = e_h^n + \frac{\tau}{2} \left(M_h^{\text{upw}} e_h^n + M_h^{\text{upw}} e_h^{n+1} \right) + \frac{\tau}{2} \left(M_h^{\text{upw}} e_\pi^n + M_h^{\text{upw}} e_\pi^{n+1} \right).$$

Applying $\langle \cdot, e_h^n + e_h^{n+1} \rangle_\mathbb{H}$ to both sides of the above equation, we obtain

$$\|e_h^{n+1}\|_\mathbb{H}^2 - \|e_h^n\|_\mathbb{H}^2 = \frac{\tau}{2}\left\langle M_h^{\text{upw}}(e_h^n + e_h^{n+1}), e_h^n + e_h^{n+1}\right\rangle_\mathbb{H}$$
$$+ \frac{\tau}{2}\left\langle M_h^{\text{upw}}(e_\pi^n + e_\pi^{n+1}), e_h^n + e_h^{n+1}\right\rangle_\mathbb{H}. \tag{5.48}$$

For the first term on the right-hand side of (5.48), it follows from Proposition 4.25 (ii) that

$$\left\langle M_h^{\text{upw}}(e_h^n + e_h^{n+1}), e_h^n + e_h^{n+1}\right\rangle_\mathbb{H} \leq 0.$$

Similar to (5.43), for the second term on the right-hand side of (5.48), we have

$$\left\langle M_h^{\text{upw}}(e_\pi^n + e_\pi^{n+1}), e_h^n + e_h^{n+1}\right\rangle_\mathbb{H} \leq -\frac{1}{2}\left\langle M_h^{\text{upw}}(e_h^n + e_h^{n+1}), e_h^n + e_h^{n+1}\right\rangle_\mathbb{H}$$
$$+ Ch^{2k-1}\|u^n + u^{n+1}\|_{H^k(D)^6}^2. \tag{5.49}$$

Plugging (5.49) into (5.48) gives

$$\mathbb{E}\left[\|e_h^{n+1}\|_\mathbb{H}^2 - \|e_h^n\|_\mathbb{H}^2\right] \leq C\tau h^{2k-1}\mathbb{E}\left[\|u^n\|_{H^k(D)^6}^2 + \|u^{n+1}\|_{H^k(D)^6}^2\right] \leq C\tau h^{2k-1}.$$

Then the Grönwall inequality yields that

$$\mathbb{E}\left[\|e_h^n\|_\mathbb{H}^2\right] \leq Ch^{2k-1}. \tag{5.50}$$

Combining (5.46) and (5.50) yields that

$$\max_{1 \leq n \leq N}\left(\mathbb{E}\left[\|u_h^n - u^n\|_\mathbb{H}^2\right]\right)^{\frac{1}{2}} \leq Ch^{k-\frac{1}{2}}.$$

The desired mean-square convergence order can be derived by adding up the temporal and spatial errors. $\qquad\square$

Remark 5.2 Using similar arguments as in the proof of Theorem 5.6, we can derive the convergence analysis for other dG full discretizations to (5.36). See also [160] for the study of the optimal spatial error estimates in the case of $d \leq 2$.

5.3 Convergence Analysis for Splitting Algorithms

This section mainly focuses on the convergence analysis of splitting algorithms discussed in Sect. 4.3 for the stochastic Maxwell equations with additive noise

$$
\begin{cases}
d\mathbf{E}(t) = \nabla \times \mathbf{H}(t)dt + \lambda_1 dW(t), & t \in (0, T], \\
d\mathbf{H}(t) = -\nabla \times \mathbf{E}(t)dt + \lambda_2 dW(t), & t \in (0, T], \\
\mathbf{E}(0) = \mathbf{E}_0, \quad \mathbf{H}(0) = \mathbf{H}_0
\end{cases}
\tag{5.51}
$$

on a cuboid $D = (a_1^-, a_1^+) \times (a_2^-, a_2^+) \times (a_3^-, a_3^+)$. We first investigate the mean-square error of the splitting process. Then the stochastic midpoint method is taken as an example to discretize each subsystem in the temporal direction and the corresponding convergence analysis is also given.

5.3.1 Splitting Error

The splitting error is introduced by solving the subproblems in a completely decoupled manner. Generally, the splitting error always exists, even when all subproblems are solved exactly. The aim of this subsection is to present an analysis of operator splitting and to provide insight into the splitting error.

Let $\{u_\tau(t), t \in [0, T]\}$ be a sequence of splitting processes. Recall that the system (4.98) is obtained based on the local one-dimensional splitting of the Maxwell operator. On the time interval $(t_n, t_{n+1}]$ with $t_n = n\tau$, $n = 0, 1, \ldots, N-1$, (4.98) reads as

$$
\begin{cases}
du_{\tau,n}^{[1]}(t) = M_x u_{\tau,n}^{[1]}(t)dt + \lambda^{[1]}dW(t), & u_{\tau,n}^{[1]}(t_n) = u_\tau(t_n), \\
du_{\tau,n}^{[2]}(t) = M_y u_{\tau,n}^{[2]}(t)dt + \lambda^{[2]}dW(t), & u_{\tau,n}^{[2]}(t_n) = u_{\tau,n}^{[1]}(t_{n+1}), \\
du_{\tau,n}^{[3]}(t) = M_z u_{\tau,n}^{[3]}(t)dt + \lambda^{[3]}dW(t), & u_{\tau,n}^{[3]}(t_n) = u_{\tau,n}^{[2]}(t_{n+1}).
\end{cases}
\tag{5.52}
$$

Let $\{\phi_{n,t-t_n}^\alpha : 0 \le t_n \le t\}$, $\alpha = x, y, z$ be the solution operators of equations in (5.52) on $(t_n, t_{n+1}]$, then the splitting process u_τ on $(t_n, t_{n+1}]$ is defined as

$$
u_\tau(t) := \left(\phi_{n,t-t_n}^z \phi_{n,\tau}^y \phi_{n,\tau}^x \right) u_\tau(t_n)
\tag{5.53}
$$

with $u_\tau(0) = u_0$. It is obvious that

$$
u_\tau(t_n) = \prod_{j=0}^{n-1} \left(\phi_{j,\tau}^z \phi_{j,\tau}^y \phi_{j,\tau}^x \right) u_\tau(0), \quad n = 1, 2, \ldots, N.
$$

Theorem 5.7 *Under conditions in Theorem 2.7, for sufficiently small τ, there exists a positive constant C such that*

$$\max_{1 \leq n \leq N} \left(\mathbb{E} \big[\| u(t_n) - u_\tau(t_n) \|_{\mathbb{H}}^2 \big] \right)^{\frac{1}{2}} \leq C\tau.$$

Proof For $n = 0, 1, \ldots, N-1$, according to (5.53) and the mild solutions of (5.52), we have

$$u_\tau(t_{n+1}) = S_z(\tau)(\phi_{n,\tau}^y \phi_{n,\tau}^x u_\tau(t_n)) + \int_{t_n}^t S_z(t_{n+1} - r)\lambda^{[3]}\mathrm{d}W(r)$$

$$= S_z(\tau)\left[S_y(\tau)\phi_{n,\tau}^x u_\tau(t_n) + \int_{t_n}^{t_{n+1}} S_y(t_{n+1} - r)\lambda^{[2]}\mathrm{d}W(r) \right]$$

$$+ \int_{t_n}^{t_{n+1}} S_z(t_{n+1} - r)\lambda^{[3]}\mathrm{d}W(r)$$

$$= S_z(\tau)S_y(\tau)\left[S_x(\tau)u_\tau(t_n) + \int_{t_n}^{t_{n+1}} S_x(t_{n+1} - r)\lambda^x\mathrm{d}W(r) \right]$$

$$+ \int_{t_n}^{t_{n+1}} S_z(\tau)S_y(t_{n+1} - r)\lambda^{[2]}\mathrm{d}W(r) + \int_{t_n}^{t_{n+1}} S_z(t_{n+1} - r)\lambda^{[3]}\mathrm{d}W(r)$$

$$= S_z(\tau)S_y(\tau)S_x(\tau)u_\tau(t_n) + \int_{t_n}^{t_{n+1}} S_z(\tau)S_y(\tau)S_x(t_{n+1} - r)\lambda^{[1]}\mathrm{d}W(r)$$

$$+ \int_{t_n}^{t_{n+1}} S_z(\tau)S_y(t_{n+1} - r)\lambda^{[2]}\mathrm{d}W(r) + \int_{t_n}^{t_{n+1}} S_z(t_{n+1} - r)\lambda^{[3]}\mathrm{d}W(r).$$

$$(5.54)$$

It follows from (5.51) and $(\lambda_1^\top, \lambda_2^\top)^\top = \lambda^{[1]} + \lambda^{[2]} + \lambda^{[3]}$ that

$$u(t_{n+1}) = S(\tau)u(t_n) + \int_{t_n}^{t_{n+1}} S(t_{n+1} - r)\lambda^{[1]}\mathrm{d}W(r)$$

$$+ \int_{t_n}^{t_{n+1}} S(t_{n+1} - r)\lambda^{[2]}\mathrm{d}W(r) + \int_{t_n}^{t_{n+1}} S(t_{n+1} - r)\lambda^{[3]}\mathrm{d}W(r).$$

$$(5.55)$$

Let $e^n := u(t_n) - u_\tau(t_n)$. Subtracting (5.54) from (5.55) leads to

$$e^{n+1} = S_z(\tau)S_y(\tau)S_x(\tau)e^n + \big[S(\tau) - S_z(\tau)S_y(\tau)S_x(\tau)\big]u(t_n)$$

$$+ \int_{t_n}^{t_{n+1}} \big[S(t_{n+1} - r) - S_z(\tau)S_y(\tau)S_x(t_{n+1} - r)\big]\lambda^{[1]}\mathrm{d}W(r)$$

$$+ \int_{t_n}^{t_{n+1}} \big[S(t_{n+1} - r) - S_z(\tau)S_y(t_{n+1} - r)\big]\lambda^{[2]}\mathrm{d}W(r) \qquad (5.56)$$

$$+ \int_{t_n}^{t_{n+1}} \big[S(t_{n+1} - r) - S_z(t_{n+1} - r)\big]\lambda^{[3]}\mathrm{d}W(r)$$

$$=: S_z(\tau)S_y(\tau)S_x(\tau)e^n + I_1 + I_2 + I_3 + I_4.$$

For the term I_1, by Lemma C.5 (iii) and the $H^2(D)^6$-regularity of u in Theorem 2.7, we have

$$\mathbb{E}\big[\,\|I_1\|_{\mathbb{H}}^2\,\big] \leq C\tau^4\mathbb{E}\big[\|u(t_n)\|_{H^2(D)^6}^2\big] \leq C\tau^4. \qquad (5.57)$$

For the term I_2, the Itô isometry yields

$$\mathbb{E}\big[\|I_2\|_{\mathbb{H}}^2\big]$$

$$= \int_{t_n}^{t_{n+1}} \left\|\Big(S(t_{n+1} - s) - S_z(\tau)S_y(\tau)S_x(t_{n+1} - s)\Big)\lambda^{[1]} \circ Q^{\frac{1}{2}}\right\|_{HS(U,\mathbb{H})}^2 \mathrm{d}s$$

$$\leq C \int_{t_n}^{t_{n+1}} \left\|S_z(\tau)S_y(\tau)\Big(S(t_{n+1} - s) - S_x(t_{n+1} - s)\Big)\lambda^{[1]} \circ Q^{\frac{1}{2}}\right\|_{HS(U,\mathbb{H})}^2 \mathrm{d}s$$

$$+ C \int_{t_n}^{t_{n+1}} \left\|\Big(S_z(\tau)\,(Id - S_y(\tau))\,S(t_{n+1} - s)\Big)\lambda^{[1]} \circ Q^{\frac{1}{2}}\right\|_{HS(U,\mathbb{H})}^2 \mathrm{d}s$$

$$+ C \int_{t_n}^{t_{n+1}} \left\|(Id - S_z(\tau))\,S(t_{n+1} - s)\lambda^{[1]} \circ Q^{\frac{1}{2}}\right\|_{HS(U,\mathbb{H})}^2 \mathrm{d}s$$

$$\leq C\tau^3,$$

where we used the unitarity of semigroups S_α ($\alpha = x, y, z$) and Lemma C.5 (i)–(ii). Similarly, one has

$$\mathbb{E}\big[\|I_3\|_{\mathbb{H}}^2\big] \leq C\tau^3, \qquad \mathbb{E}\big[\|I_4\|_{\mathbb{H}}^2\big] \leq C\tau^3.$$

Then combining the above estimates for terms $I_1, \ldots, I_4$, we obtain

$$\mathbb{E}\big[\|e^{n+1}\|_{\mathbb{H}}^2\big] \leq \mathbb{E}\Big[\|S_z(\tau)S_y(\tau)S_x(\tau)e^n\|_{\mathbb{H}}^2\Big] + 2\mathbb{E}\Big[\big\langle S_z(\tau)S_y(\tau)S_x(\tau)e^n, I_1\big\rangle_{\mathbb{H}}\Big]$$

$$+ C\mathbb{E}\Big[\|I_1\|_{\mathbb{H}}^2 + \|I_2\|_{\mathbb{H}}^2 + \|I_3\|_{\mathbb{H}}^2 + \|I_4\|_{\mathbb{H}}^2\Big]$$

$$\leq (1+C\tau)\mathbb{E}\Big[\|S_z(\tau)S_y(\tau)S_x(\tau)e^n\|_{\mathbb{H}}^2\Big] + \frac{C}{\tau}\mathbb{E}\big[\|I_1\|_{\mathbb{H}}^2\big]$$

$$+ C\mathbb{E}\Big[\|I_1\|_{\mathbb{H}}^2 + \|I_2\|_{\mathbb{H}}^2 + \|I_3\|_{\mathbb{H}}^2 + \|I_4\|_{\mathbb{H}}^2\Big]$$

$$\leq (1+C\tau)\mathbb{E}\big[\|e^n\|_{\mathbb{H}}^2\big] + C\tau^3$$

due to the Young inequality. The conclusion of the theorem follows from the Grönwall inequality. $\qquad\square$

5.3.2   Error of the Splitting Midpoint Method

In this part, we apply the stochastic midpoint method to discretize (5.52) temporally and obtain the following splitting midpoint method starting from u_0:

$$\begin{cases} u^{n+1,[1]} = u^n + \frac{1}{2}\tau M_x\big(u^n + u^{n+1,[1]}\big) + \lambda^{[1]}\Delta W^{n+1}, \\[2mm] u^{n+1,[2]} = u^{n+1,[1]} + \frac{1}{2}\tau M_y\big(u^{n+1,[1]} + u^{n+1,[2]}\big) + \lambda^{[2]}\Delta W^{n+1}, \\[2mm] u^{n+1} = u^{n+1,[2]} + \frac{1}{2}\tau M_z\big(u^{n+1,[2]} + u^{n+1}\big) + \lambda^{[3]}\Delta W^{n+1}. \end{cases} \tag{5.58}$$

The convergence analysis for (5.58) is stated below.

Theorem 5.8 *Under conditions in Theorem 2.7, for sufficiently small τ, there exists a positive constant C such that*

$$\max_{0\leq n\leq N}\Big(\mathbb{E}\big[\|u(t_n) - u^n\|_{\mathbb{H}}^2\big]\Big)^{\frac{1}{2}} \leq C\tau.$$

Proof Denoting

$$S_{\tau,\alpha} := \Big(Id - \frac{\tau}{2}M_\alpha\Big)^{-1}\Big(Id + \frac{\tau}{2}M_\alpha\Big), \quad T_{\tau,\alpha} := \Big(Id - \frac{\tau}{2}M_\alpha\Big)^{-1}$$

with $\alpha = x, y, z$, the splitting midpoint method (5.58) can be rewritten as

$$u^{n+1} = S_{\tau,z} S_{\tau,y} S_{\tau,x} u^n + S_{\tau,z} S_{\tau,y} T_{\tau,x} \lambda^{[1]} \Delta W^{n+1}$$
$$+ S_{\tau,z} T_{\tau,y} \lambda^{[2]} \Delta W^{n+1} + T_{\tau,z} \lambda^{[3]} \Delta W^{n+1}, \quad n = 0, 1, \ldots, N - 1.$$
$$(5.59)$$

Let $\hat{e}^n := u(t_n) - u^n$. Subtracting (5.59) from (5.55), we obtain

$$\hat{e}^{n+1} = S_{\tau,z} S_{\tau,y} S_{\tau,x} \hat{e}^n + \Big(S(\tau) - S_{\tau,z} S_{\tau,y} S_{\tau,x} \Big) u(t_n)$$

$$+ \int_{t_n}^{t_{n+1}} \Big[S(t_{n+1} - r) - S_{\tau,z} S_{\tau,y} T_{\tau,x} \Big] \lambda^{[1]} dW(r)$$

$$+ \int_{t_n}^{t_{n+1}} \Big[S(t_{n+1} - r) - S_{\tau,z} T_{\tau,y} \Big] \lambda^{[2]} dW(r)$$

$$+ \int_{t_n}^{t_{n+1}} \Big[S(t_{n+1} - r) - T_{\tau,z} \Big] \lambda^{[3]} dW(r)$$

$$=: S_{\tau,z} S_{\tau,y} S_{\tau,x} \hat{e}^n + \hat{I}_1 + \hat{I}_2 + \hat{I}_3 + \hat{I}_4.$$

It follows from Lemma C.6 (iii) and Theorem 2.7 that

$$\mathbb{E}\big[\|\hat{I}_1\|_{\mathbb{H}}^2\big] \le C\tau^4 \mathbb{E}\big[\|u(t_n)\|_{H^2(D)^6}^2\big] \le C\tau^4.$$

For the term $\hat{I}_2$, the Itô isometry leads to

$$\mathbb{E}\big[\|\hat{I}_2\|_{\mathbb{H}}^2\big] = \int_{t_n}^{t_{n+1}} \Big\| \Big(S(t_{n+1} - s) - S_{\tau,z} S_{\tau,y} T_{\tau,x} \Big) \lambda^{[1]} Q^{\frac{1}{2}} \Big\|_{HS(U,\mathbb{H})}^2 ds$$

$$\le C \int_{t_n}^{t_{n+1}} \Big\| S_{\tau,z} S_{\tau,y} \Big(S(t_{n+1} - s) - T_{\tau,x} \Big) \lambda^{[1]} Q^{\frac{1}{2}} \Big\|_{HS(U,\mathbb{H})}^2 ds$$

$$+ C \int_{t_n}^{t_{n+1}} \Big\| \Big(S_{\tau,z} (Id - S_{\tau,y}) S(t_{n+1} - s) \Big) \lambda^{[1]} Q^{\frac{1}{2}} \Big\|_{HS(U,\mathbb{H})}^2 ds$$

$$+ C \int_{t_n}^{t_{n+1}} \Big\| (Id - S_{\tau,z}) S(t_{n+1} - s) \lambda^{[1]} Q^{\frac{1}{2}} \Big\|_{HS(U,\mathbb{H})}^2 ds$$

$$\le C\tau^3,$$

where the last inequality follows from Lemma C.6 (i)–(ii). Similarly, we can obtain

$$\mathbb{E}\big[\|\hat{I}_3\|_{\mathbb{H}}^2\big] \le C\tau^3, \qquad \mathbb{E}\big[\|\hat{I}_4\|_{\mathbb{H}}^2\big] \le \tau^3.$$

Altogether, by the Young inequality, we obtain

$$\mathbb{E}\big[\|\hat{e}^{n+1}\|_{\mathbb{H}}^2\big] \leq \mathbb{E}\Big[\big\|S_{\tau,z}S_{\tau,y}S_{\tau,x}\hat{e}_n\big\|_{\mathbb{H}}^2\Big] + 2\mathbb{E}\Big[\big\langle S_{\tau,z}S_{\tau,y}S_{\tau,x}\hat{e}_n, \hat{I}_1\big\rangle_{\mathbb{H}}\Big]$$

$$+ C\mathbb{E}\Big[\|\hat{I}_1\|_{\mathbb{H}}^2 + \|\hat{I}_2\|_{\mathbb{H}}^2 + \|\hat{I}_3\|_{\mathbb{H}}^2 + \|\hat{I}_4\|_{\mathbb{H}}^2\Big]$$

$$\leq (1+C\tau)\mathbb{E}\Big[\big\|S_{\tau,z}S_{\tau,y}S_{\tau,x}\hat{e}_n\big\|_{\mathbb{H}}^2\Big] + \frac{C}{\tau}\mathbb{E}\big[\|\hat{I}_1\|_{\mathbb{H}}^2\big]$$

$$+ C\mathbb{E}\Big[\|\hat{I}_1\|_{\mathbb{H}}^2 + \|\hat{I}_2\|_{\mathbb{H}}^2 + \|\hat{I}_3\|_{\mathbb{H}}^2 + \|\hat{I}_4\|_{\mathbb{H}}^2\Big]$$

$$\leq (1+C\tau)\mathbb{E}\big[\|\hat{e}^n\|_{\mathbb{H}}^2\big] + C\tau^3.$$

The proof is thus completed by the Grönwall inequality.   □

Remark 5.3 If we use the exponential Euler method (4.44) to discretize (5.53) in the temporal direction, one can derive the corresponding mean-square convergence order in a similar approach as Theorem 5.8.

Summary and Outlook

With the regularity analysis of the exact solution for the stochastic Maxwell equations in Chap. 2 and that of numerical solutions in Chap. 4, we give the convergence analyses for several structure-preserving algorithms. The mean-square errors of the temporal semi-discretizations, including the stochastic symplectic Runge–Kutta methods and the exponential-type methods are estimated in Sect. 5.1. Then, a fully discrete algorithm via the dG method in space and the midpoint method in time is analyzed in Sect. 5.2. Finally, the error estimate of the splitting process is derived and the mean-square convergence order of a splitting midpoint method is obtained in the last section.

There are still some unsolved problems in the mean-square convergence analysis of structure-preserving algorithms for the stochastic Maxwell equations. For example:

- What is the mean-square convergence order of the stochastic symplectic Runge–Kutta methods for the stochastic Maxwell equations driven by multiplicative noise?
- How to give the mean-square convergence analyses of the fully discrete structure-preserving algorithms, whose spatial direction is discretized by utilizing the wavelet method or the finite difference method?

In many applications of stochastic partial differential equations, researchers sometimes are interested in simulating the average of the functional of the solution to the stochastic partial differential equation, which prompts the development of weakly convergent numerical methods. There have been plenty of works related to the weak convergence analysis of numerical algorithms for stochastic systems; see e.g., [6, 7, 54, 64, 65, 68, 86, 93, 119–121, 128, 152, 175, 176] and references therein. However, there is no work on this subject for the stochastic Maxwell equations.

Besides, we would like to mention that all the results in this chapter are derived for the stochastic Maxwell equations with globally Lipschitz continuous coeffiicents. For the case with the non-globally Lipschitz continuous drift term, such as the second harmonic generation and the Kerr-type nonlinearity, to our knowledge, there is no work on their numerical study. Generally, it is a difficult problem to obtain the mean-square convergence order of the numerical approximation of stochastic nonlinear partial differential equations. So far, one approach to deal with this problem is to study the exponential integrability of both the exact and the numerical solutions (see [17, 22, 55–57, 60, 106] and references therein). Another approach is to utilize techniques like truncating the nonlinear term [37, 63], using the tamed function [89, 108, 110], and adapting the step-size to control the stability [79, 99, 114, 115]. More efforts need to be paid to the numerical study of the stochastic Maxwell equations with non-globally Lipschitz continuous coefficients.

Chapter 6
Implementation of Numerical Experiments

The last chapter is dedicated to the numerical experiments to verify our theoretical results obtained in this monograph, particularly in Chaps. 4 and 5. The efficient numerical scientific computing environment—MATLAB—will be used in this chapter to solve numerical examples that illustrate some key concepts and theoretical properties.

We begin with MATLAB codes for the standard Brownian motion and the Q-Wiener process in Sect. 6.1. Section 6.2 turns to present some numerical experiments to confirm the structure-preserving properties and convergence order results investigated in the previous chapters. Particularly, some core MATLAB codes of numerical experiments are provided for the reader's convenience.

For more details of the MATLAB programs to the numerical simulation for the stochastic system, we refer to [95] and references therein for the stochastic ordinary differential equations, and to [130] for the stochastic partial differential equations. Readers can also find codes for the numerical implementation of the exponential Euler method (4.44) under the link www.math.chalmers.se/~cohend/Recherche/codeMaxwell@CohenCuiHongSun.zip which is provided by Cohen et al. in [53].

6.1 MATLAB Codes for Wiener Processes

This section introduces some basic MATLAB codes for the standard Brownian motion and the Q-Wiener process.

© The Author(s), under exclusive license to Springer Nature Singapore Pte Ltd. 2023   215
C. Chen et al., *Numerical Approximations of Stochastic Maxwell Equations*,
Lecture Notes in Mathematics 2341, https://doi.org/10.1007/978-981-99-6686-8_6

6.1.1   Standard Brownian Motion

A standard Brownian motion $\{W(t),\ t \in [0, T]\}$ is a stochastic process that is
continuous with respect to t. It is necessary to consider the simulation of a Brownian
motion for the computational purpose. To be specific, let $\tau = T/N$ for some positive
integer N, and denote by $W_n := W(t_n)$ the specified value at the discrete time point
$t_n = n\tau$ for $n = 0, 1, \ldots, N$.

The MATLAB M-file **brownian.m** in Listing 6.1 performs one simulation of
the Brownian motion over $[0, 1]$ with $N = 1000$; see Fig. 6.1 for the numerical
result. Here, the random number generator **randn** is used which produces a number
obeying the standard normal distribution, denoted by $\mathcal{N}(0, 1)$. In order to make
experiments repeatable, MATLAB allows the initial state of the random number
generator to be set. We set the state to be 100 with the command **randn ('state',
100)**.

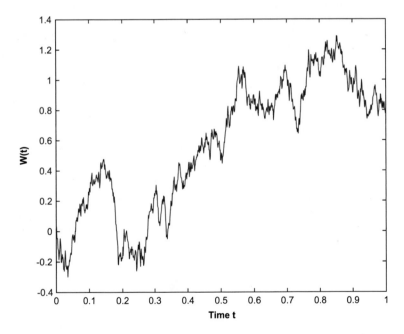

Fig. 6.1 A single sample path of $W(t)$ on $[0, 1]$, generated by using the uniform time step size
$\tau = 0.001$ and linear interpolation. The graph is produced by Listing 6.1

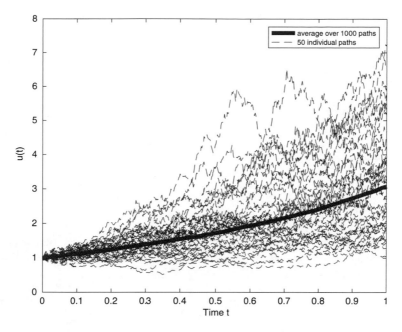

Fig. 6.2 The average of u over 1000 sample paths, and 50 individual sample paths of u. The graph is produced by Listing 6.2

Listing 6.1 *M-file* brownian.m

```
1   randn('state',100)          % set the state of randn
2   T=1; N=1000; tau=T/N; t=[0:tau:T];
3   W=zeros(1,N+1);
4   for n=2:N+1
5       dW=sqrt(tau)*randn; % increment of the Brownian motion
6       W(n)=W(n-1)+dW;
7   end
8   figure
9   plot(t,W,'r-')
10  xlabel('Time t');ylabel('W(t)');
```

The MATLAB M-file **meanbrown.m** in Lising 6.2 produces Fig. 6.2. Here, we evaluate the function $u(t) = \exp(t + 0.5W(t))$ along 1000 sample paths. The average of u over these paths is plotted with a solid blue line. Fifty individual sample paths are also plotted by a dashed red line.

Listing 6.2 *M-file* meanbrown.m

```
1   randn('state',100)          % set the state of randn
2   T=1; N=500; tau=T/N; t=[0:tau:1];
3   M=1000;                     % M Brownian paths
4   dW=sqrt(tau)*randn(M,N);
```

```
5    W=zeros(M,N+1);
6    W(:,2:end)=cumsum(dW,2);
7    u=exp(repmat(t,[M 1])+0.5*W);
8    umean=mean(u);
9    figure
10   plot(t,umean,'b-','LineWidth',5) % average over 1000 paths
11   hold on
12   plot(t,u(1:50,:),'r--') % 50 individual paths
13   hold off
14   xlabel('Time t');ylabel('u(t)');
15   legend('average over 1000 paths','50 individual paths');
```

6.1.2   *Q-Wiener Process*

Let Q be a nonnegative and symmetric operator with a finite trace on a separable Hilbert space V. A V-valued Q-Wiener process $\{W(t),\ t \geq 0\}$ has the following Karhunen–Loève expansion

$$W(t) = \sum_{j=1}^{\infty} \sqrt{\eta_j} e_j \beta_j(t), \quad t \geq 0, \tag{6.1}$$

where $\{\beta_j\}_{j\in\mathbb{N}_+}$ is a family of independent standard Brownian motions and (e_j, η_j), $j \in \mathbb{N}_+$ are the eigenpairs of Q with $\{e_j\}_{j\in\mathbb{N}_+}$ being an orthonormal basis of V.

In order to obtain the numerical approximation of $W(t)$, we first truncate (6.1) by

$$W^M(t) := \sum_{j=1}^{M} \sqrt{\eta_j} e_j \beta_j(t), \quad M \geq 1, \quad t \geq 0.$$

Note that

$$\mathbb{E}\big[\|W(t) - W^M(t)\|^2\big] = \mathbb{E}\Big[\Big\|\sum_{j=M+1}^{\infty} \sqrt{\eta_j} e_j \beta_j(t)\Big\|^2\Big] = t \sum_{j=M+1}^{\infty} \eta_j,$$

which tends to zero as $M \to \infty$, since the trace of the operator Q is finite. Hence, in the following numerical simulations, one can choose M sufficiently large to control the error induced by the truncation of the noise. For the time interval $[0, T]$, we still introduce a uniform partition with time step size $\tau = T/N$, $N \in \mathbb{N}_+$. The increment of W^M can be computed by

$$W^M(t_{n+1}) - W^M(t_n) = \sqrt{\tau} \sum_{j=1}^{M} \sqrt{\eta_j} e_j \xi_j^n,$$

where $t_n = n\tau$ and $\xi_j^n := \big(\beta_j(t_{n+1}) - \beta_j(t_n)\big)/\sqrt{\tau} \sim \mathcal{N}(0, 1)$.

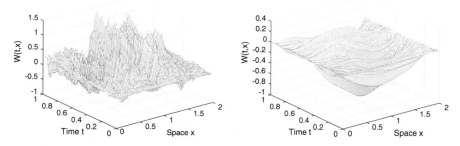

Fig. 6.3 A single sample path of the Q-Wiener process with $M = 100$, $K = 101$, and $\tau = 0.01$ for $r = 0.5$ (left) and $r = 2$ (right). The graph is produced by Listing 6.3

For example, we let $V := L^2(0, a)$ with a positive constant a, and take

$$
e_j(x) = \sqrt{\frac{2}{a}} \sin\left(\frac{j\pi x}{a}\right), \quad \eta_j = j^{-(2r+1+\varepsilon)}, \quad j \in \mathbb{N}_+ \tag{6.2}
$$

for some $\varepsilon > 0$ and $r \geq 0$. In this case, $\mathrm{Tr}(Q) = \sum_{j=1}^{\infty} \eta_j < \infty$. The value of r has an effect on the regularity of the noise. Choose K grid points $x_k = ka/K$, $k = 1, 2, \ldots, K$, then

$$
W^M(t_{n+1}, x_k) - W^M(t_n, x_k) = \sqrt{\frac{2\tau}{a}} \sum_{j=1}^{M} |j|^{-\frac{(2r+1+\varepsilon)}{2}} \sin\left(\frac{j\pi k}{K}\right) \xi_j^n. \tag{6.3}
$$

Let $T = 1$ and $a = 2$. We plot one sample path of $W^M(t)$ in Fig. 6.3 in the cases $r = 0.5$ and $r = 2$, respectively. These figures are generated with $M = 100$, $K = 101$, and $\tau = 0.01$ using M-file **Q-Wiener.m** in Listing 6.3. Here, we fix $\varepsilon = 0.001$.

Listing 6.3 *M-file Q*-Wiener.m

```
1   T=1; tau=0.01; N=T/tau;
2   a=2; epsilon=0.001; r=0.5; % parameters in (6.2)
3   M=100; K=101;dx=a/K;
4   xx=dx:dx:a; tt=0:tau:T;
5   [x,t]=meshgrid(xx,tt);
6   W=zeros(size(x)); xi=zeros(size(tt)); y=zeros(size(tt));
7   eta=[1:M].^(-(2*r+1+epsilon));
8   basis=sqrt(2/a)*sin(pi/a*repmat([1:M],K,1).*repmat(xx',1,M));
9   for i=1:N
10      dB=sqrt(tau)*randn(1,M);
11      dW=sum(repmat(sqrt(eta).*dB,K,1).*basis,2)';
12      W(i+1,:)=W(i,:)+dW;
13  end
```

```
14  % When K=M+1, one can also use the discrete sine ...
         transform 'dst' of the vector ...
         sqrt{eta}.*randn(1,M)*sqrt(2*tau/a) to generate the ...
         increment of the Wiener process. Namely, lines 4-13 ...
         can be replaced by the following code:
15  %xx=dx:dx:a-dx; tt=0:tau:T;
16  %[x,t]=meshgrid(xx,tt);
17  %W=zeros(size(x)); xi=zeros(size(tt)); y=zeros(size(tt));
18  %eta=[1:M].^(-(2*r+1+epsilon));
19  %for i=1:N
20  %      dB=sqrt(tau)*randn(1,M);
21  %      dW=sqrt(2/a)*dst(sqrt(eta).*dB);
22  %      W(i+1,:)=W(i,:)+dW;
23  %end
24  figure
25  mesh(x,t,W);
26  xlabel('Space x');ylabel('Time t');zlabel('W(t,x)');
```

We remark that it is an important problem to consider the simulation of $W(t)$ when the eigenpairs of Q are not known explicitly. Suppose that the expression of the kernel ρ of Q is known. Then one can also approximate $W(t)$, for example, by using the finite element method. Precisely, we can approximate $W(t)$ by

$$
W_h(t) := \overline{\Pi}_h W(t) = \sum_{i=1}^{N_h} \langle W(t), \psi_i \rangle_V \psi_i,
$$

where $\{\psi_i\}_{i=1}^{N_h}$ is a basis of some finite element space V_h and $\overline{\Pi}_h : V \to V_h$ is the orthogonal projection. Note that for any $\psi, \varphi \in V$, $\mathbb{E}\big[\langle W(t), \psi \rangle_V\big] = 0$ and

$$
\mathbb{E}\big[\langle W(t), \psi \rangle_V \langle W(t), \varphi \rangle_V\big] = t \int\int \psi(x)\varphi(y)\rho(x,y)\mathrm{d}x\mathrm{d}y.
$$

Therefore, $\{\langle W(t), \psi_i \rangle_V\}_{i=1}^{N_h}$ and then $W_h(t)$ can be simulated numerically. It can also be shown that $W_h(t)$ converges to $W(t)$ in the mean-square sense as h tends to 0 if some spatial regularity of W is assumed.

6.2   Numerical Experiments for Structure-Preserving Algorithms

This section provides some MATLAB codes of structure-preserving algorithms for the stochastic Maxwell equations proposed in Chap. 4.

6.2.1 Stochastic Multi-Symplectic Algorithms

We first focus on the numerical experiments of several stochastic multi-symplectic algorithms, which are proposed in Sect. 4.2.1, for the stochastic Maxwell equations. We verify the performance of the developed algorithms in the aspect of the energy evolution law and the divergence conservation laws; see Listings 6.4–6.6 for the main MATLAB codes. To give the numerical implementation, we apply the MS Method-I (4.49), the MS Method-II (4.50), and the MS Method-III (4.51) to the two-dimensional TM polarization case

$$
\begin{cases}
dE_3(t) = [\partial_x H_2(t) - \partial_y H_1(t)]dt - \lambda_1 dW, \\[4pt]
dH_1(t) = -\partial_y E_3(t)dt + \lambda_2 dW, \\[4pt]
dH_2(t) = \partial_x E_3(t)dt + \lambda_2 dW
\end{cases}
$$

for $t \in (0, 1]$, where initial data are

$$
E_{3_0}(x, y) = \sin(3\pi x)\sin(4\pi y),
$$

$$
H_{1_0}(x, y) = -0.8\cos(3\pi x)\cos(4\pi y),
$$

$$
H_{2_0}(x, y) = -0.6\sin(3\pi x)\sin(4\pi y)
$$

for $(x, y) \in D = [0, \frac{2}{3}] \times [0, \frac{1}{2}]$.

Recall the Q-Wiener process $W(t) = \sum_{j=1}^{\infty} \sqrt{\eta_j} e_j \beta_j(t)$, for $t \geq 0$. We choose the eigenvalues $\{\eta_{m,l}\}_{m,l \in \mathbb{N}_+}$ and the orthonormal basis $\{e_{m,l}\}_{m,l \in \mathbb{N}_+}$ of $U = L^2(D)$ in this subsection as

$$
e_{m,l}(x, y) = 2\sqrt{3}\sin\left(\frac{3}{2}m\pi x\right)\sin\left(2l\pi y\right), \quad \eta_{m,l} = \frac{1}{m^3 + l^3}, \quad m, l \in \mathbb{N}_+,
$$

which implies

$$
(\Delta W)_{i,j}^n := W_{i,j}^{n+1} - W_{i,j}^n = \sum_{m,l=1}^{\infty} 2\sqrt{\frac{3\tau}{m^3 + l^3}}\sin\left(\frac{3}{2}m\pi x\right)\sin\left(2l\pi y\right)\xi_{m,l}^n
$$

$$
\tag{6.4}
$$

with $\xi_{m,l}^n$ being independent $\mathcal{N}(0, 1)$-random variables.

We take the time step size $\tau = 0.001$ and the space mesh sizes $\Delta x = \Delta y = 1/150$ (i.e., $I = 100$, $J = 75$). Let $\lambda_1 = \lambda_2 = \lambda$. Now we concentrate on numerically performing the evolution of the discrete averaged energies of the MS Method-I (4.49), the MS Method-II (4.50), and the MS Method-III (4.51). It follows from the theoretical results given in Sect. 4.2.1.2 that the concrete expressions of the corresponding energies for these three algorithms have the following forms.

(i)   **MS Method-I**

$$\Phi^{[\mathrm{I}]}(t_{n+1}) = \Phi^{[\mathrm{I}]}(t_n) + 2\Delta x \Delta y \sum_{i=1}^{I}\sum_{j=1}^{J}\left(\Upsilon_{i+\frac{1}{2},j+\frac{1}{2}}^{n+\frac{1}{2}}(\Delta W)_{i,j}^{n+1}\right),$$

$$n = 0, 1, \ldots, N-1,$$

where

$$\Phi^{[\mathrm{I}]}(t_{n+1}) := \Delta x \Delta y \sum_{i=1}^{I}\sum_{j=1}^{J}\left[\left|(E_3)_{i+\frac{1}{2},j+\frac{1}{2}}^{n+1}\right|^2 + \left|(H_1)_{i+\frac{1}{2},j+\frac{1}{2}}^{n+1}\right|^2\right.$$

$$\left. + \left|(H_2)_{i+\frac{1}{2},j+\frac{1}{2}}^{n+1}\right|^2\right]$$

and

$$\Upsilon_{i+\frac{1}{2},j+\frac{1}{2}}^{n+\frac{1}{2}} := \lambda_2\left((H_1)_{i+\frac{1}{2},j+\frac{1}{2}}^{n+\frac{1}{2}} + (H_2)_{i+\frac{1}{2},j+\frac{1}{2}}^{n+\frac{1}{2}}\right) - \lambda_1(E_3)_{i+\frac{1}{2},j+\frac{1}{2}}^{n+\frac{1}{2}}.$$

(ii)   **MS Method-II**

$$\Phi^{[\mathrm{II}]}(t_{n+1}) = \Phi^{[\mathrm{II}]}(t_n) + \Delta x \Delta y \sum_{i=1}^{I}\sum_{j=1}^{J}\left(\Upsilon_{i,j}^{n}(W_{i,j}^{n+1} - W_{i,j}^{n-1})\right),$$

$$n = 1, \ldots, N-1,$$

where

$$\Phi^{[\mathrm{II}]}(t_{n+1}) := \Delta x \Delta y \sum_{i=1}^{I}\sum_{j=1}^{J}\left[(E_3)_{i,j}^{n+1}(E_3)_{i,j}^{n} + (H_1)_{i,j}^{n+1}(H_1)_{i,j}^{n}\right.$$

$$\left. + (H_2)_{i,j}^{n+1}(H_2)_{i,j}^{n}\right]$$

and

$$\Upsilon_{i,j}^{n} := \lambda_2\left((H_1)_{i,j}^{n} + (H_2)_{i,j}^{n}\right) - \lambda_1(E_3)_{i,j}^{n}.$$

(iii) **MS Method-III**

$$\Phi^{[\text{III}]}(t_{n+1}) = \Phi^{[\text{III}]}(t_n) + \Delta x \Delta y \sum_{i=1}^{I} \sum_{j=1}^{J} \left(\Upsilon_{i,j}^{n+\frac{1}{2}} (\Delta W)_{i,j}^{n+1} \right),$$

$$n = 0, 1, \ldots, N - 1,$$

where

$$\Phi^{[\text{III}]}(t_{n+1}) := \Delta x \Delta y \sum_{i=1}^{I} \sum_{j=1}^{J} \left[\left| (E_3)_{i,j}^{n+1} \right|^2 + \left| (H_1)_{i,j}^{n+1} \right|^2 + \left| (H_2)_{i,j}^{n+1} \right|^2 \right]$$

and

$$\Upsilon_{i,j}^{n+\frac{1}{2}} := \lambda_2 \left((H_1)_{i,j}^{n+\frac{1}{2}} + (H_2)_{i,j}^{n+\frac{1}{2}} \right) - \lambda_1 (E_3)_{i,j}^{n+\frac{1}{2}}.$$

Figures. 6.4, 6.5, and 6.6 present the evolution of discrete energies and averaged energies of the MS Method-I, the MS Method-II, and the MS Method-III with various sizes ($\lambda = 0.1$, 1, and 2) of the noise. Here blue lines denote discrete energies along 100 trajectories, and red lines represent discrete averaged energies over these trajectories using the Monte–Carlo method. As shown in these figures, discrete averaged energies are of linear growth with respect to time for all of the three numerical algorithms with different values of λ, which coincides with the theoretical results (see Corollaries 4.1–4.3). Note that for the MS Method-I, Figs. 6.4a, 6.5a, and 6.6a are generated by combining Listing 6.4 and the MATLAB M-file **ms-method1-energy.m** in Listing 6.7. The MATLAB codes for the evolution of discrete energies and averaged energies of the MS Method-II and the MS Method-III can be given similarly as Listing 6.7.

Next, we consider the numerical simulation for the discrete conservation law of the averaged divergence. Since the first two components of the electric field **E** are zero for the two-dimensional TM system, the averaged divergence-preserving property of the electric field holds naturally. Therefore, we focus on the divergence-preserving property of the magnetic field $\mathbf{H} = (H_1, H_2, 0)^{\top}$. Thanks to definitions of discrete divergence operators $\bar{\nabla}_{i,j}^{[\text{I}]}$ and $\bar{\nabla}_{i,j}^{[\text{III}]}$ given in (4.60), we have

(i) **MS Method-I**

$$\bar{\nabla}_{i,j}^{[\text{I}]} \cdot \mathbf{H}^{n+1} - \bar{\nabla}_{i,j}^{[\text{I}]} \cdot \mathbf{H}^n$$

$$= \frac{\lambda_2}{\Delta x} \left[(\Delta W)_{i,j}^{n+1} + (\Delta W)_{i,j-1}^{n+1} - (\Delta W)_{i-1,j}^{n+1} - (\Delta W)_{i-1,j-1}^{n+1} \right]$$

$$+ \frac{\lambda_2}{\Delta y} \left[(\Delta W)_{i,j}^{n+1} + (\Delta W)_{i-1,j}^{n+1} - (\Delta W)_{i,j-1}^{n+1} - (\Delta W)_{i-1,j-1}^{n+1} \right];$$

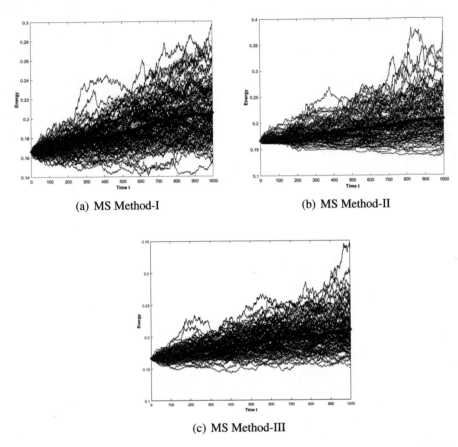

(a) MS Method-I   (b) MS Method-II

(c) MS Method-III

Fig. 6.4 The evolution of the discrete energies and averaged energies generated by (**a**) the MS Method-I, (**b**) the MS Method-II, and (**c**) the MS Method-III along 100 trajectories with $\tau = 0.001$, $\Delta x = \Delta y = 1/150$, and $\lambda = 0.1$

(ii)   **MS Method-II**

$$\bar{\nabla}_{i,j}^{[\mathrm{II}]} \cdot \mathbf{H}^{n+\frac{1}{2}} - \bar{\nabla}_{i,j}^{[\mathrm{II}]} \cdot \mathbf{H}^{n-\frac{1}{2}}$$

$$= \frac{\lambda_2}{4\Delta x} \Big[(\Delta W)_{i+1,j}^{n+1} + (\Delta W)_{i+1,j}^{n} - (\Delta W)_{i-1,j}^{n+1} - (\Delta W)_{i-1,j}^{n} \Big]$$

$$+ \frac{\lambda_2}{4\Delta y} \Big[(\Delta W)_{i,j+1}^{n+1} + (\Delta W)_{i,j+1}^{n} - (\Delta W)_{i,j-1}^{n+1} - (\Delta W)_{i,j-1}^{n} \Big];$$

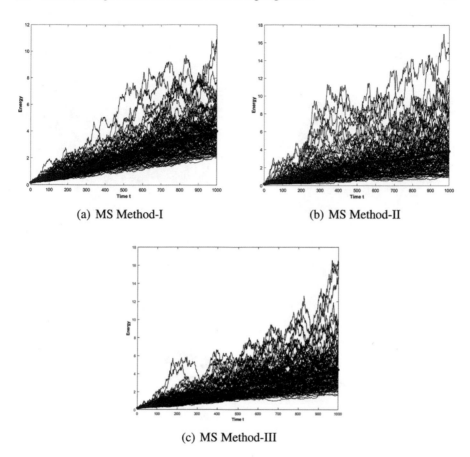

(a) MS Method-I   (b) MS Method-II

(c) MS Method-III

Fig. 6.5 The evolution of the discrete energies and averaged energies generated by (**a**) the MS Method-I, (**b**) the MS Method-II, and (**c**) the MS Method-III along 100 trajectories with $\tau = 0.001$, $\Delta x = \Delta y = 1/150$, and $\lambda = 1$

(iii) **MS Method-III**

$$\bar{\nabla}_{i,j}^{[\mathrm{II}]} \cdot \mathbf{H}^{n+1} - \bar{\nabla}_{i,j}^{[\mathrm{II}]} \cdot \mathbf{H}^n$$

$$= \frac{\lambda_2}{2\Delta x} \Big[(\Delta W)_{i+1,j}^{n+1} - (\Delta W)_{i-1,j}^{n+1} \Big] + \frac{\lambda_2}{2\Delta y} \Big[(\Delta W)_{i,j+1}^{n+1} - (\Delta W)_{i,j-1}^{n+1} \Big].$$

In the sequel, we use the Monte–Carlo method to approximate the expectation, and numerically perform the error of the discrete divergence, which is defined by

$$\mathrm{Err\text{-}Div}(n) := \Delta x \Delta y \sum_{i=1}^{I} \sum_{j=1}^{J} \left| \frac{1}{P} \sum_{s=1}^{P} \Big(\bar{\nabla}_{i,j} \cdot \mathbf{H}^{n+1}(\omega_s) - \bar{\nabla}_{i,j} \cdot \mathbf{H}^n(\omega_s) \Big) \right|$$

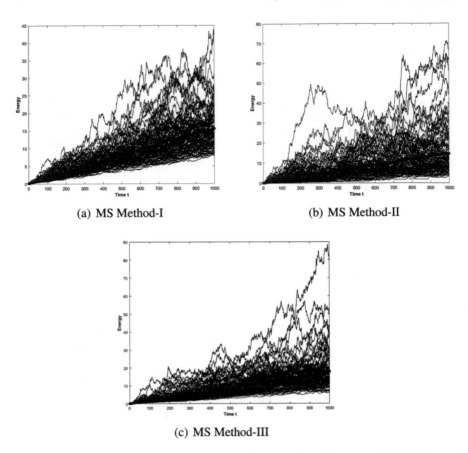

(a) MS Method-I   (b) MS Method-II

(c) MS Method-III

Fig. 6.6 The evolution of the discrete energies and averaged energies generated by (**a**) the MS Method-I, (**b**) the MS Method-II, and (**c**) the MS Method-III along 100 trajectories with $\tau = 0.001$, $\Delta x = \Delta y = 1/150$, and $\lambda = 2$

for $n = 0, 1, \ldots, N$, where $\bar{\nabla}_{i,j} \in \{\bar{\nabla}_{i,j}^{[\mathrm{I}]}, \bar{\nabla}_{i,j}^{[\mathrm{II}]}\}$. The errors of the discrete averaged divergence for the MS Method-I, the MS Method-II, and the MS Method-III are presented in Fig. 6.7. We can see that the scale of the error here is only of 10^{-2} since the number $P = 100$ of truncation is not large enough. As the value of P increases, the error of the averaged divergence decreases. This fact is also verified through numerical simulations shown in Fig. 6.8. It presents in Fig. 6.8 the error of the averaged divergence of the MS Method-III with $P = 10^s$, $s = 1, 2, \ldots, 6$.

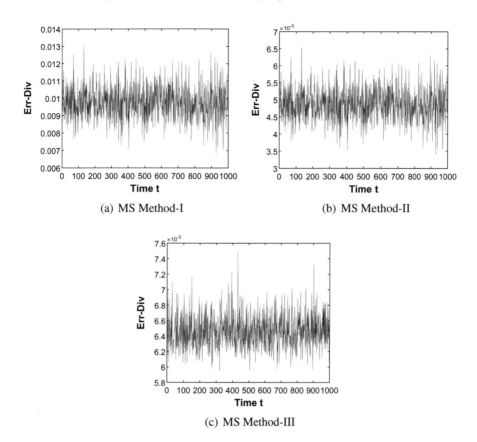

(a) MS Method-I (b) MS Method-II

(c) MS Method-III

Fig. 6.7 The errors of the discrete averaged divergence of (**a**) the MS Method-I, (**b**) the MS Method-II, and (**c**) the MS Method-III along 100 trajectories with $\tau = 0.001$, $\Delta x = \Delta y = 1/150$, and $\lambda = 0.1$

Listing 6.4 A MATLAB code for the **MS Method-I** (4.49)

```
 1 function [E_z1, H_x1, H_y1]=Method1(A, B, C, D, I, J, ...
       rrx1, rry1, E_zinitial1, H_xinitial1, H_yinitial1)
 2 % Using the splitting approach to solve Method-I, the ...
       original system is split into two subsystems: one is ...
       the system of E_z and H_y (subsystem1), and another is ...
       the system of H_x (subsystem2)
 3 % A,B,C,D: coefficient matrices of algebraic equations; ...
       rrx1: noise for subsystem1; rry1: noise for subsystem2
 4 % 1) Solve u_1^{n+1}: A*u_1^{n+1}=B*u_1^{n}+rrx1/2;
 5 % 2) Update the value of u_2^{n} by u_1^{n+1};
 6 % 3) Solve u_2^{n+1}: C*u_2^{n+1}=D*u_2^{n}+rry1;
 7 % 4) Update the value of u_1^{n} by u_2^{n+1};
 8 % 5) Solve u_1^{n+1}: A*u_1^{n+1}=B*u_1^{n}+rrx1/2;
 9
10 % I,J: numbers of the spatial grid points
```

```
11   % E_zinitial1, H_xinitial1, H_yinitial1: initial data
12
13   u1=zeros(2*I*J,1); u2=zeros(2*I*J,1);
14   u1(1:I*J)=E_zinitial1(:);
15   u1(I*J+1:2*I*J)=H_yinitial1(:);
16   u2(I*J+1:2*I*J)=H_xinitial1(:);
17   u1=A$B*u1+rrx1/2);
18   u2(1:I*J,1)=u1(1:I*J,1); u2=C$D*u2+rry1);
19   u1(1:I*J,1)=u2(1:I*J,1); u1=A$B*u1+rrx1/2);
20   E_z1colum=u1(1:I*J,1); H_x1colum=u2(I*J+1:2*I*J,1);
21   H_y1colum=u1(I*J+1:2*I*J,1);
22   E_z1=reshape(E_z1colum,[I,J]);
23   H_x1=reshape(H_x1colum,[I,J]);
24   H_y1=reshape(H_y1colum,[I,J]);
```

Listing 6.5 A MATLAB code for the **MS Method-II** (4.50)

```
1   function d=mmod2(i,I)
2   if i< =I
3        d=i;
4   elseif i> =I+2
5        d=2;
6   else
7        d=1;
```

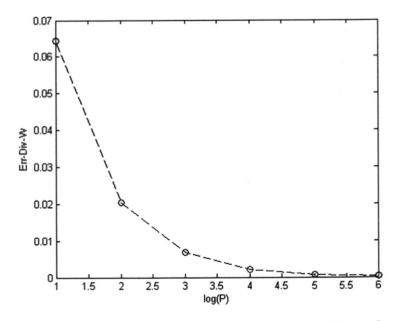

Fig. 6.8 The error of the discrete averaged divergence of the MS Method-III along $P = 10^s$, $s = 1, 2, \ldots, 6$ with $\tau = 0.001$, $\Delta x = \Delta y = 1/150$, and $\lambda = 0.1$

```
 8    end
 9    %%%%%%%%%%%%%%%%%%%%%%%%%%%%%%%%%%%%%%
10    function [E_z2 H_x2 H_y2]=Method2(I, J, lambda, c11, c12, ...
          c21, dW, E_z3, H_x3, H_y3, E_zinitial2, H_xinitial2, ...
          H_yinitial2)
11    % Method-II is a three-layer method, and we choose ...
          Method-III to initialize it
12    % I,J: numbers of the spatial grid points
13    % E_zinitial2, H_xinitial2, H_yinitial2: initial data
14    % E_z3, H_x3, H_y3: given by Listing 6.6 (i.e., Method-III)
15    % c11=tau/dx; c12=tau/dy
16    % c21=2*lamada*tau with lambda being the scale of the ...
          noise (lambda_1=lambda_2=lambda)
17
18    E_z2=zeros(I,J);H_x2=zeros(I,J);H_y2=zeros(I,J);
19    for j=2:J+1
20        for i=2:I+1
21            H_y2(mmod2(i,I),mmod2(j,J))=H_yinitial2(mmod2(i,I),...
22                mmod2(j,J))+c11*(E_z3(mmod2(i+1,I),mmod2(j,J))-...
23                E_z3(i-1,mmod2(j,J)))+c21*E_z3(mmod2(i,I),...
24                mmod2(j,J))*dW(i-1);
25            H_x2(mmod2(i,I),mmod2(j,J))=H_xinitial2(mmod2(i,I),...
26                mmod2(j,J))-c12*(E_z3(mmod2(i,I),mmod2(j+1,J))-...
27                E_z3(mmod2(i,I),j-1))+c21*E_z3(mmod2(i,I),...
28                mmod2(j,J))*dW(i-1);
29            E_z2(mmod2(i,I),mmod2(j,J))=E_zinitial2(mmod2(i,I),...
30                mmod2(j,J))+c11*(H_y3(mmod2(i+1,I),mmod2(j,J))-...
31                H_y3(i-1,mmod2(j,J)))-c12*(H_x3(mmod2(i,I),...
32                mmod2(j+1,J))-H_x3(mmod2(i,I),j-1))-c21*dW(i-1);
33        end
34    end
```

Listing 6.6 A MATLAB code for the **MS Method-III** (4.51)

```
 1    function [E_z3, H_x3, H_y3]=Method3(C, D, I, J, rrx3, ...
          E_zinitial3, H_xinitial3, H_yinitial3)
 2    % I,J: number of the spatial grid points
 3    % E_zinitial3,H_xinitial3,H_yinitial3: initial data
 4    % rrx3: noise
 5    % C,D: coefficient matrices of the algebraic equation
 6    % C*u3^{n+1}=D*u3^{n}+rrx3
 7
 8    u3=zeros(3*I*J,1);
 9    u3(1:I*J)=E_zinitial3(:);
10    u3(I*J+1:2*I*J)=H_xinitial3(:);
11    u3(2*I*J+1:3*I*J)=H_yinitial3(:);
12    u3=C$D*u3+rrx3);
13    E_z3=reshape(u3(1:I*J),[I,J]);
14    H_x3=reshape(u3(I*J+1:2*I*J),[I,J]);
15    H_y3=reshape(u3(2*I*J+1:3*I*J),[I,J]);
```

Listing 6.7 *M-file* ms-method1-energy.m

```
 1  % Generate the evolution law of the discrete energy by ...
        using Listing 6.4 (i.e., Method-I)
 2  T=1; tau=0.001; N=T/tau;
 3  xl=0;xr=2/3; yl=0;yr=1/2; % spatial domain D=[xl, ...
        xr]*[yl, yr]
 4  I=100; dx=(xr-xl)/I; J=75; dy=(yr-yl)/J;   % mesh sizes
 5  xx=xl:dx:(xr-dx); yy=yl:dy:(yr-dy);
 6  x=[xx,xr,xr+dx]; y=[yy,yr,yr+dy];
 7  t=(1:N)*tau;
 8  % covariance matrix of the noise term
 9  lambda=0.1; % scale of the noise (lambda_1=lambda_2=lambda)
10  M=I; L=J; % numbers of the trucation for the noise term
11  P=1; % number of the path
12  [NN,PP]=meshgrid([1:L],[1:M]);
13  NN=NN.^3;PP=PP.^3; %(L*M)-matrix
14  eta_ml=2*sqrt(3./(NN+PP));
15  dW=zeros(I,J);
16  EX=sin(3/2*pi*repmat([1:M],I,1).*repmat(xx',1,M));
17  EY=sin(2*pi*repmat([1:L]',1,J).*repmat(yy,L,1));
18  clear NN PP
19  % generate coefficient matrices of algebraic equations ...
        for  Method-I
20  c11=tau/dx; c12=tau/dy; c13=lambda; c14=2*lambda;
21  A_11=sparse(1:I,1:I,1*ones(1,I),I,I)+sparse(1:I-1,2:I,...
22      ones(1,I-1),I,I)+sparse(I,1,1,I,I);
23  A_12=sparse(1:I,1:I,1*ones(1,I),I,I)+sparse(1:I-1,2:I,...
24      -1*ones(1,I-1),I,I)+sparse(I,1,-1,I,I);
25  A_12x=0.5*(tau/dx)*A_12;
26  A1=kron(speye(J),A_11); A2=kron(speye(J),A_12x);
27  A=[A1,A2;A2,A1]; B=[A1,-A2;-A2,A1];
28  C1=kron(speye(J),eye(I))+kron(sparse(1:J-1,2:J,...
29      ones(1,J-1),J,J),eye(I))+kron(sparse(J,1,1,J,J),eye(I));
30  C_12x=(tau/dy)*eye(I);
31  C2=kron(speye(J),-C_12x)+kron(sparse(1:J-1,2:J,...
32      ones(1,J-1),J,J),C_12x)+kron(sparse(J,1,1,J,J),C_12x);
33  C=[C1,C2;C2,C1]; D=[C1,-C2;-C2,C1];
34  clear A_12 A_12x A1 A2 A_2 C1 C2 C_12x
35  % obtain the discrete energy using Method-I
36  Energy=zeros(P,N);
37  for p=1:P
38      % initial value
39      E_zinitial1=sin(3*pi*xx')*sin(4*pi*yy);
40      H_xinitial1=-0.8*cos(3*pi*xx')*cos(4*pi*yy);
41      H_yinitial1=-0.6*sin(3*pi*xx')*sin(4*pi*yy);
42      for n=1:N
43          Z=randn(M,L); Bcoe=sqrt(tau)*eta_ml.*Z;
44          dW=EX*Bcoe*EY; dWnewnew=reshape(dW,[I*J,1]);
45          rrx1=[-c13*dWnewnew' c14*dWnewnew']';
46          rry1=[-c13*dWnewnew' c14*dWnewnew']';
47          % solve the numerical solution by using Listing ...
                6.4 (Method-I)
```

```
48          [E_z1, H_x1, H_y1]=Method1(A, B, C, D, I, J, ...
                rrx1, rry1, E_zinitial1, H_xinitial1, ...
                H_yinitial1);
49          E11=A_11*E_zinitial1; E12=A_11*H_xinitial1;
50          E13=A_11*H_yinitial1; ...
                E14=A_11*E_zinitial1(1:I,[2:J,1]);
51          E15=A_11*H_xinitial1(1:I,[2:J,1]);
52          E16=A_11*H_yinitial1(1:I,[2:J,1]);
53          Energy(p,n)=1/16*dx*dy*sum((sum((E11+E14).^2...
54          +(E12+E15).^2+(E13+E16).^2)));
55          E_zinitial1=E_z1; H_yinitial1=H_y1; H_xinitial1=H_x1;
56          fprintf('p=%d,n=%d\n',p,n)
57      end
58  end
59  figure
60  plot(Energy','b')
61  hold on
62  plot(mean(Energy,1),'ro')
63  xlabel('Time t');ylabel('Energy')
```

6.2.2 Stochastic Multi-Symplectic Wavelet Algorithm

This section concerns the numerical experiments of the stochastic multi-symplectic wavelet algorithm developed in Sect. 4.2.2. We investigate the performance of the proposed method in the preservation of energy; see Listings 6.8–6.10 for the corresponding MATLAB codes.

We consider the stochastic Maxwell equations driven by multiplicative noise (4.61) on $D = [0, 1]^3$ with initial data

$$E_{1_0} = \cos(2\pi(x + y + z)), \ E_{2_0} = -2E_{1_0}, \ E_{3_0} = E_{1_0},$$

$$H_{1_0} = \sqrt{3}E_{1_0}, \ H_{2_0} = 0, \ H_{3_0} = -\sqrt{3}E_{1_0}$$

and the periodic boundary condition. We use the order $r = 4$ of the Daubechies scaling function ϕ (see (4.62)) to solve the problem and choose 100 realizations to approximate the expectation.

We take the eigenvalues $\{\eta_{m,l,q}\}_{m,l,q \in \mathbb{N}_+}$ and the orthonormal basis $\{e_{m,l,q}\}_{m,l,q \in \mathbb{N}_+}$ of $U = L^2(D)$ in this subsection as

$$\eta_{m,l,q} = \frac{1}{m^3 + l^3 + q^3},$$

$$e_{m,l,q}(x, y, z) = 2\sqrt{2} \sin(m\pi x) \sin(l\pi y) \sin(q\pi z)$$

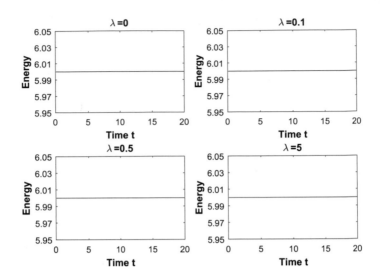

Fig. 6.9 Evolution of the discrete energy along a single trajectory until $T = 20$ with $\tau = 0.1$, $\Delta x = \Delta y = \Delta z = 1/2^4$

for $m, l, q \in \mathbb{N}_+$, which implies

$$(\Delta W)^n_{i,j,k} = 2\sqrt{2\tau} \sum_{m,l,q=1}^{\infty} \frac{1}{\sqrt{m^3 + l^3 + q^3}} \sin(m\pi x_i) \sin(l\pi y_j) \sin(q\pi z_k)\xi^n_{m,l,q}$$

(6.5)

with $\{\xi^n_{m,l,q}\}$ being independent $\mathcal{N}(0, 1)$-random variables.

Figure 6.9 shows the evolution of the discrete energy with various scales of the noise: $\lambda = 0, 0.5, 1$, and 5. Although different scales of noise are chosen, graphs of the discrete energy remain to be horizontal lines approximately. We observe the agreement with the theoretical result (see Proposition 4.24).

In Fig. 6.10, we plot the discrete energy evolution of one trajectory, and the discrete averaged energy evolution until $T = 20$ with $\lambda = 0.5$. Here the blue dotted line denotes the evolution of the discrete averaged energy over 100 trajectories and three solid lines represent three sample paths of the energy, respectively. This figure shows that the numerical result coincides with the theoretical analysis in Proposition 4.24.

Now, we further investigate the energy conservation law under various time step sizes and space mesh sizes. To this end, we define the energy error by

$$\text{Energy error} := \Upsilon^n - \Upsilon^0, \quad n = 0, 1, \ldots, N,$$

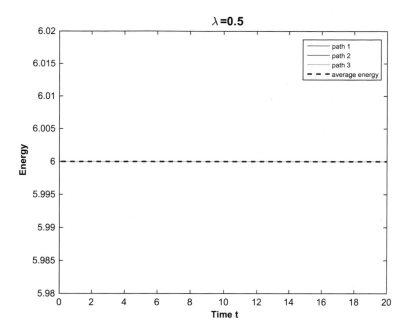

Fig. 6.10 Evolution of the discrete averaged energy over 100 trajectories and discrete energy along one trajectory until $T = 20$ with $\tau = 0.1$, $\Delta x = \Delta y = \Delta z = 1/2^4$

where Υ^n is the discrete energy along one trajectory at time t_n given by

$$\Upsilon^n = \Delta x \Delta y \Delta z \sum_{i=1}^{N_1} \sum_{j=1}^{N_2} \sum_{k=1}^{N_3} \left(|\mathbf{E}_{i,j,k}^n|^2 + |\mathbf{H}_{i,j,k}^n|^2 \right), \quad n = 0, 1, \ldots, N.$$

Figures 6.11, 6.12, 6.13, and 6.14 display the discrete energy errors with various time step sizes and spatial resolutions until $T = 20$. It can be seen that the magnitude for the errors of the discrete energy is of 10^{-11} for various parameters.

Listing 6.8 A MATLAB code for solving wavelet matrices B^x, B^y, and B^z

```
1  function W_m=wavelet_matrix(s,J,N)
2  % Solve the wavelet matrix B=F^{-1}*diag(W_m)*F
3  % s: s order derivative of autocorrelation function (s=1)
4  % J: scale (J=J_1,J_2,J_3, space mesh size 1/2^J)
5  % N: number of spatial grid points (N=N_1,N_2,N_3)
6
7  % solve Daubechies wavelet filter coefficients h_k
8  r=4; % order of the Daubechies scaling function
9  h(2*r:-1:1)=wfilters('db4'); % Daubechies wavelet db4
10 % solve s-order derivative of autocorrelation function
11 a=zeros(2*r-1,1);
```

```
12   for m=1:2:2*r-1
13       a(m)=h(1:2*r-m)*h(1+m:2*r)';
14   end
15   b=2^s*[a(2*r-1:-1:1);1;a];
16   A=zeros(4*r-3,4*r-3);
17   for l=1:2*r-2
18       A(1,1:2*l)=b(2*l:-1:1)';
19   end
20   A(2*r-1,1:4*r-3)=b(4*r-2:-1:2)';
21   for l=2*r:4*r-4
22       A(1,2*l-4*r+2:4*r-3)=b(4*r-1:-1:2*l-4*r+4)';
23   end
24   A=A-eye(4*r-3); A(4*r-3,:)=(-2*r+2:2*r-2).^s;
25   c=[zeros(4*r-4,1);(-1)^s*factorial(s)];
26   theta_s=A\c;
27   W_m=zeros(N,1);
28   W_m(1:2*r-1)=theta_s(2*r-1:4*r-3);
29   W_m(N-2*r+3:N)=(-1)^s*W_m(2*r-1:-1:2);
30   W_m=fft(2^(s*J)*W_m);
```

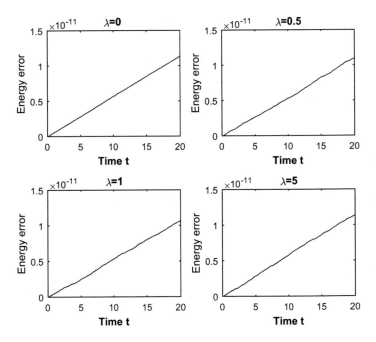

Fig. 6.11 Errors of the discrete energy for $\lambda = 0, 0.5, 1$, and 5 with $\tau = 0.1$, $\Delta x = \Delta y = \Delta z = 1/2^4$, respectively

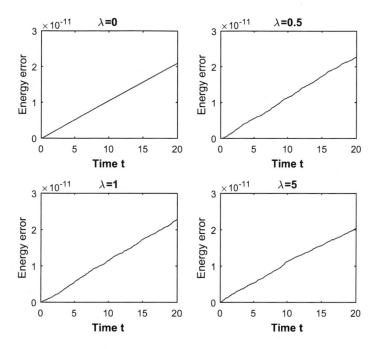

Fig. 6.12 Errors of the discrete energy for $\lambda = 0, 0.5, 1$, and 5 with $\tau = 0.1$, $\Delta x = \Delta y = \Delta z = 1/2^5$, respectively

Listing 6.9 A MATLAB code to obtain drift terms on the right-hand side of (4.65)

```
function d=dis_curl(u,J_1,J_2,J_3,N_1,N_2,N_3)
% u: electromagnetic field u=(E_1,E_2,E_3,H_1,H_2,H_3)'
% 1/2^J_1,1/2^J_2,1/2^J_3: space mesh sizes
% N_1,N_2,N_3: number of spatial grid points
d=zeros(3*N_1*N_2*N_3,1);d1=zeros(N_1*N_2*N_3,1);d2=d1;x=d1;
X=zeros(N_1*N_2,1);
Fx=u(1:N_1*N_2*N_3,1); % E_1 or H_1
Fy=u(N_1*N_2*N_3+1:2*N_1*N_2*N_3,1); % E_2 or H_2
Fz=u(2*N_1*N_2*N_3+1:3*N_1*N_2*N_3,1); % E_3 or H_3
% solve wavelet matrices by using Listing 6.8
M_1=zeros(N_1,1);M_2=zeros(N_2,1);M_3=zeros(N_3,1);
M_1=wavelet_matrix(1,J_1,N_1);
M_2=wavelet_matrix(1,J_2,N_2);
M_3=wavelet_matrix(1,J_3,N_3);
% solve drift terms on the right-hand side of (4.63)
tag=1;
for j=1:N_2
    for i=1:N_1
        x(1:N_3)=Fy(tag:N_1*N_2:N_1*N_2*N_3);
        X=real(ifft(M_3.*fft(x(1:N_3))));% Kronecker ...
            inner product in z-direction
        d1(tag:N_1*N_2:N_1*N_2*N_3)=X;
        tag=tag+1;
```

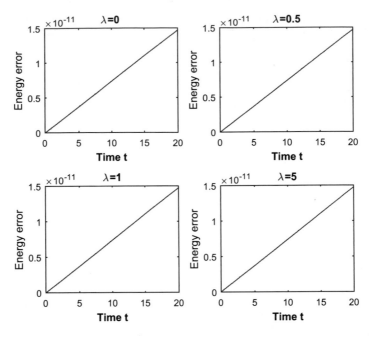

Fig. 6.13 Errors of the discrete energy for $\lambda = 0$, 0.5, 1, and 5 with $\tau = 0.005$, $\Delta x = \Delta y = \Delta z = 1/2^4$, respectively

```
23          end
24    end
25    for k=1:N_3
26          for i=1:N_1
27              x(1:N_2)=Fz((k-1)*N_1*N_2+i:N_1:k*N_1*N_2);
28              X=real(ifft(M_2.*fft(x(1:N_2))));% Kronecker ...
                    inner product in y-direction
29              d2((k-1)*N_1*N_2+i:N_1:k*N_1*N_2)=X;
30          end
31    end
32    d(1:N_1*N_2*N_3)=-d1+d2; % solve A_2H_3-A_3H_2
33
34    tag=1;
35    for j=1:N_2
36          for i=1:N_1
37              x(1:N_3)=Fx(tag:N_1*N_2:N_1*N_2*N_3);
38              X=real(ifft(M_3.*fft(x(1:N_3))));% Kronecker ...
                    inner product in z-direction
39              d1(tag:N_1*N_2:N_1*N_2*N_3)=X;
40              tag=tag+1;
41          end
42    end
43    for k=1:N_3
44          for j=1:N_2
45    x(1:N_1)=Fz((k-1)*N_1*N_2+(j-1)*N_1+1:(k-1)*N_1*N_2+j*N_1);
```

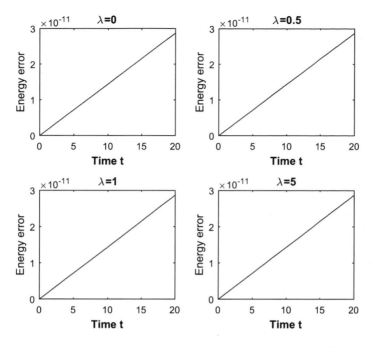

Fig. 6.14 Errors of the discrete energy for $\lambda = 0, 0.5, 1$, and 5 with $\tau = 0.005$, $\Delta x = \Delta y = \Delta z = 1/2^5$, respectively

```
46              X=real(ifft(M_1.*fft(x(1:N_1))));
47              % Kronecker inner product in x-direction
48              d2((k-1)*N_1*N_2+(j-1)*N_1+1:(k-1)*N_1*N_2+j*N_1)=X;
49         end
50   end
51   d(N_1*N_2*N_3+1:2*N_1*N_2*N_3)=d1-d2; % solve A_3H_1-A_1H_3
52
53   for k=1:N_3
54       for i=1:N_1
55           x(1:N_2)=Fx((k-1)*N_1*N_2+i:N_1:k*N_1*N_2);
56           X=real(ifft(M_2.*fft(x(1:N_2))));% Kronecker ...
                 inner product in y-direction
57           d1((k-1)*N_1*N_2+i:N_1:k*N_1*N_2)=X;
58       end
59   end
60   for k=1:N_3
61       for j=1:N_2
62       x(1:N_1)=Fy((k-1)*N_1*N_2+(j-1)*N_1+1:(k-1)*N_1*N_2+j*N_1);
63           X=real(ifft(M_1.*fft(x(1:N_1))));
64           % Kronecker inner product
65           d2((k-1)*N_1*N_2+(j-1)*N_1+1:(k-1)*N_1*N_2+j*N_1)=X;
66       end
67   end
68   d(2*N_1*N_2*N_3+1:3*N_1*N_2*N_3)=d2-d1; % solve A_1H_2-A_2H_1
```

Listing 6.10 A MATLAB code for solving the algebraic equation $Au=b$ induced by (4.65)

```
1   % this function return b
2   function b=get_b(u, mu, varepsilon, tau, J_1, J_2, J_3, ...
        N_1, N_2, N_3, xi, lambda)
3   % u: electromagnetic field u=(E_1,E_2,E_3,H_1,H_2,H_3)'
4   % mu,varepsilon: electric permittivity, magnetic permeability
5   % tau: time step size
6   % 1/2^J_1,1/2^J_2,1/2^J_3: space mesh sizes
7   % N_1,N_2,N_3: number of spatial grid points
8   % xi: noise
9   % lambda: scale of the noise
10  r_1=2*mu/tau;r_2=2*varepsilon/tau;
11  b=zeros(6*N_1*N_2*N_3,1);
12  b(1:3*N_1*N_2*N_3)=r_1*u(1:3*N_1*N_2*N_3)...
13    -dis_curl(u(3*N_1*N_2*N_3+1:6*N_1*N_2*N_3),J_1,J_2,J_3,...
14    N_1,N_2,N_3)+lambda*xi.*u(3*N_1*N_2*N_3+1:6*N_1*N_2*N_3);
15  b(3*N_1*N_2*N_3+1:6*N_1*N_2*N_3)...
16    =dis_curl(u(1:3*N_1*N_2*N_3),J_1,J_2,J_3,N_1,N_2,N_3)...
17    +r_2*u(3*N_1*N_2*N_3+1:6*N_1*N_2*N_3)...
18    -lambda*xi.*u(1:3*N_1*N_2*N_3);
19
20  % this function return A*u
21  function d=get_Au(u, mu, varepsilon, tau, J_1, J_2, J_3, ...
        N_1, N_2, N_3, xi, lambda)
22  r_1=2*mu/tau;r_2=2*varepsilon/tau;
23  d=zeros(6*N_1*N_2*N_3,1);
24  d(1:3*N_1*N_2*N_3)=r_1*u(1:3*N_1*N_2*N_3)...
25    +dis_curl(u(3*N_1*N_2*N_3+1:6*N_1*N_2*N_3),J_1,J_2,J_3,...
26    N_1,N_2,N_3)-lambda*xi.*u(3*N_1*N_2*N_3+1:6*N_1*N_2*N_3);
27  d(3*N_1*N_2*N_3+1:6*N_1*N_2*N_3)...
28    =-dis_curl(u(1:3*N_1*N_2*N_3),J_1,J_2,J_3,N_1,N_2,N_3)...
29    +r_2*u(3*N_1*N_2*N_3+1:6*N_1*N_2*N_3)...
30    +lambda*xi.*u(1:3*N_1*N_2*N_3);
31
32  % this function use to solve A*u=b
33  function u=GMRES(u, mu, varepsilon, tau, J_1, J_2, J_3, ...
        N_1, N_2, N_3, xi, lambda)
34  b=get_b(u, mu, varepsilon, tau, J_1, J_2, J_3, N_1, N_2, ...
        N_3, xi, lambda);
35  r1=b-get_Au(u, mu, varepsilon, tau, J_1, J_2, J_3, N_1, ...
        N_2, N_3, xi, lambda);
36  u1=zeros(max(size(u)),1);
37  m=20;
38  while max(abs(r1))>0.00001
39    v(:,1)=r1/norm(r1);
40    for j=1:m
41      d=get_Au(v(:,j),mu,varepsilon,tau,J_1,J_2,J_3,N_1,N_2,...
42        N_3,xi,lambda);
43      for i=1:j
44        H(i,j)=v(:,i)'*d;
45      end
46      u1(:)=0;
47      for i=1:j
```

```
48          u1=H(i,j)*v(:,i)+u1;
49       end
50    u1=d-u1;
51    H(j+1,j)=norm(u1);
52       if (H(j+1,j)<0.0001 || j==m)
53          e1=zeros(j+1,1);
54          e1(1)=norm(r1);
55          y=pinv(H(1:j+1,1:j))*e1;
56          u=u+v(:,1:j)*y;
57          r1=b-get_AX(u,mu,varepsilon,tau,J_1,J_2,J_3,N_1,N_2,...
58          N_3,xi,lambda);
59          break;
60       end
61       v(:,j+1)=u1/H(j+1,j);
62    end
63 end
```

6.2.3   Splitting Midpoint Method

This section presents some numerical experiments of the splitting algorithm developed in Sect. 4.3. To be precise, we investigate the energy evolution of the splitting midpoint method (5.58), and check the temporal accuracy by fixing the space mesh size; see Listings 6.11–6.13 for the corresponding MATLAB codes.

In the sequel, we consider the stochastic Maxwell equations (4.95) on the domain $D = [0, 1]^3$ with $\lambda_1 = \lambda_2 = (1, 1, 1)^\top$. The initial data read as

$$E_{1_0} = \frac{k_y - k_z}{r} \cos(k_x \pi x) \sin(k_y \pi y) \sin(k_z \pi z),$$

$$H_{1_0} = \sin(k_x \pi x) \cos(k_y \pi y) \cos(k_z \pi z),$$

$$E_{2_0} = \frac{k_z - k_x}{r} \sin(k_x \pi x) \cos(k_y \pi y) \sin(k_z \pi z),$$

$$H_{2_0} = \cos(k_x \pi x) \sin(k_y \pi y) \cos(k_z \pi z),$$

$$E_{3_0} = \frac{k_x - k_y}{r} \sin(k_x \pi x) \sin(k_y \pi y) \cos(k_z \pi z),$$

$$H_{3_0} = \cos(k_x \pi x) \cos(k_y \pi y) \sin(k_z \pi z)$$

with $k_x = 1$, $k_y = 2$, $k_z = -3$, and $r = \sqrt{k_x^2 + k_y^2 + k_z^2}$. The formulation of the increment of the Q-Wiener process is given by (6.5).

As stated in Proposition 4.32, the averaged energy of the splitting subsystems grows linearly. To illustrate this phenomenon, we set the time interval to be $[0, 1000]$ with $\tau = 0.001$. The discrete averaged energy over 200 trajectories is displayed in

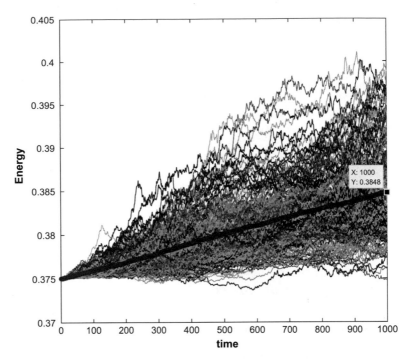

Fig. 6.15 The averaged energy over 200 trajectories with $\tau = 10^{-3}$, $\Delta x = \Delta y = \Delta z = 1/60$

Fig. 6.15, where the bold red line represents the discrete averaged energy. It can be observed the linear growth of the averaged energy of the splitting midpoint method (5.58).

Next, we illustrate the mean-square convergence order in the temporal direction of the splitting midpoint method in Theorem 5.8. To this end, we compute the error at the terminal time $T = 1/4$ and plot the error against τ on a log-log scale for the truncated Q-Wiener process with the truncation numbers $M = 1$, 4, and 8, respectively. The reference solution is computed by using the same numerical algorithm for a small time step size $\tau = 0.001$ and the expectation is realized by using the average of 100 independent paths. Fix space mesh sizes $\Delta x = \Delta y = \Delta z = 1/60$. We then compare the reference solution with solutions of the splitting midpoint method with time step sizes being $2^1\tau$, $2^2\tau$, and $2^3\tau$, respectively, to estimate the mean-square convergence order.

Figure 6.16 displays the mean-square error. It can be seen that the convergence order for the deterministic case is 1, while the observations are different from the stochastic case, where various sorts of Q-Wiener processes depending on M are used. The mean-square order of convergence drops approximately from 1 to 0.5, which results from the regularities of the noise and the exact solution of the stochastic Maxwell equations.

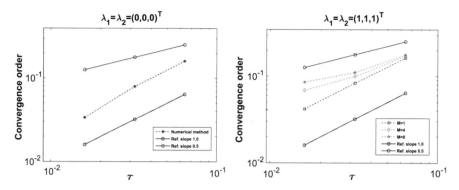

Fig. 6.16 Convergence orders of the operator splitting midpoint algorithm for the deterministic case (left) and the stochastic case for $M = 1, 4$, and 8 (right)

As mentioned in Sect. 4.3, the splitting technique can improve the computational efficiency. In order to demonstrate the efficiency and superiority of the proposed method (5.58), we compare the splitting midpoint method with the following ones:

- The stochastic multi-symplectic wavelet algorithm for (4.95) (wavelet method for short)

$$\mathbf{E}_1^{n+1} - \mathbf{E}_1^n = \tau\left(A_2\mathbf{H}_3^{n+\frac{1}{2}} - A_3\mathbf{H}_2^{n+\frac{1}{2}}\right) + \lambda_1\Delta\mathbf{W}^{n+1},$$

$$\mathbf{E}_2^{n+1} - \mathbf{E}_2^n = \tau\left(A_3\mathbf{H}_1^{n+\frac{1}{2}} - A_1\mathbf{H}_3^{n+\frac{1}{2}}\right) + \lambda_1\Delta\mathbf{W}^{n+1},$$

$$\mathbf{E}_3^{n+1} - \mathbf{E}_3^n = \tau\left(A_1\mathbf{H}_2^{n+\frac{1}{2}} - A_2\mathbf{H}_1^{n+\frac{1}{2}}\right) + \lambda_1\Delta\mathbf{W}^{n+1},$$

$$\mathbf{H}_1^{n+1} - \mathbf{H}_1^n = \tau\left(A_3\mathbf{E}_2^{n+\frac{1}{2}} - A_2\mathbf{E}_3^{n+\frac{1}{2}}\right) + \lambda_2\Delta\mathbf{W}^{n+1}, \qquad (6.6)$$

$$\mathbf{H}_2^{n+1} - \mathbf{H}_2^n = \tau\left(A_1\mathbf{E}_3^{n+\frac{1}{2}} - A_3\mathbf{E}_1^{n+\frac{1}{2}}\right) + \lambda_2\Delta\mathbf{W}^{n+1},$$

$$\mathbf{H}_3^{n+1} - \mathbf{H}_3^n = \tau\left(A_2\mathbf{E}_1^{n+\frac{1}{2}} - A_1\mathbf{E}_2^{n+\frac{1}{2}}\right) + \lambda_2\Delta\mathbf{W}^{n+1},$$

where A_s ($s = 1, 2, 3$) and $\Delta\mathbf{W}^{n+1}$ are given in (4.66).
- The splitting wavelet method (obtained by applying the wavelet method (6.6) to each subsystem (5.52)).

Figure 6.17 displays the total CPU time of each numerical method with various mesh sizes for 2000 samples. We summarize that the splitting methods have higher efficiency than non-splitting ones.

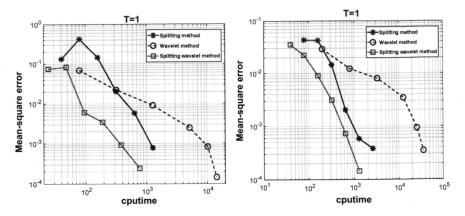

Fig. 6.17 Efficiency for the splitting midpoint method, wavelet method, and splitting wavelet method applied to three-dimensional stochastic Maxwell equations (4.95). Left: mesh sizes $\Delta x = \Delta y = \Delta z = 1/2^4$; Right: mesh sizes $\Delta x = \Delta y = \Delta z = 1/2^7$

Listing 6.11 A MATLAB code for the first subsystem (5.58)

```
 1  function [E_1,E_2,E_3,H_1,H_2,H_3]=sto_subsystem1( I, J, ...
        K, tau, dx, lambda_1, lambda_2, noise, E_10, E_20, ...
        E_30, H_10, H_20, H_30)
 2  % I,J,K: numbers of spatial grid points
 3  % tau: time step size
 4  % dx: mesh size in x-direction
 5  % E_10,E_20,E_30,H_10,H_20,H_30: initial data
 6  % lambda_1, lambda_2: scales of the noise (lambda_1= ...
        lambda_2)
 7  % noise: the increment of the Wiener process
 8  E_1=zeros(I,J,K);E_2=zeros(I,J,K);E_3=zeros(I,J,K);
 9  H_1=zeros(I,J,K);H_2=zeros(I,J,K);H_3=zeros(I,J,K);
10  X1=zeros(2*I,J);X2=zeros(2*I,J);
11  Y1=zeros(2*I,J);Y2=zeros(2*I,J);
12  % solve coefficient matrices of the algebraic equation
13  rx=tau/dx;
14  Ax1=sparse(1:I,1:I,1*ones(1,I),I,I)+sparse(1:I-1,2:I,...
15      1*ones(1,I-1),I,I)+sparse(I,1,1,I,I);
16  Ax2=sparse(1:I,1:I,(-1)*ones(1,I),I,I)+sparse(1:I-1,2:I,...
17      1*ones(1,I-1),I,I)+sparse(I,1,1,I,I);
18  Ax2=rx*Ax2;
19  A=[Ax1,Ax2;Ax2,Ax1];B=[Ax1,-Ax2;-Ax2,Ax1];
20  % solve the numerical solution
21  % A(E_2^{n+1},[1]},H_3^{n+1},[1]})=B(E_2^{n,[1]},H_3^{n,[1]})
22  % B(E_3^{n+1},[1]},H_2^{n+1},[1]})=A(E_3^{n,[1]},H_2^{n,[1]})
23  % E_1^{n+1},[1]}=E_1^{n,[1]}+\lambda_1*\Delta W
24  % H_1^{n+1},[1]}=H_1^{n,[1]}+\lambda_2*\Delta W
25  E_1=E_10+lambda_1*noise;H_1=H_10+lambda_2*noise;
26  for k=1:K
27      X1=[E_20(:,:,k);H_30(:,:,k)]; Y1=A$B*X1);
28      E_2(:,:,k)=Y1(1:I,:);H_3(:,:,k)=Y1(I+1:2*I,:);
```

```
29    X2=[E_30(:,:,k);H_20(:,:,k)];  Y2=B$A*X2);
30    E_3(:,:,k)=Y2(1:I,:);H_2(:,:,k)=Y2(I+1:2*I,:);
31  end
```

Listing 6.12 A MATLAB code for the second subsystem (5.58)

```
1  function [E_1,E_2,E_3,H_1,H_2,H_3]=sto_subsystem2(I, J, ...
      K, tau, dy, lambda_1, lambda_2, noise, E_10, E_20, ...
      E_30, H_10, H_20, H_30)
2  % I,J,K: numbers of spatial grid points
3  % tau: time step size
4  % dy: mesh size in y-direction
5  % E_10,E_20,E_30,H_10,H_20,H_30: initial data
6  % lambda_1, lambda_2: scales of the noise (lambda_1= ...
      lambda_2)
7  % noise: the increment of the Wiener process
8  E_1=zeros(I,J,K);E_2=zeros(I,J,K);E_3=zeros(I,J,K);
9  H_1=zeros(I,J,K);H_2=zeros(I,J,K);H_3=zeros(I,J,K);
10 X1=zeros(2*J,I);X2=zeros(2*J,I);
11 Y1=zeros(2*J,I);Y2=zeros(2*J,I);
12 % solve coefficient matrices of the algebraic equation
13 ry=tau/dy;
14 Ay1=sparse(1:J,1:J,1*ones(1,J),J,J)+sparse(1:J-1,2:J,...
15     1*ones(1,J-1),J,J)+sparse(J,1,1,J,J);
16 Ay2=sparse(1:J,1:J,(-1)*ones(1,J),J,J)+sparse(1:J-1,2:J,...
17     1*ones(1,J-1),J,J)+sparse(J,1,1,J,J);
18 Ay2=ry*Ay2;
19 A=[Ay1,Ay2;Ay2,Ay1];B=[Ay1,-Ay2;-Ay2,Ay1];
20 % solve the numerical solution
21 % A(E_3^{n+1,[2]},H_1^{n+1,[2]})=B(E_3^{n,[2]},H_1^{n,[2]})
22 % B(E_1^{n+1,[2]},H_3^{n+1,[2]})=A(E_1^{n,[2]},H_3^{n,[2]})
23 % E_2^{n+1,[2]}=E_2^{n,[2]}+\lambda_1*\Delta W
24 % H_2^{n+1,[2]}=H_2^{n,[2]}+\lambda_2*\Delta W
25 E_2=E_20+lambda_1*noise; H_2=H_20+lambda_2*noise;
26 for k=1:K
27     X1=[E_30(:,:,k)';H_10(:,:,k)'];  Y1=A$B*X1);
28     E_3(:,:,k)=Y1(1:J,:)';H_1(:,:,k)=Y1(J+1:2*J,:)';
29     X2=[E_10(:,:,k)';H_30(:,:,k)'];  Y2=B$A*X2);
30     E_1(:,:,k)=Y2(1:J,:)';H_3(:,:,k)=Y2(J+1:2*J,:)';
31 end
```

Listing 6.13 A MATLAB code for the third subsystem (5.58)

```
1  function [E_1,E_2,E_3,H_1,H_2,H_3]=sto_subsystem3(I, J, ...
      K, tau, dz, lambda_1, lambda_2, noise, E_10, E_20, ...
      E_30, H_10, H_20, H_30)
2  % I,J,K: numbers of spatial grid points
3  % tau: time step size
4  % dz: mesh size in z-direction
5  % E_10,E_20,E_30,H_10,H_20,H_30: initial data
```

```
 6  % lambda_1, lambda_2: scales of the noise (lambda_1= ...
       lambda_2)
 7  % noise: the increment of the Wiener process
 8  E_1=zeros(I,J,K);E_2=zeros(I,J,K);E_3=zeros(I,J,K);
 9  H_1=zeros(I,J,K);H_2=zeros(I,J,K);H_3=zeros(I,J,K);
10  X1=zeros(2*K,J);X2=zeros(2*K,J);
11  Y1=zeros(2*K,J);Y2=zeros(2*K,J);
12  % solve coefficient matrices of the algebraic equation
13  rz=tau/dz;
14  Az1=sparse(1:K,1:K,1*ones(1,K),K,K)+sparse(1:K-1,2:K,...
15      1*ones(1,K-1),K,K)+sparse(K,1,1,K,K);
16  Az2=sparse(1:K,1:K,(-1)*ones(1,K),K,K)+sparse(1:K-1,2:K,...
17      1*ones(1,K-1),K,K)+sparse(K,1,1,K,K);
18  Az2=rz*Az2;
19  A=[Az1,Az2;Az2,Az1];B=[Az1,-Az2;-Az2,Az1];
20  % solve the numerical solution
21  % A(E_1^{n+1,[3]},H_2^{n+1,[3]})=B(E_1^{n,[3]},H_2^{n,[3]})
22  % B(E_2^{n+1,[3]},H_1^{n+1,[3]})=A(E_2^{n,[3]},H_1^{n,[3]})
23  % E_3^{n+1,[3]}=E_3^{n,[3]}+\lambda_1*\Delta W
24  % H_3^{n+1,[3]}=H_3^{n,[3]}+\lambda_2*\Delta W
25  E_3=E_30+lambda_1*noise;H_3=H_30+lambda_2*noise;
26  for i=1:I
27      v=reshape(E_10(i,:,:),[J,K]); ...
             w=reshape(H_20(i,:,:),[J,K]);
28      X1=[v';w'];Y1 =A$B*X1);
29      E_1(i,:,:)=reshape(Y1(1:K,:)',[1,J,K]);
30      H_2(i,:,:)=reshape(Y1(K+1:2*K,:)',[1,J,K]);
31      v1=reshape(E_20(i,:,:),[J,K]);
32      w1=reshape(H_10(i,:,:),[J,K]);
33      X2=[v1';w1'];Y2=B$A*X2);
34      E_2(i,:,:)=reshape(Y2(1:K,:)',[1,J,K]);
35      H_1(i,:,:)=reshape(Y2(K+1:2*K,:)',[1,J,K]);
36  end
```

Appendix A
Basic Identities and Inequalities

This chapter is devoted to giving some basic identities and inequalities which are frequently used throughout this monograph. For the identities, the divergence theorem, the Stokes theorem, and the Green formulae are presented. And for the inequalities, the Young inequality, the Hölder inequality, the Minkowski inequality, several versions of the Grönwall inequality, and the algebraic inequality are introduced.

A.1 Basic Identities

Let u be a scalar function, and let $\mathbf{u} = (u_1, u_2, u_3)^\top$, $\mathbf{v} = (v_1, v_2, v_3)^\top$, and $\mathbf{w} = (w_1, w_2, w_3)^\top$ be vector functions. Denote the gradient, divergence, and curl operators, respectively by

$$\nabla u := (\partial_x u, \partial_y u, \partial_z u)^\top,$$

$$\nabla \cdot \mathbf{u} := \partial_x u_1 + \partial_y u_2 + \partial_z u_3,$$

$$\nabla \times \mathbf{u} := \begin{bmatrix} \partial_y u_3 - \partial_z u_2 \\ \partial_z u_1 - \partial_x u_3 \\ \partial_x u_2 - \partial_y u_1 \end{bmatrix}.$$

The cross product of vector functions $\mathbf{u}$ and $\mathbf{v}$ is denoted by

$$\mathbf{u} \times \mathbf{v} := \begin{bmatrix} u_2 v_3 - u_3 v_2 \\ u_3 v_1 - u_1 v_3 \\ u_1 v_2 - u_2 v_1 \end{bmatrix}.$$

The commonly used vector identities include:

$$\nabla \times (u\mathbf{v}) = u(\nabla \times \mathbf{v}) - \mathbf{v} \times (\nabla u),$$

$$\nabla \times (\nabla \times \mathbf{u}) = -\Delta\mathbf{u} + \nabla(\nabla \cdot \mathbf{u}),$$

$$\nabla \cdot (\nabla u) = \Delta u,$$

$$\nabla \times (\nabla u) = 0,$$

$$\nabla \cdot (\nabla \times \mathbf{u}) = 0,$$

$$\nabla \cdot (\mathbf{u} \times \mathbf{v}) = \mathbf{v} \cdot (\nabla \times \mathbf{u}) - \mathbf{u} \cdot (\nabla \times \mathbf{v}),$$

$$\mathbf{u} \cdot (\mathbf{v} \times \mathbf{w}) = \mathbf{v} \cdot (\mathbf{w} \times \mathbf{u}) = \mathbf{w} \cdot (\mathbf{u} \times \mathbf{v}).$$

Let $D \subset \mathbb{R}^3$ be an open, bounded, and Lipschitz domain with boundary ∂D. Let $S \subset \mathbb{R}^3$ be an open surface with boundary ∂S. Denote by $\mathbf{n}$ the unit outward normal vector on ∂D or S, and by $\hat{\mathbf{n}}$ the unit tangential vector on the contour ∂S that encloses S. The commonly used integral identities include:

The divergence theorem:

$$\int_D \nabla \cdot \mathbf{u}\, d\mathbf{x} = \oint_{\partial D} \mathbf{u} \cdot \mathbf{n}\, ds.$$

The Stokes theorem:

$$\int_S (\nabla \times \mathbf{u}) \cdot \mathbf{n}\, ds = \oint_{\partial S} \mathbf{u} \cdot \hat{\mathbf{n}}\, dl.$$

The Green formulae:

$$\int_D \mathbf{u} \cdot \nabla\phi\, d\mathbf{x} + \int_D (\nabla \cdot \mathbf{u})\phi\, d\mathbf{x} = \oint_{\partial D} \mathbf{n} \cdot \mathbf{u}\phi\, ds,$$

$$\int_D \nabla \times \mathbf{u} \cdot \mathbf{w}\, d\mathbf{x} - \int_D \mathbf{u} \cdot \nabla \times \mathbf{w}\, d\mathbf{x} = \oint_{\partial D} (\mathbf{n} \times \mathbf{u}) \cdot \mathbf{w}\, ds.$$

A.2 Basic Inequalities

Below we present the Young inequality, the Hölder inequality, the Minkowski inequality, several versions of the Grönwall inequality, and the algebraic inequality. We refer to e.g. [77, 87] for the proof of different versions of the Grönwall inequality, and to [70] for the algebraic inequality.

Proposition A.1 (Young Inequality) *Let γ be a positive real number, and $p, q >$ 1 such that $\frac{1}{p} + \frac{1}{q} = 1$. Then*

$$|ab| \leq \gamma \frac{|a|^p}{p} + \gamma^{-\frac{q}{p}} \frac{|b|^q}{q}, \qquad a, b \in \mathbb{R}.$$

Proposition A.2 (Hölder Inequality) *Let $p, q > 1$ such that $\frac{1}{p} + \frac{1}{q} = 1$. Then*

$$\|fg\|_{L^1(D)} \leq \|f\|_{L^p(D)} \|g\|_{L^q(D)}, \qquad f \in L^p(D), \ g \in L^q(D).$$

Proposition A.3 (Minkowski Inequality) *Let $p > 1$ and $f, g \in L^p(D)$. Then*

$$\|f + g\|_{L^p(D)} \leq \|f\|_{L^p(D)} + \|g\|_{L^p(D)}.$$

Proposition A.4 (Continuous Grönwall Inequality) *Let $T > 0$, $f, g \in L^\infty(0, T)$, the constant $c \geq 0$, and g be a monotonically increasing and continuous function. If f satisfies*

$$f(t) \leq g(t) + c \int_0^t f(s) ds \quad a.e. \ in \ t \in [0, T],$$

then

$$f(t) \leq e^{ct} g(t) \quad a.e. \ in \ t \in [0, T].$$

Proposition A.5 (Discrete Grönwall Inequality) *Let $\{a_n\}_{n \in \mathbb{N}}$ and $\{b_n\}_{n \in \mathbb{N}}$ be two sequences. Let $\theta \in [0, 1]$, $c_0 > 0$, $\tau > 0$, and $1 - \theta c_0 \tau > 0$. If $a_0 \leq b_0$ and*

$$a_{n+1} \leq b_{n+1} + c_0 \tau \sum_{j=0}^n \left((1 - \theta)a_j + \theta a_{j+1}\right) \quad \forall n \in \mathbb{N},$$

then for all $n \in \mathbb{N}$,

$$a_{n+1} \leq b_{n+1} + \frac{c_0 \tau}{1 - \theta c_0 \tau} \sum_{j=0}^n \left(\frac{1 + (1 - \theta)c_0 \tau}{1 - \theta c_0 \tau}\right)^{n-j} \left((1 - \theta)b_j + \theta b_{j+1}\right).$$

Moreover, if $\{b_n\}_{n \in \mathbb{N}}$ is monotonically increasing, it follows that for all $n \in \mathbb{N}$,

$$a_n \leq b_n \left(\frac{1 + (1 - \theta)c_0 \tau}{1 - \theta c_0 \tau}\right)^n.$$

Proposition A.6 (Grönwall Inequality: Differential Form) *Let $T > 0$, $f \in W^{1,1}(0, T)$, and g, $\lambda \in L^1(0, T)$. Then*

$$f'(t) \leq g(t) + \lambda(t) f(t) \quad a.e. \text{ in } t \in [0, T]$$

implies that for almost all $t \in [0, T]$,

$$f(t) \leq e^{\Lambda(t)} f(0) + \int_0^t e^{\Lambda(t) - \Lambda(s)} g(s) ds,$$

where $\Lambda(t) := \int_0^t \lambda(s) ds$.

Proposition A.7 (Algebraic Inequality) *Let $p \geq 2$. Then for any $a, b \in \mathbb{R}^m$ with $m \in \mathbb{N}_+$, it holds that*

$$\left(|a|^{p-2} a - |b|^{p-2} b \right) \cdot (a - b) \geq \gamma_0 |a - b|^p.$$

Moreover, if $1 < p < 2$, then

$$\left(|a|^{p-2} a - |b|^{p-2} b \right) \cdot (a - b) \leq \gamma_1 |a - b|^p.$$

Here, the positive constants γ_0, γ_1 depend on p and m.

Appendix B
Semigroup, Sobolev Space, and Differential Calculus

B.1 Operator Semigroup

We refer readers to [78, 142] for more details about the theory of the operator semigroup.

Let $(V, \langle \cdot, \cdot \rangle_V)$ be a Hilbert space with corresponding norm $\| \cdot \|_V = \sqrt{\langle \cdot, \cdot \rangle_V}$. By $\mathscr{L}(V)$ we denote the space of all bounded linear operators on V. For any $A \in \mathscr{L}(V)$, the operator norm is given by

$$\| A \|_{\mathscr{L}(V)} := \sup_{x \neq 0} \frac{\| Ax \|_V}{\| x \|_V}.$$

Some fundamentals of semigroup theory needed in this monograph are shown below.

Definition B.1 A mapping $S(\cdot) : [0, +\infty) \to \mathscr{L}(V)$ is called a strongly continuous semigroup or C_0-semigroup if the following conditions are satisfied:

(i) $S(0) = Id$ and $S(t + s) = S(t)S(s)$ for all $t, s \geq 0$;
(ii) For each $u \in V$, the mapping $S(\cdot)u : [0, \infty) \to V, t \mapsto S(t)u$ is continuous.

In particular, a C_0-semigroup $S(\cdot)$ is called a unitary semigroup if $\| S(t)u \|_V = \| u \|_V$ for all $u \in V$ and $t \geq 0$.

The operator A defined by

$$Au := \lim_{t \to 0^+} \frac{S(t)u - u}{t} \quad \forall\, u \in \mathscr{D}(A)$$

© The Author(s), under exclusive license to Springer Nature Singapore Pte Ltd. 2023
C. Chen et al., *Numerical Approximations of Stochastic Maxwell Equations*,
Lecture Notes in Mathematics 2341, https://doi.org/10.1007/978-981-99-6686-8

with domain

$$\mathscr{D}(A) := \left\{ u \in V \;\middle|\; \lim_{t \to 0^+} \frac{S(t)u - u}{t} \quad \text{exists in } V \right\}$$

is called the infinitesimal generator of the strongly continuous semigroup $S(\cdot)$.

Definition B.2 Let $(A, \mathscr{D}(A))$ be a densely defined linear operator on V, i.e., $\overline{\mathscr{D}(A)} = V$. The adjoint operator A^* of A is defined by

$$\mathscr{D}(A^*) := \left\{ u \in V : \exists v \in V \ s.t. \ \langle v, w \rangle_V = \langle u, Aw \rangle_V \ \text{for all } w \in \mathscr{D}(A) \right\},$$

and $A^* u := v$ for $u \in \mathscr{D}(A^*)$.

Definition B.3 Let $A : \mathscr{D}(A) \to V$ be a densely defined linear operator. The operator A is called

(i)   symmetric if $Au = A^*u$ for all $u \in \mathscr{D}(A) \subset \mathscr{D}(A^*)$;
(ii)   skew-symmetric if $Au = -A^*u$ for all $u \in \mathscr{D}(A) \subset \mathscr{D}(A^*)$;
(iii) self-adjoint if $A = A^*$ (i.e., A is symmetric and $\mathscr{D}(A) = \mathscr{D}(A^*)$);
(iv)   skew-adjoint if $A^* = -A$ (i.e., A is skew-symmetric and $\mathscr{D}(A) = \mathscr{D}(A^*)$).

The following lemma states the criterion for a skew-symmetric operator to be skew-adjoint; see for instance [147].

Lemma B.1 *Let $A : \mathscr{D}(A) \to V$ be skew-symmetric. Then A is skew-adjoint if $Id \pm A$ have dense range, that is,*

$$\overline{\operatorname{ran}(Id \pm A)} = V.$$

The skew-adjointness of the densely defined operator forms a sufficient and necessary condition for the unitary C_0-semigroup as stated in [78, Theorem 3.24].

Theorem B.1 *Let $A : \mathscr{D}(A) \to V$ be a densely defined linear operator. Then A generates a unitary C_0-semigroup $S(\cdot)$ if and only if A is skew-adjoint.*

B.2 Sobolev Space

In this section, we state some fundamental definitions and results in the context of the Sobolev space. We refer to [165] and [23, Chap. 8] for more details.

Let $D \subset \mathbb{R}^d$ be an open set with a Lipschitz boundary ∂D. For each multi-index $\alpha = (\alpha_1, \ldots, \alpha_d)$, we define

$$|\alpha| := \sum_{m=1}^d \alpha_m, \quad D^\alpha := \frac{\partial^{\alpha_1}}{\partial x_1^{\alpha_1}} \cdots \frac{\partial^{\alpha_d}}{\partial x_d^{\alpha_d}}.$$

For an integer $k \geq 0$ and a real number $p \in [1, \infty]$, we define the Sobolev space $W^{k,p}(D)$ as

$$W^{k,p}(D) := \{u \in L^p(D) : D^\alpha u \in L^p(D) \text{ for all } \alpha \text{ with } |\alpha| \leq k\},$$

where the derivatives are understood in the weak sense. When endowed with the norm

$$\|u\|_{W^{k,p}(D)} := \Big(\sum_{|\alpha| \leq k} \|D^\alpha u\|^p_{L^p(D)} \Big)^{\frac{1}{p}},$$

$W^{k,p}(D)$ is a Banach space. In the special case $p = 2$, we denote $H^k(D) := W^{k,2}(D)$. With the inner product

$$\langle u, v \rangle_{H^k(D)} := \sum_{|\alpha| \leq k} \int_D D^\alpha u(\mathbf{x}) D^\alpha v(\mathbf{x}) d\mathbf{x},$$

the space $H^k(D)$ becomes a Hilbert space. If D is a bounded domain with a Lipschitz boundary, then there exists a trace operator

$$\gamma^{\partial D} : H^1(D) \to H^{1/2}(\partial D)$$

such that $\gamma^{\partial D} u = u|_{\partial D}$ for all $u \in C^\infty(\overline{D})$. The kernel of $\gamma^{\partial D}$ is denoted by $H^1_0(D)$.

The following Sobolev embedding theorem is very useful in applications; see for instance [148, Theorem 3.2.2]. Denote by $H \hookrightarrow V$ (resp. $H \overset{c}{\hookrightarrow} V$) the continuous (resp. compact) embedding of H into V.

Theorem B.2 (Sobolev Embedding Theorem) *Let D be a d-dimensional domain with $C^{0,1}$ boundary, and $j, k \in \mathbb{N}$ and $p \in [1, \infty)$.*

(i) *If $kp < d$ and $1 \leq p \leq q < \infty$ satisfy $q(d - kp) \leq dp$, then*

$$W^{j+k,p}(D) \hookrightarrow W^{j,q}(D).$$

In particular, if D is bounded, then the above embedding holds additionally for $1 \leq q < p$.

(ii) *If $kp = d$ and $1 \leq p \leq q < \infty$, then*

$$W^{j+k,p}(D) \hookrightarrow W^{j,q}(D).$$

In particular, if D is bounded, then the above embedding holds additionally for $1 \leq q < p$. If $p = 1$, the embedding holds for $q = \infty$ as well.

(iii) *If $kp > d$ and one of the following cases holds*

   (a) $d > (k-1)p$ and $0 < \alpha \le (k-d/p)$;
   (b) $d = (k-1)p$ and $0 < \alpha < 1$;
   (c) $d = k-1$, $p = 1$ and $0 < \alpha \le 1$,

   then

$$W^{j+k,p}(D) \hookrightarrow C^{j,\alpha}(\bar{D}).$$

To end this section, we give several lemmas which will be used to analyze the H^k-regularity of the solution of the stochastic Maxwell equations on a cuboid $D \subset \mathbb{R}^3$.

Lemma B.2 ([5, Theorem 2.17]) *The spaces*

$$\left\{ u \in H(\mathrm{curl}, D) \cap H(\mathrm{div}, D) : \mathbf{n} \times u|_{\partial D} = 0 \right\}$$

and

$$\left\{ u \in H(\mathrm{curl}, D) \cap H(\mathrm{div}, D) : \mathbf{n} \cdot u|_{\partial D} = 0 \right\}$$

are both continuously imbedded in $H^1(D)^3$. Moreover,

$$\|u\|_{H^1(D)^3} \le C\left(\|u\|_{L^2(D)^3} + \|\nabla \times u\|_{L^2(D)^3} + \|\nabla \cdot u\|_U \right), \tag{B.1}$$

where the positive constant C depends on $|D|$.

Lemma B.3 ([5, Theorem 2.8]) *The following compact embeddings hold:*

   (i) $\{u \in H(\mathrm{curl}, D) \cap H(\mathrm{div}, D) : \mathbf{n} \times u|_{\partial D} = 0\} \overset{c}{\hookrightarrow} L^2(D)^3$,
   (ii) $\{u \in H(\mathrm{curl}, D) \cap H(\mathrm{div}, D) : \mathbf{n} \cdot u|_{\partial D} = 0\} \overset{c}{\hookrightarrow} L^2(D)^3$.

The following lemma is about the homogeneous boundary value problem for the Laplacian on a cuboid.

Lemma B.4 ([98, Lemma 3.6]) *Let $\overline{\Gamma}$ be a union of faces of a cuboid D in $\mathbb{R}^3$, and $\widetilde{\Gamma}$ be the union of the remaining open faces. Let $f \in U$. Then there is a unique function $v \in H^1_{\overline{\Gamma}}(D)$ such that*

$$\int_D v\varphi \mathbf{dx} + \int_D \nabla v \cdot \nabla \varphi \mathbf{dx} = \int_D f\varphi \mathbf{dx}$$

for all $\varphi \in H^1_{\overline{\Gamma}}(D)$. Moreover, the solution v belongs to $H^2(D) \cap H^1_{\overline{\Gamma}}(D)$ and satisfies $v - \Delta v = f$ on D, $\partial_{\mathbf{n}} v := \nabla v \cdot \mathbf{n} = 0$ on $\widetilde{\Gamma}$, and $\|v\|_{H^2(D)} \le C\|f\|_{L^2(D)}$ with the positive constant C depending only on $|D|$.

The case for the mixed inhomogeneous boundary value problem for the Laplacian on a cuboid is stated below. We denote

$$\Gamma_1^+ := \{\mathbf{x} \in \overline{D} : x = a_1^+\}, \quad \Gamma_2^+ := \{\mathbf{x} \in \overline{D} : y = a_2^+\},$$

$$\Gamma_3^+ := \{\mathbf{x} \in \overline{D} : z = a_3^+\},$$

$$\Gamma_1^- := \{\mathbf{x} \in \overline{D} : x = a_1^-\}, \quad \Gamma_2^- := \{\mathbf{x} \in \overline{D} : y = a_2^-\},$$

$$\Gamma_3^- := \{\mathbf{x} \in \overline{D} : z = a_3^-\},$$

and $\Gamma_j := \Gamma_j^- \cup \Gamma_j^+$ for $j = 1, 2, 3$. For $p \geq 1$, $s \in (0, \infty) \backslash \mathbb{N}$, $s < k \in \mathbb{N}$, and some open set $D_1 \subseteq \mathbb{R}^d$, we define the Sobolev–Slobodeckij space $W^{s,p}(D_1) := \left(L^p(D_1), W^{k,p}(D_1)\right)_{s/k,p}$ by the real interpolation, see for instance [2].

Lemma B.5 ([73, Lemma 3.1]) *Let $j \in \{1, 2, 3\}$ and $\Gamma^* := \partial D \backslash \Gamma_j$. Take $f \in U$ and $g \in H_0^{1/2}(\Gamma_j) := \left(L^2(\Gamma_j), H_0^1(\Gamma_j)\right)_{1/2,2}$. If there exists a unique function $v \in H_{\Gamma^*}^1(D)$ solving*

$$\int_D v\varphi d\mathbf{x} + \int_D \nabla v \cdot \nabla \varphi d\mathbf{x} = \int_D f\varphi d\mathbf{x} + \int_{\Gamma_j^+} g\varphi d\sigma - \int_{\Gamma_j^-} g\varphi d\sigma$$

for all $\varphi \in H_{\Gamma^*}^1(D)$, then the solution v belongs to $H^2(D) \cap H_{\Gamma^*}^1(D)$ and satisfies $v - \Delta v = f$ on D, $\partial_{\mathbf{n}} v = g$ on Γ_j, and $\|v\|_{H^2(D)} \leq C\left(\|f\|_U + \|g\|_{H_0^{1/2}(\Gamma_j)}\right)$ with the positive constant C depending only on $|D|$.

To check the boundary conditions, we often use the following lemma.

Lemma B.6 ([74, Lemma 2.1]) *For some $j, k \in \{1, 2, 3\}$ with $k \neq j$, let $f \in U$ satisfy $\partial_j f$, $\partial_k f$, $\partial_{jk} f \in U$ and $f = 0$ on Γ_j. Then $\partial_k f = 0$ on Γ_j.*

B.3 Fréchet and Gâteaux Derivatives

Now we recall some basic facts on the Fréchet differentiability and Gâteaux differentiability of a mapping. For more details, see e.g., [9, Sect. 5.3] and [14, Sect. 2.1].

Let E and V be two Banach spaces. Let $\mathscr{L}(E, V)$ be the Banach space of all bounded linear operators from E into V endowed with the usual operator norm.

Definition B.4 A mapping $\psi : E \to V$ is Fréchet differentiable at $x \in E$ if there exists a continuous linear map $A \in \mathscr{L}(E, V)$ such that

$$\lim_{h \to 0} \frac{\psi(x + h) - \psi(x) - Ah}{\|h\|_E} = 0. \tag{B.2}$$

The linear map A will be denoted by $\psi_x(x)$ and is called the Fréchet derivative of ψ at x.

Definition B.5 A mapping $\phi : E \to V$ is Gâteaux differentiable at $x \in E$ if there exists a continuous linear map $A \in \mathscr{L}(E, V)$ such that

$$\lim_{\rho \to 0} \frac{\phi(x + \rho h) - \phi(x)}{\rho} = Ah \quad \forall\, h \in E. \tag{B.3}$$

The linear map A will be denoted by $D_x \phi(x)$ and is called the Gâteaux derivative of ϕ at x.

Evidently, (B.3) is equivalent to

$$\psi(x + \rho h) = \psi(x) + \rho Ah + o(|\rho|) \quad \forall\, h \in E.$$

Thus Fréchet differentiability implies Gâteaux differentiability, while the converse is not true. We shall often use the following characterization of Fréchet differentiability (see e.g., [134, Lemma 2.1]).

Lemma B.7 *A mapping $\psi : E \to V$ is Fréchet differentiable at $x \in E$ with $\psi_x(x) = A$ if and only if for each bounded set $\widehat{E} \subset E$,*

$$\lim_{\rho \to 0} \frac{\psi(x + \rho h) - \psi(x) - \rho Ah}{\rho} = 0$$

uniformly with respect to $h \in \widehat{E}$.

Appendix C
Estimates Related to Maxwell Operators

This chapter presents some estimates related to the continuous and discrete Maxwell operators, which are frequently used in the convergence analysis of the structure-preserving algorithms for the stochastic Maxwell equations. We refer to [29, 41–43] for more details.

Recall the Maxwell operator

$$M = \begin{bmatrix} 0 & \varepsilon^{-1}\nabla\times \\ -\mu^{-1}\nabla\times & 0 \end{bmatrix}$$

with domain

$$\mathcal{D}(M) = \left\{ \begin{bmatrix} \mathbf{E} \\ \mathbf{H} \end{bmatrix} \in \mathbb{H} : M\begin{bmatrix} \mathbf{E} \\ \mathbf{H} \end{bmatrix} = \begin{bmatrix} \varepsilon^{-1}\nabla\times\mathbf{H} \\ -\mu^{-1}\nabla\times\mathbf{E} \end{bmatrix} \in \mathbb{H}, \ \mathbf{n}\times\mathbf{E}\big|_{\partial D} = 0 \right\},$$

where $D \subset \mathbb{R}^3$ is an open, bounded, and Lipschitz domain with boundary ∂D. The corresponding norm is defined as

$$\|u\|_{\mathcal{D}(M)} := \left(\|u\|_{\mathbb{H}}^2 + \|Mu\|_{\mathbb{H}}^2 \right)^{\frac{1}{2}}.$$

The following lemma presents the estimate for the C_0-semigroup

$$\left\{ S(t) = e^{tM}, t \geq 0 \right\}$$

generated by the Maxwell operator M.

© The Author(s), under exclusive license to Springer Nature Singapore Pte Ltd. 2023 255
C. Chen et al., *Numerical Approximations of Stochastic Maxwell Equations*,
Lecture Notes in Mathematics 2341, https://doi.org/10.1007/978-981-99-6686-8

Lemma C.1 *For the semigroup* $\{S(t) = e^{tM}, t \geq 0\}$ *on* $\mathbb{H}$*, there exists a positive constant C such that*

$$\|S(t) - Id\|_{\mathscr{L}(\mathscr{D}(M),\mathbb{H})} \leq Ct$$

for all $t \geq 0$.

Proof We start from the deterministic system

$$\frac{du(t)}{dt} = Mu(t), \quad t \in (0, T]; \quad u(0) = u_0. \tag{C.1}$$

Thus

$$\frac{d}{dt}\|u(t)\|_{\mathbb{H}}^2 = 2\left\langle \frac{du(t)}{dt}, u(t) \right\rangle_{\mathbb{H}} = 2\left\langle Mu(t), u(t) \right\rangle_{\mathbb{H}} = 0,$$

which leads to

$$\|u(t)\|_{\mathbb{H}} = \|S(t)u_0\|_{\mathbb{H}} = \|u_0\|_{\mathbb{H}},$$

that is, $\|S(t)\|_{\mathscr{L}(\mathbb{H})} = 1$.

Similarly, consider

$$\frac{d}{dt}\|Mu(t)\|_{\mathbb{H}}^2 = 2\left\langle M\frac{du(t)}{dt}, Mu(t) \right\rangle_{\mathbb{H}} = 2\left\langle M^2u(t), Mu(t) \right\rangle_{\mathbb{H}} = 0,$$

from which we have $\|S(t)\|_{\mathscr{L}(\mathscr{D}(M))} = 1$.

The assertion in this lemma is equivalent to

$$\|u(t) - u_0\|_{\mathbb{H}} = \|(S(t) - Id)u_0\|_{\mathbb{H}} \leq Ct\|u_0\|_{\mathscr{D}(M)}.$$

In fact, we can conclude from (C.1) that

$$\langle u(t) - u_0, u(t) \rangle_{\mathbb{H}} = \left\langle \int_0^t Mu(s)ds, u(t) \right\rangle_{\mathbb{H}},$$

where the term on the left-hand side is

$$\frac{1}{2}\left(\|u(t)\|_{\mathbb{H}}^2 - \|u_0\|_{\mathbb{H}}^2 + \|u(t) - u_0\|_{\mathbb{H}}^2\right) = \frac{1}{2}\|u(t) - u_0\|_{\mathbb{H}}^2$$

and the term on the right-hand side can be estimated by

$$\left\langle \int_0^t Mu(s)\,ds,\; u(t) \right\rangle_{\mathbb{H}} \le \int_0^t \|Mu(s)\|_{\mathbb{H}} \|u(t)\|_{\mathbb{H}}\,ds$$

$$\le \|u_0\|_{\mathbb{H}} \int_0^t \|u(s)\|_{\mathscr{D}(M)}\,ds \le Ct\|u_0\|^2_{\mathscr{D}(M)}.$$

The proof is thus finished. $\square$

The following lemma gives the estimates of the operator

$$\left[Id - \tau(A \otimes M) \right]^{-1},$$

which is related to the stochastic symplectic Runge–Kutta method (4.5)–(4.6). Recall that the matrix $A = (a_{ij})^s_{i,j=1}$ is said to satisfy the coercivity condition if it is invertible, and there exists a diagonal positive definite matrix $\mathscr{K} = \mathrm{diag}(k_1, k_2, \dots, k_s)$ and a constant $\alpha > 0$ such that

$$u^\top \mathscr{K} A^{-1} u \ge \alpha u^\top \mathscr{K} u \quad \forall u \in \mathbb{R}^s.$$

Lemma C.2 *Suppose that the matrix* $A = (a_{ij})^s_{i,j=1}$ *satisfies the coercivity condition. Then there exist positive constants* C *such that*

(i) $\left\| \left[Id - \tau(A \otimes M) \right]^{-1} \right\|_{\mathscr{L}(\mathbb{H}^{\otimes s})} \le C$;

(ii) $\left\| Id - \left[Id - \tau(A \otimes M) \right]^{-1} \right\|_{\mathscr{L}(\mathscr{D}(M)^{\otimes s}, \mathbb{H}^{\otimes s})} \le C\tau$,

where $\mathbb{H}^{\otimes s} := \underbrace{\mathbb{H} \times \mathbb{H} \times \cdots \times \mathbb{H}}_{s}$ *and* $\mathscr{D}(M)^{\otimes s} = \underbrace{\mathscr{D}(M) \times \mathscr{D}(M) \times \cdots \times \mathscr{D}(M)}_{s}$.

Proof Denote $u = \left[Id - \tau(A \otimes M) \right]^{-1} v$, where $v = \left((v^1)^\top, (v^2)^\top, \dots, (v^s)^\top \right)^\top$ with $v^i \in \mathbb{H}$ for each $i = 1, 2, \dots, s$. Then we have

$$u = v + \tau(A \otimes M)u.$$

Since A satisfies the coercivity condition, we apply $\langle u, (\mathscr{K} A^{-1} \otimes Id) \cdot \rangle_{\mathbb{H}^{\otimes s}}$ to both sides of the above equation and obtain

$$\left\langle u, (\mathscr{K} A^{-1} \otimes Id)u \right\rangle_{\mathbb{H}^{\otimes s}} = \left\langle u, (\mathscr{K} A^{-1} \otimes Id)v \right\rangle_{\mathbb{H}^{\otimes s}}$$
$$+ \tau \left\langle u, (\mathscr{K} A^{-1} \otimes Id)(A \otimes M)u \right\rangle_{\mathbb{H}^{\otimes s}}. \tag{C.2}$$

Note that

$$\left\langle u, \left(\mathscr{K} A^{-1} \otimes Id\right)u\right\rangle_{\mathbb{H}^{\otimes s}} \geq \alpha \sum_{i=1}^{s} k_i \|u^i\|_{\mathbb{H}}^2 \geq \alpha \min_{1 \leq i \leq s} \{k_i\} \|u\|_{\mathbb{H}^{\otimes s}}^2$$

and

$$\left\langle u, \left(\mathscr{K} A^{-1} \otimes Id\right)\left(A \otimes M\right)u\right\rangle_{\mathbb{H}^{\otimes s}} = \left\langle u, \left(\mathscr{K} \otimes M\right)u\right\rangle_{\mathbb{H}^{\otimes s}}$$

$$= \sum_{i=1}^{s} k_i \left\langle u^i, M u^i\right\rangle_{\mathbb{H}} = 0.$$

Thus, it follows from (C.2) and the Young inequality that

$$\alpha \min_{1 \leq i \leq s} \{k_i\} \|u\|_{\mathbb{H}^{\otimes s}}^2 \leq \langle u, \left(\mathscr{K} A^{-1} \otimes Id\right)v\rangle_{\mathbb{H}^{\otimes s}} \leq \gamma \|u\|_{\mathbb{H}^{\otimes s}}^2 + \frac{C}{\gamma}\|v\|_{\mathbb{H}^{\otimes s}}^2.$$

Taking $\gamma = \alpha \min_{1 \leq i \leq s} \{k_i\}/2$ leads to

$$\|u\|_{\mathbb{H}^{\otimes s}}^2 \leq C\|v\|_{\mathbb{H}^{\otimes s}}^2,$$

where the positive constant C depends on α, $\min_{1 \leq i \leq s}\{k_i\}$, $|\mathscr{K}|$, and $|A^{-1}|$. It means that

$$\left\|\left[Id - \tau\left(A \otimes M\right)\right]^{-1}v\right\|_{\mathbb{H}^{\otimes s}}^2 \leq C\|v\|_{\mathbb{H}^{\otimes s}}^2. \tag{C.3}$$

Therefore, we prove the first assertion.

Similarly, we can show that

$$\|\left(A \otimes M\right)u\|_{\mathbb{H}^{\otimes s}}^2 \leq C\|\left(A \otimes M\right)v\|_{\mathbb{H}^{\otimes s}}^2.$$

It follows from

$$\left(\left[Id - \tau(A \otimes M)\right]^{-1} - Id\right)v = u - v = \tau(A \otimes M)u$$

that

$$\left\|\left(\left[Id - \tau(A \otimes M)\right]^{-1} - Id\right)v\right\|_{\mathbb{H}^{\otimes s}} = \tau\|\left(A \otimes M\right)u\|_{\mathbb{H}^{\otimes s}}$$

$$\leq C\tau\|\left(A \otimes M\right)v\|_{\mathbb{H}^{\otimes s}} \leq C\tau\|v\|_{\mathscr{D}(M)^{\otimes s}},$$

which leads to the second assertion. $\square$

The estimates of operators

$$S_\tau = \left(Id - \frac{\tau}{2}M\right)^{-1}\left(Id + \frac{\tau}{2}M\right) \quad \text{and} \quad T_\tau = \left(Id - \frac{\tau}{2}M\right)^{-1}$$

related to the stochastic midpoint method (4.15) are stated in the following lemma.

Lemma C.3 *There exist positive constants C such that*

(i) $\|T_\tau\|_{\mathscr{L}(\mathbb{H})} \le 1$, $\quad \|Id - T_\tau\|_{\mathscr{L}(\mathscr{D}(M),\mathbb{H})} \le C\tau$;

(ii) $\|S_\tau\|_{\mathscr{L}(\mathbb{H})} = 1$, $\quad \|Id - S_\tau\|_{\mathscr{L}(\mathscr{D}(M),\mathbb{H})} \le C\tau$;

(iii) $\displaystyle\max_{1 \le n \le N} \|S(t_n) - (S_\tau)^n\|_{\mathscr{L}(\mathscr{D}(M^k),\mathbb{H})} \le C\tau^{\frac{k}{2}}$ *with $k \in \{1, 2\}$;*

(iv) *for $n = 1, 2, \ldots, N$ and $s \in [t_j, t_{j+1}]$ with $j = 0, 1, \ldots, n - 1$, it holds that*

$$\|S(t_n - s) - (S_\tau)^{n-j-1}T_\tau\|_{\mathscr{L}(\mathscr{D}(M^k),\mathbb{H})} \le C\tau^{\frac{k}{2}}, \quad k \in \{1, 2\}.$$

Proof

(i) Define $\widetilde{v} := T_\tau v$ for any $v \in \mathbb{H}$, which means that

$$\widetilde{v} = v + \frac{\tau}{2}M\widetilde{v}.$$

Taking the inner product with $\widetilde{v}$ yields

$$\frac{1}{2}\left[\|\widetilde{v}\|_{\mathbb{H}}^2 - \|v\|_{\mathbb{H}}^2 + \|\widetilde{v} - v\|_{\mathbb{H}}^2\right] = \frac{\tau}{2}\langle M\widetilde{v}, \widetilde{v}\rangle_{\mathbb{H}} = 0.$$

Hence, $\|\widetilde{v}\|_{\mathbb{H}} = \|T_\tau v\|_{\mathbb{H}} \le \|v\|_{\mathbb{H}}$ leads to $\|T_\tau\|_{\mathscr{L}(\mathbb{H},\mathbb{H})} \le 1$. Notice that for $v \in \mathscr{D}(M)$,

$$\|(Id - T_\tau)v\|_{\mathbb{H}} = \|v - \widetilde{v}\|_{\mathbb{H}} = \frac{\tau}{2}\|M\widetilde{v}\|_{\mathbb{H}} = \frac{\tau}{2}\|MT_\tau v\|_{\mathbb{H}} \le \frac{\tau}{2}\|v\|_{\mathscr{D}(M)},$$

which completes the proof of (i).

(ii) Let $\widetilde{u} := S_\tau u$ for any $u \in \mathbb{H}$. Then

$$\widetilde{u} = u + \frac{\tau}{2}M(u + \widetilde{u}).$$

Taking the inner product with $u + \widetilde{u}$ yields

$$\|\widetilde{u}\|_{\mathbb{H}}^2 - \|u\|_{\mathbb{H}}^2 = \frac{\tau}{2}\langle M(u + \widetilde{u}), u + \widetilde{u}\rangle_{\mathbb{H}} = 0.$$

Hence, $\|\widetilde{u}\|_{\mathbb{H}} = \|S_\tau u\|_{\mathbb{H}} = \|u\|_{\mathbb{H}}$ leads to $\|S_\tau\|_{\mathscr{L}(\mathbb{H},\mathbb{H})} = 1$. For $u \in \mathscr{D}(M)$, we have

$$\|(Id - S_\tau)u\|_{\mathbb{H}} = \|u - \widetilde{u}\|_{\mathbb{H}} = \frac{\tau}{2}\|M(u + \widetilde{u})\|_{\mathbb{H}} \leq \frac{\tau}{2}\big(\|Mu\|_{\mathbb{H}} + \|MS_\tau u\|_{\mathbb{H}}\big)$$

$$\leq \tau \|u\|_{\mathscr{D}(M)},$$

which completes the proof of (ii).

(iii) Denote

$$v(t) := S(t)v_0 \quad \text{and} \quad v^j := S_\tau^j v_0, \quad j = 0, 1, \ldots, N,$$

then $v(t)$ is the exact solution of

$$\begin{cases} \frac{dv(t)}{dt} = Mv(t), & t \in (0, T], \\ v(0) = v_0, \end{cases}$$

while v^j is the solution of

$$\begin{cases} v^j = v^{j-1} + \frac{\tau}{2}\big(Mv^{j-1} + Mv^j\big), & j = 1, 2, \ldots, N, \\ v^0 = v_0. \end{cases}$$

Let $e^j := v(t_j) - v^j, j = 0, 1, \ldots, N$. Note that $v(t_j) = v(t_{j-1}) + \int_{t_{j-1}}^{t_j} Mv(s)ds$, we have

$$e^j = e^{j-1} + \frac{\tau}{2}\big(Me^{j-1} + Me^j\big) + \int_{t_{j-1}}^{t_j} \Big(Mv(s) - \frac{1}{2}Mv(t_{j-1}) - \frac{1}{2}Mv(t_j)\Big)ds.$$

Applying $\langle \,\cdot\,, e^j + e^{j-1}\rangle_{\mathbb{H}}$ to both sides of the above equation, and using the skew-adjointness of the Maxwell operator M lead to

$$\|e^j\|_{\mathbb{H}}^2 = \|e^{j-1}\|_{\mathbb{H}}^2 + \frac{1}{2}\int_{t_{j-1}}^{t_j} \Big\langle 2Mv(s) - Mv(t_{j-1}) - Mv(t_j),\ e^j + e^{j-1}\Big\rangle_{\mathbb{H}} ds$$

$$= \|e^{j-1}\|_{\mathbb{H}}^2 - \frac{1}{2}\int_{t_{j-1}}^{t_j} \Big\langle \int_{t_{j-1}}^s Mv(r)dr - \int_s^{t_j} Mv(r)dr,\ Me^j + Me^{j-1}\Big\rangle_{\mathbb{H}} ds$$

$$\leq \|e^{j-1}\|_{\mathbb{H}}^2 + C\tau^2\Big(\sup_{0 \leq t \leq T} \|v(t)\|_{\mathscr{D}(M)}^2 + \max_{0 \leq j \leq N} \|v^j\|_{\mathscr{D}(M)}^2\Big)$$

$$\leq \|e^{j-1}\|_{\mathbb{H}}^2 + C\tau^2\|v_0\|_{\mathscr{D}(M)}^2,$$

$$\tag{C.4}$$

which implies

$$\max_{0\le j\le N}\|e^j\|_{\mathbb{H}} = \max_{0\le j\le N}\|(S(t_j)-(S_\tau)^j)v_0\|_{\mathbb{H}} \le C\tau^{\frac{1}{2}}\|v_0\|_{\mathscr{D}(M)}.$$

Furthermore, it follows from (C.4) that

$$\|e^j\|_{\mathbb{H}}^2 = \|e^{j-1}\|_{\mathbb{H}}^2 - \frac{1}{2}\int_{t_{j-1}}^{t_j}\Big\langle \int_{t_{j-1}}^{s} Mv(r)dr$$

$$- \int_s^{t_j} Mv(r)dr,\ Me^j + Me^{j-1}\Big\rangle_{\mathbb{H}} ds$$

$$= \|e^{j-1}\|_{\mathbb{H}}^2$$

$$+ \frac{1}{2}\int_{t_{j-1}}^{t_j}\Big\langle\Big(\int_{t_{j-1}}^{s}\int_{t_{j-1}}^{r} - \int_s^{t_j}\int_{t_{j-1}}^{r}\Big)Mv(\xi)d\xi\,dr,$$

$$M^2(e^j+e^{j-1})\Big\rangle_{\mathbb{H}} ds$$

$$\le \|e^{j-1}\|_{\mathbb{H}}^2 + C\tau^3\|v_0\|_{\mathscr{D}(M^2)}^2,$$

from which we obtain

$$\max_{0\le j\le N}\|e^j\|_{\mathbb{H}} = \max_{0\le j\le N}\|(S(t_j)-(S_\tau)^j)v_0\|_{\mathbb{H}} \le C\tau\|v_0\|_{\mathscr{D}(M^2)}.$$

(iv) For $s \in [t_j, t_{j+1}]$, $j = 0, 1, \ldots, n-1$, we have

$$S(t_n-s) - (S_\tau)^{n-j-1}T_\tau = S(t_n-t_{j+1})\big(S(t_{j+1}-s)-Id\big)$$

$$+ \Big(S(t_n-t_{j+1})-(S_\tau)^{n-j-1}\Big)$$

$$+ \Big((S_\tau)^{n-j-1}(Id-T_\tau)\Big).$$

Then assertion (iv) follows from combining results in (i)–(iii).

$\square$

Below, we give estimates of operators

$$S_{h,\tau} = \Big(Id - \frac{\tau}{2}M_h^{\text{upw}}\Big)^{-1}\Big(Id + \frac{\tau}{2}M_h^{\text{upw}}\Big) \quad\text{and}\quad T_{h,\tau} = \Big(Id - \frac{\tau}{2}M_h^{\text{upw}}\Big)^{-1}$$

related to the stochastic symplectic dG midpoint algorithm (5.37).

Lemma C.4 *For any $v \in \mathbb{H}_{h,r}$, it holds*

(i) $\|T_{h,\tau} v\|_{\mathbb{H}} \leq \|v\|_{\mathbb{H}}$;

(ii) $\|(S_{h,\tau})^n v\|_{\mathbb{H}} \leq \|v\|_{\mathbb{H}}$, $n = 0, 1, \ldots, N$.

Proof

(i) Define $\tilde{v} := T_{h,\tau} v$ for any $v \in \mathbb{H}_{h,r}$, then

$$\tilde{v} = v + \frac{\tau}{2} M_h^{\mathrm{upw}} \tilde{v}. \tag{C.5}$$

Taking the $\mathbb{H}$-inner product with $\tilde{v}$ on both sides of (C.5), we have

$$\frac{1}{2}\left[\|\tilde{v}\|_{\mathbb{H}}^2 - \|v\|_{\mathbb{H}}^2 + \|\tilde{v} - v\|_{\mathbb{H}}^2 \right] = \frac{\tau}{2} \langle M_h^{\mathrm{upw}} \tilde{v}, \tilde{v} \rangle_{\mathbb{H}} \leq 0$$

due to Proposition 4.25 (ii). Hence, for any $v \in \mathbb{H}_{h,r}$, we have

$$\|T_{h,\tau} v\|_{\mathbb{H}} = \|\tilde{v}\|_{\mathbb{H}} \leq \|v\|_{\mathbb{H}},$$

which implies assertion (i).

(ii) Similarly, to prove assertion (ii), define $v_h^n := (S_{h,\tau})^n v$ for any $v \in \mathbb{H}_{h,r}$, which means that

$$v_h^\ell = v_h^{\ell-1} + \frac{\tau}{2}\left(M_h^{\mathrm{upw}} v_h^{\ell-1} + M_h^{\mathrm{upw}} v_h^\ell \right), \quad \ell = 1, 2, \ldots, n \tag{C.6}$$

with $v_h^0 = v$.

Taking the $\mathbb{H}$-inner product with $v_h^{\ell-1} + v_h^\ell$ on both sides of (C.6) yields

$$\|v_h^\ell\|_{\mathbb{H}}^2 - \|v_h^{\ell-1}\|_{\mathbb{H}}^2 \leq 0,$$

and thus $\|v_h^\ell\|_{\mathbb{H}} \leq \|v_h^{\ell-1}\|_{\mathbb{H}} \leq \cdots \leq \|v_h^0\|_{\mathbb{H}} = \|v\|_{\mathbb{H}}$. This leads to assertion (ii).

$\square$

Finally, we present estimates of operators $S_\alpha(t) := e^{t M_\alpha}$, $t \geq 0$,

$$S_{\tau,\alpha} = \left(Id - \frac{\tau}{2} M_\alpha \right)^{-1} \left(Id + \frac{\tau}{2} M_\alpha \right) \quad \text{and} \quad T_{\tau,\alpha} = \left(Id - \frac{\tau}{2} M_\alpha \right)^{-1}, \quad \alpha = x, y, z$$

related to the splitting midpoint method (5.58).

Lemma C.5 *Let assumptions $\mathbf{n} \times \mathbf{E}|_{\partial D} = 0$, $\mathbf{n} \cdot \mathbf{H}|_{\partial D} = 0$, and $\nabla \cdot \mathbf{E} \in L^2(\Omega, H_{00}^1(D))$ hold. Then for any $v = (\mathbf{E}^\top, \mathbf{H}^\top)^\top \in H^2(D)^6$, it holds that*

(i) $\|S_\alpha(t) - Id\|_{\mathscr{L}(\mathscr{D}(M_\alpha), \mathbb{H})} \leq Ct$;

(ii) $\|\left(S(\tau) - S_z(\tau) S_y(\tau) S_x(\tau) \right) v \|_{\mathbb{H}} \leq C\tau^2 \|v\|_{H^2(D)^6}$.

Proof

(i) Following the approach in [92], for the generator A of a C_0-semigroup and $\tau \geq 0$, we define the bounded operators $\beta_0(\tau A) = e^{\tau A}$ and

$$\beta_k(\tau A) = \int_0^1 e^{(1-\xi)\tau A} \frac{\xi^{k-1}}{(k-1)!} d\xi$$

for any $k \geq 1$. These operators satisfy the following recurrence relation

$$\beta_k(\tau A) = \frac{1}{k!} Id + \tau A \beta_{k+1}(\tau A), \quad k \geq 0.$$

Then, assertion (i) follows from the fact $S_\alpha(t) - Id = t\beta_1(t M_\alpha) M_\alpha$ for all $t \in [0, T]$.

(ii) Note that $S(\tau)v$ is the solution $u(t_{n+1})$ of the problem

$$\begin{cases} \frac{d}{dt} u(t) = (M_z + M_y + M_x)u(t), & t \in (t_n, t_{n+1}], \\ u(t_n) = v. \end{cases}$$

By the variation of constants formula and the integration by parts formula, we have

$$u(t_{n+1}) = S(\tau)v$$

$$= S_z(\tau)v + \int_0^\tau S_z(s)M_y S(\tau - s)v ds + \int_0^\tau S_z(s)M_x S(\tau - s)v ds$$

$$= S_z(\tau)v + \tau S_z(\tau)M_y v - \int_0^\tau s S_z(s)(M_z M_y - M_y M)S(\tau - s)v ds$$

$$+ \tau S_z(\tau)M_x v - \int_0^\tau s S_z(s)(M_z M_x - M_x M)S(\tau - s)v ds.$$

$$(C.7)$$

For the term $S_z(\tau)v$, we use relations

$$S_y(\tau) = \beta_0(\tau M_y) = Id + \tau M_y + \tau^2 \beta_2(\tau M_y)M_y^2$$

and

$$S_x(\tau) = \beta_0(\tau M_x) = Id + \tau M_x + \tau^2 \beta_2(\tau M_x)M_x^2$$

to obtain

$$S_z(\tau)v = S_z(\tau)\Big[S_y(\tau) - \tau M_y - \tau^2 \beta_2(\tau M_y) M_y^2 \Big] v$$

$$= S_z(\tau) S_y(\tau) v - \tau S_z(\tau) M_y v - \tau^2 S_z(\tau) \beta_2(\tau M_y) M_y^2 v$$

$$= S_z(\tau) S_y(\tau)\Big[S_x(\tau) - \tau M_x - \tau^2 \beta_2(\tau M_x) M_x^2 \Big] v \qquad (C.8)$$

$$- \tau S_z(\tau) M_y v - \tau^2 S_z(\tau) \beta_2(\tau M_y) M_y^2 v$$

$$= S_z(\tau) S_y(\tau) S_x(\tau) v - \tau S_z(\tau) S_y(\tau) M_x v - \tau S_z(\tau) M_y v$$

$$- \tau^2 S_z(\tau) S_y(\tau) \beta_2(\tau M_x) M_x^2 v - \tau^2 S_z(\tau) \beta_2(\tau M_y) M_y^2 v.$$

Plugging (C.8) into (C.7) and using $S_y(\tau) - Id = \tau M_y \beta_1(\tau M_y)$ yield

$$\big(S(\tau) - S_z(\tau) S_y(\tau) S_x(\tau) \big) v$$

$$= -\tau^2 S_z(\tau) \beta_1(\tau M_y) M_y M_x v - \tau^2 S_z(\tau) S_y(\tau) \beta_2(\tau M_x) M_x^2 v$$

$$- \tau^2 S_z(\tau) \beta_2(\tau M_y) M_y^2 v - \int_{t_n}^{t_{n+1}} s S_z(s) \big(M_z M_y - M_y M \big) S(t_{n+1} - s) v ds$$

$$- \int_{t_n}^{t_{n+1}} s S_z(s) \big(M_z M_x - M_x M \big) S(t_{n+1} - s) v ds.$$

Hence,

$$\big\| \big(S(\tau) - S_z(\tau) S_y(\tau) S_x(\tau) \big) v \big\|_{\mathbb{H}} \le C \|v\|_{H^2(D)^6} \tau^2,$$

which completes the proof of assertion (ii).

$\square$

Lemma C.6 *Let conditions in Lemma C.5 hold. Then*

(i) $\big\| S_{\tau,\alpha} \big\|_{\mathscr{L}(\mathbb{H})} = 1, \quad \big\| S_{\tau,\alpha} - Id \big\|_{\mathscr{L}(\mathscr{D}(M_\alpha),\mathbb{H})} \le C\tau;$

(ii) $\big\| T_{\tau,\alpha} \big\|_{\mathscr{L}(\mathbb{H})} \le 1, \quad \big\| T_{\tau,\alpha} - Id \big\|_{\mathscr{L}(\mathscr{D}(M_\alpha),\mathbb{H})} \le C\tau;$

(iii) $\big\| \big(S(\tau) - S_{\tau,z} S_{\tau,y} S_{\tau,x} \big) v \big\|_{\mathbb{H}} \le C\tau^2 \|v\|_{H^2(D)^6}.$

Proof Proofs of assertions (i) and (ii) are similar to those of Lemma C.3 (i)–(ii) and Lemma C.5 (i), and hence are omitted here. Below we focus on the proof of assertion (iii). From Lemma C.5 (ii), it suffices to prove

$$\|S_z(\tau)S_y(\tau)S_x(\tau)v - S_{\tau,z}S_{\tau,y}S_{\tau,x}v\|_{\mathbb{H}} \leq C\tau^2\|v\|_{H^2(D)^6} \quad \forall\, v \in H^2(D)^6.$$

For $\alpha = x, y, z$, denote $a = \frac{\tau}{2}M_\alpha$ and $\xi = (Id - a)^{-1}$. Then

$$\xi = Id + \xi a = Id + \frac{\tau}{2}\xi M_\alpha,$$

which gives

$$
\begin{aligned}
S_{\tau,\alpha} &= \xi(Id + \frac{\tau}{2}M_\alpha) = Id + \frac{\tau}{2}M_\alpha + \frac{\tau}{2}\xi M_\alpha + \frac{\tau^2}{4}\xi M_\alpha^2 \\
&= Id + \frac{\tau}{2}M_\alpha + \frac{\tau}{2}(Id + \frac{\tau}{2}\xi M_\alpha)M_\alpha + \frac{\tau^2}{4}\xi M_\alpha^2 \\
&= Id + \tau M_\alpha + \frac{\tau^2}{2}\xi M_\alpha^2.
\end{aligned}
$$

Based on the relation

$$S_\alpha(\tau) = Id + \tau M_\alpha + \tau^2\beta_2(\tau M_\alpha)M_\alpha^2,$$

we have

$$\|(S_\alpha(\tau) - S_{\tau,\alpha})v\|_{\mathbb{H}} = \left\| \tau^2\beta_2(\tau M_\alpha)M_\alpha^2 v - \frac{\tau^2}{2}\xi M_\alpha^2 v \right\|_{\mathbb{H}} \leq C\tau^2\|v\|_{\mathscr{D}(M_\alpha^2)}.$$

Therefore,

$$\|S_z(\tau)S_y(\tau)S_x(\tau)v - S_{\tau,z}S_{\tau,y}S_{\tau,x}v\|_{\mathbb{H}}$$

$$\leq \|(S_z(\tau) - S_{\tau,z})S_y(\tau)S_x(\tau)v\|_{\mathbb{H}} + \|S_{\tau,z}(S_y(\tau) - S_{\tau,y})S_x(\tau)v\|_{\mathbb{H}}$$

$$+ \|S_{\tau,z}S_{\tau,y}(S_x(\tau) - S_{\tau,x})v\|_{\mathbb{H}} \leq C\tau^2\|v\|_{H^2(D)^6}.$$

The proof is thus finished. $\square$

Appendix D
Some Results of Stochastic Partial Differential Equations

This appendix is devoted to the introduction of the Wiener process and some properties of the stochastic integral, including the Itô isometry, the Itô formula, and the stochastic Fubini theorem. Furthermore, the Burkholder–Davis–Gundy type inequalities are also presented. Finally, a concise introduction to the strong, weak, and mild solutions of stochastic partial differential equations is given. We refer readers to [62, 129, 145] for more details.

D.1 Hilbert Space Valued Wiener Process

We give the definition of the Q-Wiener process. Let Q be a nonnegative and symmetric operator with a finite trace on a separable Hilbert space $(V, \langle \cdot, \cdot \rangle_V, \|\cdot\|_V)$.

Definition D.1 Let $T > 0$. A V-valued stochastic process $\{W(t), t \in [0, T]\}$, on a probability space $(\Omega, \mathscr{F}, \mathbb{P})$ is called a Q-Wiener process if

(i) $W(0) = 0$, $\mathbb{P}$-a.s.;
(ii) W has $\mathbb{P}$-a.s. continuous trajectories;
(iii) W has independent increments, that is, for all $n \in \mathbb{N}_+$ and all partitions $0 \le t_1 < \cdots < t_n \le T$, the random variables

$$W(t_1), \ W(t_2) - W(t_1), \ \ldots, \ W(t_n) - W(t_{n-1})$$

are independent;
(iv) for all $0 \le s < t \le T$, the increment $W(t) - W(s)$ is a Gaussian random variable with mean zero and covariance operator $(t - s)Q$, that is,

$$\mathbb{P} \circ \big(W(t) - W(s)\big)^{-1} = N(0, \ (t - s)Q).$$

Note that there exists an orthonormal basis $\{e_k\}_{k\in\mathbb{N}}$ of V and a sequence of nonnegative real numbers $\{\eta_k\}_{k\in\mathbb{N}}$ such that $Qe_k = \eta_k e_k$, $k \in \mathbb{N}$.

Proposition D.1 *Assume that W is a Q-Wiener process. Then the following statements hold*:

(i) *W is a Gaussian process on V and*

$$\mathbb{E}[W(t)] = 0, \quad \mathbb{E}[\langle W(t), u\rangle_V \langle W(s), v\rangle_V] = \min\{t, s\}\langle Qu, v\rangle_V$$

for $u, v \in V$, $t, s \geq 0$;

(ii) *for $t \geq 0$, $W(t)$ has the expansion*

$$W(t) = \sum_{k\in\mathbb{N}} \sqrt{\eta_k}\beta_k(t)e_k, \qquad (D.1)$$

where $\{\beta_k\}_{k\in\mathbb{N}}$ is a family of independent standard Brownian motions on $(\Omega, \mathscr{F}, \mathbb{P})$ and the series in (D.1) is convergent in $L^2(\Omega, V)$.

Definition D.2 A Q-Wiener process $\{W(t), t \geq 0\}$, is called a Q-Wiener process with respect to a filtration $\{\mathscr{F}_t\}_{t\geq0}$, if

(i) $W(t)$ is $\mathscr{F}_t$-measurable;

(ii) $W(t) - W(s)$ is independent of $\mathscr{F}_s$ for all $0 \leq s \leq t$.

D.2   Hilbert–Schmidt Operator

Let $(H, \langle\cdot, \cdot\rangle_H, \|\cdot\|_H)$ be a separable Hilbert space.

Definition D.3 Let $A \in \mathscr{L}(V)$ and $\{e_k\}_{k\in\mathbb{N}}$ be an orthonormal basis of V. The trace of the operator A is defined as

$$\mathrm{Tr}(A) := \sum_{k\in\mathbb{N}}\langle Ae_k, e_k\rangle_V,$$

if the series is convergent.

Definition D.4 (Hilbert–Schmidt Operator) A bounded linear operator $A : V \to H$ is called Hilbert–Schmidt if

$$\sum_{k\in\mathbb{N}} \|Ae_k\|_H^2 < \infty,$$

where $\{e_k\}_{k\in\mathbb{N}}$ is an orthonormal basis of V.

Denote by $HS(V, H)$ the space of all Hilbert–Schmidt operators from V to H. One can check that $HS(V, H)$ equipped with the inner product $\langle S, A \rangle_{HS(V,H)} := \sum_{k \in \mathbb{N}} \langle Se_k, Ae_k \rangle_H$ and the norm $\|A\|_{HS(V,H)} := \left(\sum_{k \in \mathbb{N}} \|Ae_k\|_H^2 \right)^{1/2}$ is a separable Hilbert space.

Proposition D.2 *Let G be a separable Hilbert space, and let $S_1 \in \mathcal{L}(H, G)$, $S_2 \in \mathcal{L}(G, V)$, $A \in HS(V, H)$. Then $S_1 A \in HS(V, G)$ and $A S_2 \in HS(G, H)$ satisfying*

$$\|S_1 A\|_{HS(V,G)} \leq \|S_1\|_{\mathcal{L}(H,G)} \|A\|_{HS(V,H)},$$

$$\|A S_2\|_{HS(G,H)} \leq \|A\|_{HS(V,H)} \|S_2\|_{\mathcal{L}(G,V)}.$$

D.3   Properties of the Stochastic Integral

Let $Q \in \mathcal{L}(V)$ be symmetric, nonnegative, and of finite trace, and $\{W(t), t \in [0, T]\}$ be a Q-Wiener process with respect to the filtration $\{\mathscr{F}_t\}_{t \in [0,T]}$. This section introduces some properties of the stochastic integral with respect to W.

Introduce the subspace $V_0 := Q^{\frac{1}{2}} V$ of V with the inner product given by

$$\langle u, v \rangle_{V_0} := \langle Q^{-\frac{1}{2}} u, Q^{-\frac{1}{2}} v \rangle_V$$

for all $u, v \in V_0$, where $Q^{-\frac{1}{2}}$ is the pseudo-inverse of $Q^{\frac{1}{2}}$. We define the following predictable σ-algebra:

$$\mathscr{P}_T := \sigma \Big(\big\{ (s, t] \times G_s, 0 \leq s < t \leq T, G_s \in \mathscr{F}_s \big\} \cup \big\{ \{0\} \times G_0, G_0 \in \mathscr{F}_0 \big\} \Big)$$

$$\sigma \Big(Y : [0, T] \times \Omega \to \mathbb{R} \,\Big|\, Y \text{ is left-continuous and adapted to } \{\mathscr{F}_t\}_{t \in [0,T]} \Big).$$

Let $\Phi : [0, T] \times \Omega \to H$ be an $\mathscr{P}_T / \mathscr{B}(H)$-measurable mapping. Then Φ is called an H-predictable stochastic process.

Proposition D.3 (Itô Isometry) *Let $T \in (0, \infty)$ and $[a, b] \subset [0, T]$. Let the stochastic process $\Phi : [0, T] \times \Omega \to HS(V_0, H)$ be $HS(V_0, H)$-predictable with*

$$\int_a^b \mathbb{E}\big[\|\Phi(s)\|_{HS(V_0,H)}^2 \big] ds < \infty.$$

Then

$$\mathbb{E}\left[\left\|\int_a^b \Phi(s)dW(s)\right\|_H^2\right] = \mathbb{E}\left[\int_a^b \|\Phi(s)\|_{HS(V_0,H)}^2 ds\right].$$

Theorem D.1 (Stochastic Fubini Theorem) *Assume that $(A, \mathscr{A}, \mu)$ is a measurable space, where μ is finite. Let $\Phi(\cdot, \cdot, x)$ be an $HS(V_0, H)$-predictable stochastic process for all $x \in A$ and satisfy*

$$\int_A \left[\mathbb{E}\left(\int_0^T \|\Phi(s, \cdot, x)\|_{HS(V_0,H)}^2 ds\right)\right]^{\frac{1}{2}} \mu(dx) < \infty.$$

Then

$$\int_A \left[\int_0^T \Phi(s, x)dW(s)\right]\mu(dx) = \int_0^T \left[\int_A \Phi(s, x)\mu(dx)\right]dW(s), \quad \mathbb{P}\text{-}a.s.$$

Assume that Φ is an $HS(V_0, H)$-valued predictable stochastic process with

$$\mathbb{E}\left[\int_0^T \|\Phi(s)\|_{HS(V_0,H)}^2 ds\right] < \infty,$$

and φ is a Bochner integrable H-valued predictable process. Let $X(0)$ be an $\mathscr{F}_0$-measurable H-valued random variable. Then for the stochastic process

$$X(t) = X(0) + \int_0^t \varphi(s)ds + \int_0^t \Phi(s)dW(s), \quad t \in [0, T],$$

we have the following Itô formula.

Theorem D.2 (Itô Formula) *Assume that a functional $F : [0, T] \times H \to \mathbb{R}$ and its Fréchet derivatives F_t, F_x, and F_{xx} are uniformly continuous on bounded subsets of $[0, T] \times H$. Then, for all $t \in [0, T]$,*

$$F(t, X(t)) = F(0, X(0)) + \int_0^t \left\langle F_x(s, X(s)), \Phi(s)dW(s)\right\rangle$$

$$+ \int_0^t \left(F_t(s, X(s)) + \left\langle F_x(s, X(s)), \varphi(s)\right\rangle\right)ds$$

$$+ \frac{1}{2}\int_0^t \text{Tr}\left[F_{xx}(s, X(s))(\Phi(s)Q^{1/2})(\Phi(s)Q^{1/2})^*\right]ds, \quad \mathbb{P}\text{-}a.s.$$

$$(D.2)$$

We conclude this section by presenting two propositions which are widely used to show the well-posedness and the regularity of solutions of stochastic partial differential equations and their numerical approximations. The first proposition is about the Burkholder–Davis–Gundy type inequalities; see for instance [109, Corollary A.2] for more details about proofs and applications.

Proposition D.4 (Burkholder–Davis–Gundy Type Inequalities) *Let Φ be an $HS(V_0, H)$-valued predictable stochastic process with*

$$\mathbb{E}\Big[\int_0^t \|\Phi(s)\|_{HS(V_0,H)}^2 ds\Big] < \infty.$$

Then

(i) *for all $p \geq 2$ and $t \geq 0$,*

$$\Big(\mathbb{E}\Big[\sup_{t\in[0,T]} \Big\|\int_0^t \Phi(s)dW(s)\Big\|_H^p\Big]\Big)^{\frac{1}{p}}$$
$$\leq \Big(\frac{p(p-1)}{2}\Big)^{\frac{1}{2}}\Big(\int_0^t \Big(\mathbb{E}\big[\|\Phi(s)\|_{HS(V_0,H)}^p\big]\Big)^{\frac{2}{p}} ds\Big)^{\frac{1}{2}};$$

(ii) *for all $p \geq 2$, there exists a constant $C_p > 0$ such that for all $t \geq 0$,*

$$\mathbb{E}\Big[\sup_{0\leq s\leq t} \Big\|\int_0^s \Phi(\tau)dW(\tau)\Big\|_H^p\Big] \leq C_p\mathbb{E}\Big[\Big(\int_0^t \|\Phi(s)\|_{HS(V_0,H)}^2 ds\Big)^{\frac{p}{2}}\Big].$$

The second proposition is the estimate of the stochastic convolution; see e.g., [62, Proposition 7.3] for the case $p > 2$ and [94] for $p > 0$.

Proposition D.5 *Suppose that $\{S(t), t \geq 0\}$ is a C_0-semigroup on H satisfying $\|S(t)\|_{\mathscr{L}(H)} \leq e^{\zeta t}$ for some $\zeta \geq 0$ and all $t \geq 0$. Let Φ be an $HS(V_0, \mathbb{H})$-predictable stochastic process with*

$$\mathbb{E}\Big[\int_0^t \|\Phi(s)\|_{HS(V_0,H)}^2 ds\Big] < \infty.$$

Then for every $p > 0$, there exists a positive constant C_p such that

$$\mathbb{E}\Big[\sup_{0\leq t\leq T} \Big\|\int_0^t S(t-s)\Phi(s)dW(s)\Big\|_H^p\Big] \leq C_p e^{\zeta pT}\mathbb{E}\Big[\Big(\int_0^T \|\Phi(s)\|_{HS(V_0,H)}^2 ds\Big)^{\frac{p}{2}}\Big].$$

D.4   Solutions of Stochastic Partial Differential Equations

Consider the following stochastic partial differential equation

$$\begin{cases} dX(t) = [AX(t) + F(t, X(t))]dt + B(t, X(t))dW(t), & t \in (0, T], \\ X(0) = X_0, \end{cases}$$

(D.3)

where $A : \mathscr{D}(A) \to H$ is the infinitesimal generator of a C_0-semigroup $\{S(t) = e^{tA}, t \geq 0\}$, $F : [0, T] \times \Omega \times H \to H$ is $\mathscr{P}_T \times \mathscr{B}(H)/\mathscr{B}(H)$-measurable, $B : [0, T] \times \Omega \times H \to HS(V_0, H)$ is $\mathscr{P}_T \times \mathscr{B}(H)/\mathscr{B}(HS(V_0, H))$-measurable, X_0 is an H-valued $\mathscr{F}_0$-measurable random variable, and W is the Q-Wiener process.

Below we give definitions of the strong, mild, and weak solutions for (D.3).

Definition D.5 (Strong Solution) A $\mathscr{D}(A)$-valued predictable process $\{X(t), t \in [0, T]\}$ is called a strong solution of (D.3) if

$$X(t) = X_0 + \int_0^t \left[AX(s) + F(s, X(s)) \right]ds + \int_0^t B(s, X(s))dW(s), \quad \mathbb{P}\text{-a.s.}$$

for $t \in [0, T]$.

Definition D.6 (Mild Solution)   An H-valued predictable stochastic process $\{X(t), t \in [0, T]\}$ is called a mild solution of (D.3) if

$$X(t) = S(t)X_0 + \int_0^t S(t-s)F(s, X(s))ds + \int_0^t S(t-s)B(s, X(s))dW(s), \quad \mathbb{P}\text{-a.s.}$$

for $t \in [0, T]$.

Definition D.7 (Weak Solution) An H-valued predictable process $\{X(t), t \in [0, T]\}$ is called a weak solution of (D.3) if

$$\langle X(t), \eta \rangle_H = \langle X_0, \eta \rangle_H + \int_0^t \langle X(s), A^*\eta \rangle_H + \langle F(s, X(s)), \eta \rangle_H ds$$

$$+ \int_0^t \langle \eta, B(s, X(s))dW(s) \rangle_H, \quad \mathbb{P}\text{-a.s.}$$

for $t \in [0, T]$ and $\eta \in \mathscr{D}(A^*)$. Here, $(A^*, \mathscr{D}(A^*))$ is the adjoint of $(A, \mathscr{D}(A))$ on H.

The following proposition states that a strong solution is also a weak solution; conversely, under certain conditions, a weak solution can also be a strong one; see e.g., [62, 129].

Proposition D.6 (Weak Versus Strong Solutions)

(i) *Every strong solution of (D.3) is also a weak solution.*
(ii) *Let $\{X(t), t \in [0, T]\}$ be a weak solution of (D.3) with values in $\mathscr{D}(A)$ such that $B(t, X(t))$ takes values in $HS(V_0, H)$ for all $t \in [0, T]$. In addition, assume that*

$$\mathbb{P}\left(\int_0^T \|AX(t)\|_H dt < \infty \right) = 1,$$

$$\mathbb{P}\left(\int_0^T \|F(t, X(t))\|_H dt < \infty \right) = 1,$$

$$\mathbb{P}\left(\int_0^T \|B(t, X(t))\|_{HS(V_0,H)}^2 dt < \infty \right) = 1.$$

Then $\{X(t), t \in [0, T]\}$ is also a strong solution.

The following proposition gives the relationship between the weak solution and the mild solution; see [62, 129],

Proposition D.7 (Weak Versus Mild Solutions)

(i) *Let $\{X(t), t \in [0, T]\}$ be a weak solution of (D.3) such that $B(t, X(t))$ takes values in $HS(V_0, H)$ for all $t \in [0, T]$. In addition, assume that*

$$\mathbb{P}\left(\int_0^T \|X(t)\|_H dt < \infty \right) = 1,$$

$$\mathbb{P}\left(\int_0^T \|F(t, X(t))\|_H dt < \infty \right) = 1,$$

$$\mathbb{P}\left(\int_0^T \|B(t, X(t))\|_{HS(V_0,H)}^2 dt < \infty \right) = 1.$$

Then $\{X(t), t \in [0, T]\}$ is also a mild solution.
(ii) *Let $\{X(t), t \in [0, T]\}$ be a mild solution of (D.3) such that*

$$\int_0^t S(t - s)F(s, X(s))ds, \quad \int_0^t S(t - s)B(s, X(s))dW(s), \quad t \in [0, T]$$

have predictable versions. In addition, assume that

$$\mathbb{P}\left(\int_0^T \|F(t, X(t))\|_H dt < \infty \right) = 1,$$

$$\mathbb{E}\left[\int_0^T \int_0^t \|\langle S(t - s)B(s, X(s)), A^*\eta \rangle_H\|_{HS(V_0,\mathbb{R})}^2 ds dt \right] < \infty \quad \forall \eta \in \mathscr{D}(A^*).$$

Then $\{X(t), t \in [0, T]\}$ is also a weak solution.

References

1. A. Abdulle, G. Vilmart, K. Zygalakis, High order numerical approximation of the invariant measure of ergodic SDEs. SIAM J. Numer. Anal. **52**, 1600–1622 (2014)
2. R. Adams, J. Fournier, *Sobolev spaces*. Pure and Applied Mathematics, vol. 140, 2nd edn. (Elsevier/Academic Press, Amsterdam, 2003)
3. R. Alonso, L. Borcea, Electromagnetic wave propagation in random waveguides. Multiscale Model. Simul. **13**, 847–889 (2015)
4. R. Alonso, L. Borcea, J. Garnier, Wave propagation in waveguides with random boundaries. Commun. Math. Sci. **11**, 233–267 (2013)
5. C. Amrouche, M. Bernardi, M. Dauge, V. Girault, Vector potentials in three-dimensional nonsmooth domains. Math. Methods Appl. Sci. **21**, 823–864 (1998)
6. A. Andersson, S. Larsson, Weak convergence for a spatial approximation of the nonlinear stochastic heat equation. Math. Comput. **85**, 1335–1358 (2016)
7. A. Andersson, R. Kruse, S. Larsson, Duality in refined Sobolev–Malliavin spaces and weak approximation of SPDE. Stoch. Partial Differ. Equ. Anal. Comput. **4**, 113–149 (2016)
8. C. Anton, Weak backward error analysis for stochastic Hamiltonian systems. BIT **59**, 613–646 (2019)
9. K. Atkinson, W. Han, *Theoretical numerical analysis: a functional analysis framework*, vol. 39, 3rd edn. Texts in Applied Mathematics (Springer, Dordrecht, 2009)
10. M. Badieirostami, A. Adibi, H. Zhou, S. Chow, Wiener chaos expansion and simulation of electromagnetic wave propagation excited by a spatially incoherent source. Multiscale Model. Simul. **8**, 591–604 (2010)
11. G. Bao, P. Li, *Maxwell's equations in periodic structures*, vol. 208. Applied Mathematical Sciences (Singapore/Science Press, Springer/Beijing, 2022)
12. G. Bao, C. Chen, P. Li, Inverse random source scattering problems in several dimensions. SIAM/ASA J. Uncertain. Quantif. **4**, 1263–1287 (2016)
13. P. Benner, J. Schneider, Uncertainty quantification for Maxwell's equations using stochastic collocation and model order reduction. Int. J. Uncertain. Quantif. **5**, 195–208 (2015)
14. M. Berger, *Nonlinearity and functional analysis: lectures on nonlinear problems in mathematical analysis*. Pure and Applied Mathematics (Academic Press, New York-London, 1977)
15. S. Bertoluzza, G. Naldi, A wavelet collocation method for the numerical solution of partial differential equations. Appl. Comput. Harmon. Anal. **3**, 1–9 (1996)
16. G. Beylkin, On the representation of operators in bases of compactly supported wavelets. SIAM J. Numer. Anal. **6**, 1716–1740 (1992)
17. S. Bobkov, F. Götze, Exponential integrability and transportation cost related to logarithmic Sobolev inequalities. J. Funct. Anal. **163**, 1–28 (1999)

18. A. Bork, Maxwell, displacement current, and symmetry. AM J. Phys. **31**, 854–859 (1963)
19. M. Botchev, J. Verwer, Numerical integration of damped Maxwell equations. SIAM J. Sci. Comput. **31**, 1322–1346 (2008/2009)
20. C. Bréhier, Approximation of the invariant measure with an Euler scheme for stochastic PDEs driven by space-time white noise. Potential Anal. **40**, 1–40 (2014)
21. C. Bréhier, G. Vilmart, High order integrator for sampling the invariant distribution of a class of parabolic stochastic PDEs with additive space-time noise. SIAM J. Sci. Comput. **38**, A2283–A2306 (2016)
22. C. Bréhier, J. Cui, J. Hong, Strong convergence rates of semidiscrete splitting approximations for the stochastic Allen–Cahn equation. IMA J. Numer. Anal. **39**, 2096–2134 (2019)
23. H. Brezis, *Functional analysis, sobolev spaces and partial differential equations*. Universitext (Springer, New York, 2011)
24. Z. Brzeźniak, E. Motyl, M. Ondrejat, Invariant measure for the stochastic Navier–Stokes equations in unbounded 2D domains. Ann. Probab. **45**, 3145–3201 (2017)
25. J. Cai, J. Hong, Y. Wang, Y. Gong, Two energy-conserved splitting methods for three-dimensional time-domian Maxwell's equations and the convergence analysis. SIAM J. Numer. Anal. **53**, 1918–1940 (2015)
26. J. Cai, Y. Wang, Y. Gong, Convergence of time-splitting energy-conserved symplectic schemes for 3D Maxwell's equations. Appl. Math. Comput. **265**, 51–67 (2015)
27. J. Cai, Y. Wang, Y. Gong, Numerical analysis of AVF methods for three-dimensional time-domain Maxwell's equations. J. Sci. Comput. **66**, 141–176 (2016)
28. E. Celledoni, V. Grimm, R. McLachlan, D. McLaren, D. O'Neale, B. Owren, G. Quispel, Preserving energy resp. dissipation in numerical PDEs using the "Averaged Vector Field" method. J. Comput. Phys. **231**, 6770–6789 (2012)
29. C. Chen, A symplectic discontinuous Galerkin full discretization for stochastic Maxwell equations. SIAM J. Numer. Anal. **59**, 2197–2217 (2021)
30. J. Chen, K. Yee, The finite-difference time-domain and the finite-volume time-domain methods in solving Maxwell's equations. IEEE Trans. Antennas Propagat. **45**, 354–363 (1997)
31. C. Chen, J. Hong, Symplectic Runge–Kutta semidiscretization for stochastic Schrödinger equation. SIAM J. Numer. Anal. **54**, 2569–2593 (2016)
32. Z. Chen, Q. Du, J. Zou, Finite element methods with matching and nonmatching meshes for Maxwell equations with discontinuous coefficients. SIAM J. Numer. Anal. **37**, 1542–1570 (2000)
33. D. Chan, M. Soljačić, J. Joannopoulos, Direct calculation of thermal emission for three-dimensionally periodic photonic crystal slabs. Phys. Rev. E **74**, 036615 (2006)
34. W. Chen, X. Li, D. Liang, Energy-conserved splitting FDTD methods for Maxwell's equations. Numer. Math. **108**, 445–485 (2008)
35. W. Chen, X. Li, D. Liang, Symmetric energy-conserved splitting FDTD scheme for the Maxwell's equations. Commun. Comput. Phys. **6**, 804–825 (2009)
36. W. Chen, X. Li, D. Liang, Energy-conserved splitting finite-difference time-domain methods for Maxwell's equations in three dimensions. SIAM J. Numer. Anal. **48**, 1530–1554 (2010)
37. C. Chen, J. Hong, A. Prohl, Convergence of a θ-scheme to solve the stochastic nonlinear Schrödinger equation with Stratonovich noise. Stoch. Partial Differ. Equ. Anal. Comput. **4**, 274–318 (2016)
38. C. Chen, J. Hong, L. Zhang, Preservation of physical properties of stochastic Maxwell equations with additive noise via stochastic multi-symplectic methods. J. Comput. Phys. **306**, 500–519 (2016)
39. C. Chen, J. Hong, X. Wang, Approximation of invariant measure for damped stochastic nonlinear Schrödinger equation via an ergodic numerical scheme. Potential Anal. **46**, 323–367 (2017)
40. C. Chen, J. Hong, L. Ji, Mean-square convergence of a symplectic local discontinuous Galerkin method applied to stochastic linear Schrödinger equation. IMA J. Numer. Anal. **37**, 1041–1065 (2017)

41. C. Chen, J. Hong, L. Ji, Mean-square convergence of a semidiscrete scheme for stochastic Maxwell equations. SIAM J. Numer. Anal. **57**, 728–750 (2019)
42. C. Chen, J. Hong, L. Ji, Runge–Kutta semidiscretizations for stochastic Maxwell equations with additive noise. SIAM J. Numer. Anal. **57**, 702–727 (2019)
43. C. Chen, J. Hong, L. Ji, A new efficient operator splitting method for stochastic Maxwell equations (2021). arXiv:2102.10547
44. C. Chen, J. Hong, D. Jin, L. Sun, Asymptotically-preserving large deviations principles by stochastic symplectic methods for a linear stochastic oscillator. SIAM J. Numer. Anal. **59**, 32–59 (2021)
45. C. Chen, J. Hong, L. Ji, G. Liang, Ergodic numerical approximations for stochastic Maxwell equations (2022). arXiv:2210.06092
46. C. Chen, J. Hong, D. Jin, L. Sun, Large deviations principles for symplectic discretizations of stochastic linear Schrödinger equation. Potential Anal. **59**(3), 971–1011 (2023).
47. E. Chung, B. Engquist, Convergence analysis of fully discrete finite volume methods for Maxwell's equations in nonhomogeneous media. SIAM J. Numer. Anal. **43**, 303–317 (2005)
48. T. Chung, J. Zou, A finite volume method for Maxwell's equations with discontinuous physical coefficients. Int. J. Appl. Math. **7**, 201–223 (2001)
49. E. Chung, Q. Du, J. Zou, Convergence analysis on a finite volume method for Maxwell's equations in nonhomogeneous media. SIAM J. Numer. Anal. **41**, 37–63 (2003)
50. P. Ciarlet, J. Zou, Fully discrete finite element approaches for time-dependent Maxwell equations. Numer. Math. **82**, 193–219 (1999)
51. B. Cockburn, G. Karniadakis, C. Shu, *The development of discontinuous Galerkin methods*, vol. 11. Lecture Notes in Computational Science and Engineering (Springer, Berlin, 2000)
52. B. Cockburn, F. Li, C. Shu, Locally divergence-free discontinuous Galerkin methods for the Maxwell equations. J. Comput. Phys. **194**, 588–610 (2004)
53. D. Cohen, J. Cui, J. Hong, L. Sun, Exponential integrators for stochastic Maxwell's equations driven by Itô noise. J. Comput. Phys. **410**, 109382 (2020)
54. D. Conus, A. Jentzen, R. Kurniawan, Weak convergence rates of spectral Galerkin approximations for SPDEs with nonlinear diffusion coefficients. Ann. Appl. Probab. **29**, 653–716 (2019)
55. J. Cui, J. Hong, Analysis of a splitting scheme for damped stochastic nonlinear Schrödinger equation with multiplicative noise. SIAM J. Numer. Anal. **56**, 2045–2069 (2018)
56. J. Cui, J. Hong, Strong and weak convergence rates of finite element method for stochastic partial differential equation with non-globally Lipschitz coefficients. SIAM J. Numer. Anal. **57**, 1815–1841 (2019)
57. J. Cui, J. Hong, Z. Liu, Strong convergence rate of finite difference approximations for stochastic cubic Schrödinger equations. J. Differ. Equ. **263**, 3687–3713 (2017)
58. J. Cui, J. Hong, Z. Liu, W. Zhou, Numerical analysis on ergodic limit of approximations for stochastic NLS equation via multi-symplectic scheme. SIAM. J. Numer. Anal. **55**, 305–327 (2017)
59. J. Cui, J. Hong, Z. Liu, W. Zhou, Stochastic symplectic and multi-symplectic methods for nonlinear Schrödinger equation with white noise dispersion. J. Comput. Phys. **342**, 267–285 (2017)
60. J. Cui, J. Hong, L. Sun, On global existence and blow-up for damped stochastic nonlinear Schrödinger equation. Discrete Contin. Dyn. Syst. Ser. B **24**, 6837–6854 (2019)
61. G. Da Prato, *An introduction to infinite-dimensional analysis*. Universitext (Springer-Verlag, Berlin, 2006)
62. G. Da Prato, J. Zabczyk, Stochastic equations in infinite dimensions, vol. 152, 2nd edn. Encyclopedia of Mathematics and Its Applications (Cambridge University Press, Cambridge, 2014)
63. A. De Bouard, A. Debussche, A semi-discrete scheme for the stochastic nonlinear Schrödinger equation. Numer. Math. **96**, 733–770 (2004)

64. A. De Bouard, A. Debussche, Weak and strong order of convergence of a semidiscrete scheme for the stochastic nonlinear Schrödinger equation. Appl. Math. Optim. **54**, 369–399 (2006)

65. A. Debussche, Weak approximation of stochastic partial differential equations: the nonlinear case. Math. Comput. **80**, 89–117 (2011)

66. A. Debussche, E. Faou, Weak backward error analysis for SDEs. SIAM J. Numer. Anal. **50**, 1735–1752 (2012)

67. A. Debussche, C. Odasso, Ergodicity for a weakly damped stochastic non-linear Schrödinger equation. J. Evol. Equ. **5**, 317–356 (2005)

68. A. Debussche, J. Printems, Weak order for the discretization of the stochastic heat equation. Math. Comput. **78**, 845–863 (2009)

69. A. Dembo, O. Zeitouni, Large deviations techniques and applications, vol. 38. Stochastic Modelling and Applied Probability (Springer-Verlag, Berlin, 2010)

70. E. Di Benedetto, *Degenerate parabolic equations*. Universitext (Springer-Verlag, New York, 1993)

71. Z. Dong, L. Xu, X. Zhang, Invariant measures of stochastic 2D Navier–Stokes equations driven by α-stable processes. Electron. Commun. Probab. **16**, 678–688 (2011)

72. J. Douglas, S. Kim, Improved accuracy for locally one-dimensional methods for parabolic equations. Math. Models Methods Appl. Sci. **11**, 1563–1579 (2001)

73. J. Eilinghoff, R. Schnaubelt, Error estimates in L^2 of an ADI splitting scheme for the Maxwell equations. Technical Report 2017/32, CRC 1173. Karlsruhe Institute of Technology, Karlsruhe, Germany, 2017. https://doi.org/10.5445/IR/1000077909

74. J. Eilinghoff, R. Schnaubelt, Error analysis of an ADI splitting scheme for the inhomogeneous Maxwell equations. Discrete Contin. Dyn. Syst. **38**, 5685–5709 (2018)

75. J. Eilinghoff, T. Jahnke, R. Schnaubelt, Error analysis of an energy preserving ADI splitting scheme for the Maxwell equations. SIAM J. Numer. Anal. **57**, 1036–1057 (2019)

76. M. El Bouajaji, V. Dolean, M. Gander, S. Lanteri, Optimized Schwarz methods for the time-harmonic Maxwell equations with damping. SIAM J. Sci. Comput. **34**, A2048–A2071 (2012)

77. E. Emmrich, *Discrete versions of Gronwall's lemma and their application to the numerical analysis of parabolic problems*. Preprint series of the Institute of Mathematics, Technische Universität Berlin, 1999. https://doi.org/10.14279/depositonce-14714

78. K. Engel, R. Nagel, *One-parameter semigroups for linear evolution equations*. Graduate Texts in Mathematics, vol. 194 (Springer-Verlag, New York, 2000)

79. W. Fang, M. Giles, Adaptive Euler-Maruyama method for SDEs with nonglobally Lipschitz drift. Ann. Appl. Probab. **30**, 526–560 (2020)

80. K. Feng, M. Qin, *Symplectic geometric algorithms for hamiltonian systems* (Zhejiang Science and Technology Publishing House/Springer, Hangzhou/Heidelberg, 2010). Translated and revised from the Chinese original

81. M. Francoeur, M. Mengüç, Role of fluctuational electrodynamics in near-field radiative heat transfer. J. Quant. Spectrosc. RA **109**, 280–293 (2008)

82. M. Ganesh, S. Hawkins, D. Volkov, An efficient algorithm for a class of stochastic forward and inverse Maxwell models in $\mathbb{R}^3$. J. Comput. Phys. **398**, 108881 (2019)

83. L. Gao, D. Liang, New energy-conserved identities and super-convergence of the symmetric EC-S-FDTD scheme for Maxwell's equations in 2D. Commun. Comput. Phys. **11**, 1673–1696 (2012)

84. L. Gao, M. Cao, R. Shi, H. Guo, Energy conservation and super convergence analysis of the EC-S-FDTD schemes for Maxwell equations with periodic boundaries. Numer. Methods Partial Differ. Equ. **35**, 1562–1587 (2019)

85. S. Garcia, T. Lee, S. Hagness, On the accuracy of the ADI-FDTD method. IEEE Antennas Wirel. Propag. Lett. **1**, 31–34 (2002)

86. M. Geissert, M. Kovács, S. Larsson, Rate of weak convergence of the finite element method for the stochastic heat equation with additive noise. BIT **49**, 343–356 (2009)

87. T. Gronwall, Note on the derivatives with respect to a parameter of the solutions of a system of differential equations. Ann. Math. (2) **20**, 292–296 (1919)

88. M. Gunzburger, W. Zhao, Descriptions, discretizations, and comparisons of time/space colored and white noise forcings of the Navier-Stokes equations. SIAM J. Sci. Comput. **41**, A2579–A2602 (2019)

89. I. Gyöngy, S. Sabanis, D. Šiška, Convergence of tamed Euler schemes for a class of stochastic evolution equations. Stoch. Partial Differ. Equ. Anal. Comput. **4**, 225–245 (2016)

90. E. Hairer, C. Lubich, G. Wanner, *Geometric numerical integration: structure-preserving algorithms for ordinary differential equations*. Springer Series in Computational Mathematics (Springer, Heidelberg, 2010). Reprint of the second edition (2006)

91. M. Hairer, J. Mattingly, M. Scheutzow, Asymptotic coupling and a general form of Harris' theorem with applications to stochastic delay equations. Probab. Theory Related Fields **149**, 223–259 (2011)

92. E. Hansen, A. Ostermann, Dimension splitting for evolution equations. Numer. Math. **108**, 557–570 (2008)

93. E. Hausenblas, Weak approximation of the stochastic wave equation. J. Comput. Appl. Math. **235**, 33–58 (2010)

94. E. Hausenblas, J. Seidler, A note on maximal inequality for stochastic convolutions. Czechoslovak Math. J. **51**(126), 785–790 (2001)

95. D. Higham, An algorithmic introduction to numerical simulation of stochastic differential equations. SIAM Rev. **43**, 525–546 (2001)

96. R. Hiptmair, Finite elements in computational electromagnetism. Acta Numer. **11**, 237–339 (2002)

97. M. Hochbruck, T. Pažur, Implicit Runge–Kutta methods and discontinuous Galerkin discretizations for linear Maxwell's equations. SIAM J. Numer. Anal. **53**, 485–507 (2015)

98. M. Hochbruck, T. Jahnke, R. Schnaubelt, Convergence of an ADI splitting for Maxwell's equations. Numer. Math. **129**, 535–561 (2015)

99. N. Hofmann, T. Müller-Gronbach, K. Ritter, Optimal approximation of stochastic differential equations by adaptive step-size control. Math. Comput. **69**, 1017–1034 (2000)

100. J. Hong, L. Sun, *Symplectic integration of stochastic hamiltonian systems*. Lecture Notes in Mathematics, vol. 2314 (Springer Singapore, 2023)

101. J. Hong, X. Wang, Invariant Measures for Stochastic Nonlinear Schrödinger equations: Numerical approximations and symplectic structures. Lecture Notes in Mathematics, vol. 2251 (Springer Singapore, 2019)

102. J. Hong, L. Ji, L. Zhang, A stochastic multi-symplectic scheme for stochastic Maxwell equations with additive noise. J. Comput. Phys. **268**, 255–268 (2014)

103. J. Hong, L. Sun, X. Wang, High order conformal symplectic and ergodic schemes for the stochastic Langevin equation via generating functions. SIAM J. Numer. Anal. **55**, 3006–3029 (2017)

104. J. Hong, B. Hou, Q. Li, L. Sun, Three kinds of novel multi-symplectic methods for stochastic Hamiltonian partial differential equations. J. Comput. Phys. **467**, 111453 (2022)

105. L. Hornung, Strong solutions to a nonlinear stochastic Maxwell equation with a retarded material law. J. Evol. Equ. **18**, 1427–1469 (2018)

106. Y. Hu, G. Kallianpur, Exponential integrability and application to stochastic quantization. Appl. Math. Optim. **37**, 295–353 (1998)

107. D. Hundertmark, M. Meyries, L. Machinek, R. Schnaubelt, Operator semigroups and dispersive equations, in *16th internet seminar on evolution equations* (2013). https://www.math.tecnico.ulisboa.pt/~czaja/ISEM/16internetseminar201213.pdf

108. M. Hutzenthaler, A. Jentzen, P. Kloeden, Strong convergence of an explicit numerical method for SDEs with nonglobally Lipschitz continuous coefficients. Ann. Appl. Probab. **22**, 1611–1641 (2012)

109. A. Jentzen, P. Kloeden, *Taylor approximations for stochastic partial differential equations*. CBMS-NSF regional conference series in applied mathematics, vol. 83 (Society for Industrial and Applied Mathematics (SIAM), Philadelphia, 2011)

110. A. Jentzen, P. Pušnik, Strong convergence rates for an explicit numerical approximation method for stochastic evolution equations with non-globally Lipschitz continuous nonlinearities. IMA J. Numer. Anal. **40**, 1005–1050 (2020)

111. S. Jiang, L. Wang, J. Hong, Stochastic multisymplectic integrator for stochastic KdV equation. AIP Conf. Proc. **1479**, 1757–1760 (2012)

112. S. Jiang, L. Wang, J. Hong, Stochastic multi-symplectic integrator for stochastic nonlinear Schrödinger equation. Commun. Comput. Phys. **14**, 393–411 (2013)

113. J. Jin, *The finite element method in electromagnetics*. Wiley-Interscience, 2nd edn. (John Wiley & Sons, New York, 2002)

114. C. Kelly, G. Lord, Adaptive time-stepping strategies for nonlinear stochastic systems. IMA J. Numer. Anal. **38**, 1523–1549 (2018)

115. C. Kelly, G. Lord, Adaptive Euler methods for stochastic systems with non-globally Lipschitz coefficients. Numer. Algorithms **89**, 721–747 (2022)

116. A. Klenke, *Probability theory: a comprehensive course*. Universitext, 3rd edn. (Springer, Cham, 2020)

117. L. Kong, J. Hong, J. Zhang, Splitting multisymplectic integrators for Maxwell's equations. J. Comput. Phys. **229**, 4259–4278 (2010)

118. L. Kong, Y. Hong, N. Tian, W. Zhou, Stable and efficient numerical schemes for two-dimensional Maxwell equations in lossy medium. J. Comput. Phys. **397**, 108703, 21 (2019)

119. M. Kovács, S. Larsson, F. Lindgren, Weak convergence of finite element approximations of linear stochastic evolution equations with additive noise. BIT **52**, 85–108 (2012)

120. M. Kovács, S. Larsson, F. Lindgren, Weak convergence of finite element approximations of linear stochastic evolution equations with additive noise II. Fully discrete schemes. BIT **53**, 497–525 (2013)

121. R. Kruse, *Strong and weak approximation of semilinear stochastic evolution equations*. Lecture Notes in Mathematics, vol. 2093 (Springer, Cham, 2014)

122. L. Kurt, T. Schäfer, Propagation of ultra-short solitons in stochastic Maxwell's equations. J. Math. Phys. **55**, 011503, 11 (2014)

123. J. Li, Optimal L^2 error estimates for the interior penalty DG method for Maxwell's equations in cold plasma. Commun. Comput. Phys. **11**, 319–334 (2012)

124. P. Li, X. Wang, An inverse random source problem for Maxwell's equations. Multiscale Model. Simul. **19**, 25–45 (2021)

125. P. Li, X. Wang, Inverse random source scattering for the Helmholtz equation with attenuation. SIAM J. Appl. Math. **81**, 485–506 (2021)

126. J. Li, Z. Fang, G. Lin, Regularity analysis of metamaterial Maxwell's equations with random coefficients and initial conditions. Comput. Methods Appl. Mech. Eng. **335**, 24–51 (2018)

127. D. Liang, Q. Yuan, The spatial fourth-order energy-conserved S-FDTD scheme for Maxwell's equations. J. Comput. Phys. **243**, 344–364 (2013)

128. F. Lindgren, R. Schilling, Weak order for the discretization of the stochastic heat equation driven by impulsive noise. Potential Anal. **38**, 345–379 (2013)

129. W. Liu, M. Röckner, *Stochastic partial differential equations: an introduction*. Universitext (Springer, Cham, 2015)

130. G. Lord, C. Powell, T. Shardlow, *An introduction to computational stochastic PDEs*. Cambridge texts in applied mathematics (Cambridge University Press, New York, 2014)

131. T. Lu, P. Zhang, W. Cai, Discontinuous Galerkin methods for dispersive and lossy Maxwell's equations and PML boundary conditions. J. Comput. Phys. **200**, 549–580 (2004)

132. A. Macho, C. Meca, F. Peláez, F. Cortés, R. Llorente, Ultra-short pulse propagation model for multi-core fibers based on local modes. Sci. Rep. **7**, 1–14 (2017)

133. X. Mao, *Stochastic differential equations and applications*, 2nd edn. (Horwood Publishing Limited, Chichester, 2008)

134. C. Marinelli, L. Scarpa, Fréchet differentiability of mild solutions to SPDEs with respect to the initial datum. J. Evol. Equ. **20**, 1093–1130 (2020)

135. J. Mattingly, A. Stuart, D. Higham, Ergodicity for SDEs and approximations: Locally Lipschitz vector fields and degenerate noise. Stoch. Process. Appl. **101**, 185–232 (2002)

136. G. Milstein, M. Tretyakov, Computing ergodic limits for Langevin equations. Phys. D **229**, 81–95 (2007)
137. G. Milstein, Y. Repin, M. Tretyakov, Numerical methods for stochastic systems preserving symplectic structures. SIAM J. Numer. Anal. **40**, 1583–1604 (2002)
138. H. Minh, P. Niyogi, Y. Yao, Mercer's theorem, feature maps, and smoothing, in *International conference on computational learning theory*, vol. 4005 (2006), pp. 154–168
139. P. Monk, A mixed method for approximating Maxwell's equations, SIAM J. Numer. Anal. **28**, 1610–1634 (1991)
140. P. Monk, *Finite element methods for Maxwell's equations*. Numerical analysis and scientific computation (Oxford University Press, New York, 2003)
141. T. Namiki, A new FDTD algorithm based on alternating-direction implicit method. IEEE Trans. Microw. Theory Tech. **47**, 2003–2007 (1999)
142. A. Pazy, *Semigroups of linear operators and applications to partial differential equations*. Applied Mathematical Sciences, vol. 44 (Springer-Verlag, New York, 1992)
143. D. Pietro, A. Ern, *Mathematical aspects of discontinuous Galerkin methods*. Mathématiques & Applications, vol. 69 (Springer, Heidelberg/Berlin, 2012)
144. D. Polder, M. Van Hove, Nonlinear optics of intense few-cycle pulses: A overview of recent theoretical and experimental developments. Rom. J. Phys. **59**, 767–784 (2014)
145. C. Prèvôt, M. Röckner, *A concise course on stochastic partial differential equations*. Lecture notes in mathematics, vol. 1905 (Springer, Berlin, 2007)
146. M. Rao, R. Swift, *Probability theory with applications*. Mathematics and its applications, vol. 582, 2nd edn. (Springer, New York, 2006)
147. M. Reed, B. Simon, *Methods of modern mathematical physics i. functional analysis*, 2nd edn. (Academic Press/Harcourt Brace Jovanovich, Publishers, New York, 1980)
148. G. Roach, I. Stratis, A. Yannacopoulos, *Mathematical analysis of deterministic and stochastic problems in complex media electromagnetics*. Princeton series in applied mathematics (Princeton University Press, Princeton, 2012)
149. S. Rytov, Y. Kravtsov, V. Tatarskii, *Principles of statistical radiophysics 3: elements of random fields* (Springer-Verlag, Berlin, 1989)
150. M. Salins, Smoluchowski–Kramers approximation for the damped stochastic wave equation with multiplicative noise in any spatial dimension. Stoch. Partial Differ. Equ. Anal. Comput. **7**, 86–122 (2019)
151. V. Shankar, A. Mohammadian, W. Hall, A time-domain, finite-volume treatment for the Maxwell equations. Electromagnetics, **10**, 127–145 (1990)
152. T. Shardlow, Weak convergence of a numerical method for a stochastic heat equation. BIT **43**, 179–193 (2003)
153. T. Shardlow, Modified equations for stochastic differential equations. BIT **46**, 111–125 (2006)
154. Z. Shen, S. Zhou, W. Shen, One-dimensional random attractor and rotation number of the stochastic damped sine-Gordon equation. J. Differ. Equ. **248**, 1432–1457 (2010)
155. B. Song, A. Fiorina, E. Meyhofer, P. Reddy, Near-field radiative thermal transport: from theory to experiment. AIP Adv. **5**, 053503 (2015)
156. M. Song, X. Qian, T. Shen, S. Song, Stochastic conformal schemes for damped stochastic Klein-Gordon equation with additive noise. J. Comput. Phys. **411**, 109300 (2020)
157. G. Strang, On the construction and comparison of difference schemes. SIAM J. Numer. Anal. **5**, 506–517 (1968)
158. Y. Sun, P. Tse, Symplectic and multisymplectic numerical methods for Maxwell's equations. J. Comput. Phys. **230**, 2076–2094 (2011)
159. Z. Sun, Y. Xing, On structure-preserving discontinuous Galerkin methods for Hamiltonian partial differential equations: energy conservation and multi-symplecticity. J. Comput. Phys. **419**, 109662, 25 (2020)
160. J. Sun, C. Shu, Y. Xing, Multi-symplectic discontinuous Galerkin methods for the stochastic Maxwell equations with additive noise. J. Comput. Phys. **461**, 111199 (2022)
161. J. Sun, C. Shu, Y. Xing, Discontinuous Galerkin methods for stochastic Maxwell equations with multiplicative noise. ESAIM Math. Model. Numer. Anal. **57**, 841–864 (2023)

162. A. Taflove, S. Hagness, *Computational electrodynamics: the finite-difference time-domain method*, 2nd edn. (Artech House, Boston, 2000)
163. D. Talay, Second-order discretization schemes of stochastic differential systems for the computation of the invariant law. Stoch. Stoch. Rep. **29**, 13–36 (1990)
164. D. Talay, Stochastic Hamiltonian systems: Exponential convergence to the invariant measure, and discretization by the implicit Euler scheme. Markov Process. Relat. Fields **8**, 163–198 (2002)
165. L. Tartar, *An introduction to Sobolev spaces and interpolation spaces*. Lecture notes of the unione matematica Italiana, vol. 3 (Springer/UMI, Berlin/Bologna, 2007)
166. I. Tsantili, M. Cho, W. Cai, G. Karniadakis, A computational stochastic methodology for the design of random meta-materials under geometric constraints. SIAM J. Sci. Comput. **40**, B353–B378 (2018)
167. F. Tveter, Deriving the Hamilton equations of motion for a nonconservative system using a variational principle. J. Math. Phys. **39**, 1495–1500 (1998)
168. T. Van, A. Wood, A time-domain finite element method for Maxwell equations. SIAM J. Numer. Anal. **42**, 1592–1609 (2004)
169. S. Varadhan, Stochastic analysis and applications. Bull. Am. Math. Soc. (N.S.) **40**, 89–97 (2003)
170. O. Vasilyev, S. Paolucci, A fast adaptive wavelet collocation algorithm for multidimensional PDEs. J. Comput. Phys. **138**, 16–56 (1997)
171. C. Villani, *Optimal transport: old and new*. Grundlehren der mathematischen Wissenschaften, vol. 338 (Springer-Verlag, Berlin, 2009)
172. J. Volakis, A. Chatterjee, L. Kempel, *Finite element method for electromagnetics: antennas, microwave circuits, and scattering applications*. IEEE/OUP Series on Electromagnetic Wave Theory (IEEE Press/Oxford University Press, New York/Oxford, 1998)
173. A. Volokitin, B. Persson, Radiative heat transfer and noncontact friction between nanostructures. Physics-Uspekhi **50**, 879–906 (2007)
174. L. Wang, *Variational integrators and generating functions for stochastic Hamiltonian systems*. Ph.D. Thesis, Universitätsverlag Karlsruhe, 2007
175. X. Wang, Weak error estimates of the exponential Euler scheme for semi-linear SPDEs without Malliavin calculus. Discrete Contin. Dyn. Syst. **36**, 481–497 (2016)
176. X. Wang, S. Gan, Weak convergence analysis of the linear implicit Euler method for semilinear stochastic partial differential equations with additive noise. J. Math. Anal. Appl. **398**, 151–169 (2013)
177. S. Yan, J. Jin, A continuity-preserving and divergence-cleaning algorithm based on purely and damped hyperbolic Maxwell equations in inhomogeneous media. J. Comput. Phys. **334**, 392–418 (2017)
178. K. Yee, Numerical solution of initial boundary value problems involving Maxwell's equations in isotropic media. IEEE Trans. Antennas Propag. **14**, 302–307 (1966)
179. K. Zhang, Numerical studies of some stochastic partial differential equations, ProQuest LLC, Ann Arbor, MI. Ph.D. Thesis, The Chinese University of Hong Kong, 2008
180. L. Zhang, L. Ji, Stochastic multi-symplectic Runge–Kutta methods for stochastic Hamiltonian PDEs. Appl. Numer. Math. **135**, 396–406 (2019)
181. Y. Zhou, F. Chen, J. Cai, H. Liang, Optimal error estimate for energy-preserving splitting schemes for Maxwell's equations. Appl. Math. Comput. **333**, 32–41 (2018)
182. H. Zhu, S. Song, Y. Chen, Multi-symplectic wavelet collocation method for Maxwell's equations. Adv. Appl. Math. Mech. **3**, 663–688 (2011)

LECTURE NOTES IN MATHEMATICS   🐎 Springer

Editors in Chief: J.-M. Morel, B. Teissier;

Editorial Policy

1. Lecture Notes aim to report new developments in all areas of mathematics and their applications – quickly, informally and at a high level. Mathematical texts analysing new developments in modelling and numerical simulation are welcome.

   Manuscripts should be reasonably self-contained and rounded off. Thus they may, and often will, present not only results of the author but also related work by other people. They may be based on specialised lecture courses. Furthermore, the manuscripts should provide sufficient motivation, examples and applications. This clearly distinguishes Lecture Notes from journal articles or technical reports which normally are very concise. Articles intended for a journal but too long to be accepted by most journals, usually do not have this "lecture notes" character. For similar reasons it is unusual for doctoral theses to be accepted for the Lecture Notes series, though habilitation theses may be appropriate.

2. Besides monographs, multi-author manuscripts resulting from SUMMER SCHOOLS or similar INTENSIVE COURSES are welcome, provided their objective was held to present an active mathematical topic to an audience at the beginning or intermediate graduate level (a list of participants should be provided).

   The resulting manuscript should not be just a collection of course notes, but should require advance planning and coordination among the main lecturers. The subject matter should dictate the structure of the book. This structure should be motivated and explained in a scientific introduction, and the notation, references, index and formulation of results should be, if possible, unified by the editors. Each contribution should have an abstract and an introduction referring to the other contributions. In other words, more preparatory work must go into a multi-authored volume than simply assembling a disparate collection of papers, communicated at the event.

3. Manuscripts should be submitted either online at www.editorialmanager.com/lnm to Springer's mathematics editorial in Heidelberg, or electronically to one of the series editors. Authors should be aware that incomplete or insufficiently close-to-final manuscripts almost always result in longer refereeing times and nevertheless unclear referees' recommendations, making further refereeing of a final draft necessary. The strict minimum amount of material that will be considered should include a detailed outline describing the planned contents of each chapter, a bibliography and several sample chapters. Parallel submission of a manuscript to another publisher while under consideration for LNM is not acceptable and can lead to rejection.

4. In general, **monographs** will be sent out to at least 2 external referees for evaluation.

   A final decision to publish can be made only on the basis of the complete manuscript, however a refereeing process leading to a preliminary decision can be based on a pre-final or incomplete manuscript.

   Volume Editors of **multi-author works** are expected to arrange for the refereeing, to the usual scientific standards, of the individual contributions. If the resulting reports can be forwarded to the LNM Editorial Board, this is very helpful. If no reports are forwarded or if other questions remain unclear in respect of homogeneity etc, the series editors may wish to consult external referees for an overall evaluation of the volume.

5. Manuscripts should in general be submitted in English. Final manuscripts should contain at least 100 pages of mathematical text and should always include

   - a table of contents;
   - an informative introduction, with adequate motivation and perhaps some historical remarks: it should be accessible to a reader not intimately familiar with the topic treated;
   - a subject index: as a rule this is genuinely helpful for the reader.
   - For evaluation purposes, manuscripts should be submitted as pdf files.

6. Careful preparation of the manuscripts will help keep production time short besides ensuring satisfactory appearance of the finished book in print and online. After acceptance of the manuscript authors will be asked to prepare the final LaTeX source files (see LaTeX templates online: https://www.springer.com/gb/authors-editors/book-authors-editors/manuscriptpreparation/5636) plus the corresponding pdf- or zipped ps-file. The LaTeX source files are essential for producing the full-text online version of the book, see http://link.springer.com/bookseries/304 for the existing online volumes of LNM). The technical production of a Lecture Notes volume takes approximately 12 weeks. Additional instructions, if necessary, are available on request from lnm@springer.com.

7. Authors receive a total of 30 free copies of their volume and free access to their book on SpringerLink, but no royalties. They are entitled to a discount of 33.3 % on the price of Springer books purchased for their personal use, if ordering directly from Springer.

8. Commitment to publish is made by a *Publishing Agreement*; contributing authors of multiauthor books are requested to sign a *Consent to Publish form*. Springer-Verlag registers the copyright for each volume. Authors are free to reuse material contained in their LNM volumes in later publications: a brief written (or e-mail) request for formal permission is sufficient.

Addresses:
Professor Jean-Michel Morel, CMLA, École Normale Supérieure de Cachan, France
E-mail: moreljeanmichel@gmail.com

Professor Bernard Teissier, Equipe Géométrie et Dynamique,
Institut de Mathématiques de Jussieu – Paris Rive Gauche, Paris, France
E-mail: bernard.teissier@imj-prg.fr

Springer: Ute McCrory, Mathematics, Heidelberg, Germany,
E-mail: lnm@springer.com